Boundary Element Methods in Structural Analysis

Prepared under the auspices of the
Committee on Electronic Computation
of the Structural Division of the
American Society of Civil Engineers

Edited by D.E. Beskos

Published by the American Society of Civil Engineers
345 East 47th Street
New York, New York 10017-2398

ABSTRACT

Tutorial and state-of-art aspects of the Boundary Element Method (BEM) are combined with applications in structural analysis in order to provide information about the advantages of this numerical method, its most recent developments and the pertinent literature. In these 12 chapters, experts cover all the important applications of the BEM in structural analysis. More specifically, chapter one provides a historical introduction, while chapter two deals with torsion of elastic bars. The elastostatic analysis of beams plates and shells as well as general two- and three-dimensional bodies is described in chapter three and four, respectively. Stability and dynamic analysis of beams, plates and shells are discussed in chapters five and six, respectively, while general two- and three-dimensional elastodynamics is discussed in chapter seven. Chapter eight deals with material and geometric nonlinearities, while chapter nine with fracture analysis and chapter ten with dynamic soil-structure interaction. Computer programs are described in chapter eleven, while chapter twelve provides an overview of the method and its future developments.

Library of Congress Cataloging-in-Publication Data

Boundary element methods in structural analysis/prepared under the auspices of the Committee on Electronic Computation of the Structural Division of the American Society of Civil Engineers; edited by D.E. Beskos.

p. cm.

Bibliography: p.

Includes indexes.

ISBN 0-87262-694-6

1. Boundary element method—Data processing. 2. Structural analysis (Engineering)—Data processing. I. Beskos, D.E. II. American Society of Civil Engineers. Structural Division. Committee on Electronic Computation.

TA347.B69B673 1989 89-6565

624.1'71—dc20 CIP

Library of Congress Catalog Card No: 89-6565

ISBN 0-87262-694-6

Manufactured in the United States of America.

PREFACE

The use of numerical methods in conjunction with digital computers has enabled structural engineers during the last 30 years to solve very complicated problems in their area of interest. The most widely used numerical techniques are the Finite Difference Method (FMD) and especially the Finite Element Method (FEM).

During the last 15 years the Boundary Element Method (BEM) has emerged as a powerful tool for numerically solving various complex problems. This method has some distinct advantages over "domain" type techniques, such as the FDM and FEM, especially for certain classes of linear problems characterized by infinite or semi-infinite domains, stress concentrations, three-dimensionality or a combination of those characteristics. The BEM has become very popular lately for its effectiveness to deal with problems of current interest and importance such as, structural dynamics, soil-structure interaction, earthquake engineering, offshore structures and fracture mechanics.

Recently, some textbooks have appeared in the market dealing with BEM applications either in specialized subjects (potential theory, elastostatics, plasticity, porous media flow) or in the general area of engineering science. In addition some volumes in the "Developments" and "Progress-Research Topics" series have appeared in the literature dealing with various BEM applications in engineering science. However, no specific book designed primarily for structural engineers and emphasizing BEM applications in structural engineering exists in the literature.

The book at hand represents an attempt to fill this gap by combining tutorial and state-of-the-art aspects of the BEM and its applications in structural engineering. It aims at informing civil engineers, engaged in the analysis of complex structures, about the advantages of this relatively new technique, the most recent developments in that area and the pertinent literature for further study. In order to obtain a clear understanding of the contents of this volume, the reader is expected to have a good knowledge of advanced courses in engineering mechanics and mathematics. The beginner can understand the basics of the BEM in five different ways: historically in chapter one, by means of torsion in chapter two, beam bending in chapter three, elastostatics in chapter four and computer implementation in chapter eleven.

This book has been divided into 12 chapters, written by persons very well known for their contributions in the field, which cover almost all the important applications of the BEM in structural analysis. More specifically, chapter one provides a historical introduction to the method, while chapter two deals with the simple problem of torsion of elastic bars. The static analysis of beams, plates and shells is taken up in chapter three, while two -and three-dimensional problems of general elastostatics are discussed in chapter four. Stability and dynamic analysis of beams, plates and shells are described in chapters five and six, respectively, while two -and three- dimensional problems of general elastodynamics are discussed in chapter seven. Chapter eight deals with nonlinear structural behavior due to material and geometric nonlinearities, while chapter nine discusses fracture mechanics applications in conjunction with the powerful tool of computer graphics. The special but very important problem of dynamic soil-structure interaction is the subject of chapter ten. The description of the structure of BEM computer programs as well as of special and general purpose existing computer programs is given in chapter eleven. Finally, an overview of the method and its future developments are briefly discussed in chapter twelve.

The editor would like to express his appreciation and thanks to all those persons who helped him in one way or another during the various stages of the preparation of this book and in particular, all the authors of the chapters of this book, Professor C.W. Martin, Chairman of the Electronic Computation Committee of the ASCE, the Structural Division Executive Committee of the ASCE and Ms. S. Menaker of the ASCE Book Production Department.

Patras, Greece 1987

D.E. Beskos
Editor

CONTENTS

The Boundary Element Method
Some Early History - A Personal View

by Frank J. Rizzo[1]

Introduction

The aim of this chapter, as originally intended, was to provide an introduction to the boundary element method (BEM), trace the history of its development, and include a catalogue of areas of its application to engineering problems in general and structural analysis problems in particular. A discussion of advantages and disadvantages of the BEM compared with other numerical methods was also intended, along with an assessment of the present status and likely future developments with the BEM. I agreed to this plan, but after some consideration and several false starts, it occurred to me that in following this plan I would, in essence, be duplicating much of the information already provided in the early chapters of several recent books and monographs (2,4,6) and at least one recent comprehensive review article (41). In these works, the authors have taken some pains to search the literature and at least indicate, however briefly, the subject matter of each paper. In some cases there is an author's commentary on the significance of a given contribution. A certain chronology of development is indicated in each (e.g. (2,41)) of these works as well. Finally, few of the authors seem to have been hesitant to offer their views of 'the essence of the BEM' or 'how it really differs from' or is 'essentially the same as' some other method when 'viewed properly'. Invariably, the method is discussed as having many advantages, growing in applicability, efficiency, and overall popularity, and as generally having a bright future. The papers to follow in this volume will surely give additional substance to these latter views.

I have no quarrel with most of the above except to worry a bit about the accuracy of any blanket claims of clear superiority of the BEM. Also, I question the validity of the view that it really is or should be thought of as a weighted residual method. Regardless of such issues, what I could add or provide here, that doesn't duplicate what already exists elsewhere, had been my main concern.

After much consideration I decided to depart from my original plan. Instead, I would try to share with the reader the evolution of my own personal experience and thinking with the BEM, long before it had that or any other acronym, going back to my days as a research student at the University of Illinois in 1963. I would share also, as best I can

---

[1] Professor and Chairman, Department of Engineering Science and Mechanics, Iowa State University, Ames, Iowa 50011.

remember it, some thinking of certain colleagues, associates, and friends, involved with or contributors to the BEM, principally over say the five years through 1968. The result of this altered plan, while not as archival as the works cited above, may not be entirely subjective, and would have, I hope, a certain kind of interest and appeal if not historical value. In any case, the difference in perspective and sense of what is important about the BEM that emerges here, as contrasted with say the first three cited references above, will probably be quite noticeable.

It became all too clear to me over the years, and I wish to make it clear to the reader, that there were many researchers I never met and whose work I never knew about during the years mentioned or earlier (and later), and whose work was also clearly BEM research. Failure to have known about them then, or to mention them in this chapter, should in no way be interpreted as a judgment, explicit or implied, on the relative merits of anyone's contributions. Such failure is merely indicative of the narrowness of my earlier experience, if not my interest, and the scope of this article as I have chosen to define it.

The Illinois Days - Jaswon's Influence

From 1961 to 1964 I was a research student under Marvin Stippes at the University of Illinois in the Department of Theoretical and Applied Mechanics studying solid mechanics, in general, and theoretical elasticity, in particular. Marvin was very fond of classical work such as to be found in Kellogg's Potential Theory (19) and Love's treatise (25) on elasticity theory, and he had the highest regard for the original contributions of Cauchy to elasticity. He wanted his students to be acquainted with the fundamental contributions of the early Italian elasticians such as Betti, Lauricella, and Somigliana, as well as be familiar with work of more modern masters such as Mindlin and Sternberg.

It was a stimulating, if occasionally intimidating, atmosphere. Marvin was such a superb example himself of all he admired, that we students were committed to go wherever he wanted to lead us. But finding suitable and worthwhile thesis topics in what we were studying often proved to be a problem. In spring 1963 Marvin suggested that "something with integral equations" might be fruitful for me. He pointed me in the direction of Lovitt (26), Fredholm (13), a recent paper by Jaswon (17) and promptly left to go on sabbatical for a year. I was left to formulate and hopefully to write a Ph.D. thesis while he was gone.

My copy of Jaswon (17) is now yellow, tattered and dog-eared from frequent handling that year and copying over 22 years hence. That work on potential theory was the model, motivation, and springboard for everything I did that year for elasticity theory and, indeed, for any other topic that I have subsequently worked on. That paper (17), Jaswon's companion paper with Ponter (18), and the related one by Jaswon's colleague George Symm (40) were all related and relevant. Indeed, in retrospect, all three of those papers represent at once the birth and quintessence of what has become known as the 'direct' boundary element or boundary integral equation method for problems of every description.

The Direct BEM.--The key issues involved in the direct BEM[2], as contained in these 1963 papers, are as follows. Imagine a region of the plane D bounded by a closed curve or curves L; e.g., the interior of an ellipse or square, or perhaps the annular region between two concentric circles. If ω is a function satisfying Laplace's equation

$$\nabla^2\omega = 0 \quad (1)$$

throughout D, one can speak of different boundary value problems for Eq. (1) wherein we seek ω throughout D subject to (a) knowledge of ω on L, or (b) knowledge of $\partial\omega/\partial n$ on L, i.e., the rate of change of ω in the direction of the (outer) normal n to D on L, or (c) knowledge of a linear combination of ω and $\partial\omega/\partial n$ on L, or (d) knowledge of ω on a part L_1 of L and knowledge of $\partial\omega/\partial n$ on the remaining part L_2 of L. Such problems have significance in a variety of physical contexts, e.g., ω may be the warping function for the cross section of an elastic bar in torsion, ω may be the electrostatic potential in a plate, or ω may be the steady-state temperature in a solid conducting heat according to Fourier's Law. An important point for our purposes here is that regardless of the physical context and particular boundary conditions, both ω and $\partial\omega/\partial n$ cannot be simultaneously prescribed *a priori* at a given point on L in a well posed problem. For example, the temperature (ω) and the heat flux ($\sim \partial\omega/\partial n$) are not simultaneously known on L, in advance, in a well posed heat conduction problem.

The solution procedure for the above class of problems adopted in the cited 1963 papers is to analytically recast the boundary value problems, in all of their generality as stated, from a differential form to an integral form. This is done using i) Green's reciprocal identity (in the plane) and ii) the so-called fundamental solution of Laplace's equation (in two dimensions), namely log r(p,q) which is the natural logarithm of the distance r(p,q) between a fixed point p in D or on L and a variable point q on L. Specifically, Green's reciprocal identity for two sufficiently smooth functions ϕ and ψ has the familiar form

$$\int_D (\phi\nabla^2\psi - \psi\nabla^2\phi)\,dD = \int_L \left(\phi \frac{\partial\psi}{\partial n} - \psi \frac{\partial\phi}{\partial n}\right) dL \quad (2)$$

The choices $\phi \equiv \omega$ and $\psi = \log r$ in Eq. (2), with due attention to the singularity in log r, for r=0, leads, using a limiting process, to the integral relation

$$c(p)\omega(p) = \int_L \left[\omega(q) \frac{\partial}{\partial n} \log r(p,q) - \frac{\partial\omega}{\partial n}(q) \log r(p,q)\right] dL(q) \quad (3)$$

[2]The direct BEM as discussed in this paper refers, for the most part, to time-independent problems. Some modifications in parts of the analysis and subsequent numerical procedures, as discussed here, are necessary for time-dependent problems (see e.g. (2)).

wherein the normal derivatives are taken at q and $c(p) \equiv 2\pi$ if p is in D, and $c(p) \equiv \pi$ if p is on L (assuming there is a unique tangent at p on L).

The boundary value problem solution strategy now, in light of Eq. (3), is very straightforward. With p on L and thus $c(p) = \pi$, specify that part of the pair $\{\omega, \partial\omega/\partial n\}$ which is *a priori* known on L according to the type (a) through (d) of boundary value problem at hand. Then 'solve' Eq. (3), regarded as an integral equation, for that part of the pair $\{\omega, \partial\omega/\partial n\}$ which is not *a priori* known. Once that is done, put p inside, i.e., p in D, and, with $c(p) \equiv 2\pi$, generate ω at any chosen p using Eq. (3) as a simple quadrature. The original problem of obtaining ω at any p in D, subject to the prescribed boundary data is thereby solved.

A number of observations about the extraordinarily powerful but simple process just outlined is in order. The analytical procedure of going from Eq. (1) (with attendant boundary conditions) to Eq. (3) reduces, in effect, the dimension of the problems at hand. That is, a problem posed for a two-dimensional region D is converted to a problem involving the one-dimensional region L of D. (A similar reduction to a two-dimensional problem is available for a problem originally posed for a three-dimensional region). Also, Eq. (3) with p on L provides a constraint between ω and $\partial\omega/\partial n$ on L which must hold so that they pertain to one and the same harmonic function throughout D. Moreover, and most importantly, this constraint is the very mechanism for finding 'one from knowledge of the other' on L. In a very real sense, a given boundary value problem through Eq. (3) is 'solved on the boundary', and one and the same formula is applicable regardless of the type ((a through (d)) of boundary value problem at issue.

Of course it is usually impossible to carry out the above strategy for concrete problems by purely analytical means. Nevertheless, the need for numerical approximations in solving the integral equations and in performing necessary quadratures should in no way cloud 1) the simplicity and elegance of the solution strategy as outlined above or 2) the fact that all steps leading to formula (3) are entirely analytical and classical. These steps may be taken, indeed, I submit were taken in 1963, with no regard to any specific type of numerical procedures to be actually used for particular problems. These steps were and are independent of any concept of approximate solutions. It should be noted further that the analytical essence of what is now called the 'direct BEM' is contained either in formula (3) or vector extensions of it. Although some authors now introduce them early in the presentation of the BEM, such concepts as trial functions, weighted residuals, least squares or variational principles, for good or ill, played no part in the thinking of the cited 1963 authors.

As for the specific method of numerical attack on formula (3), Dr. Jaswon and his colleagues did the simplest, most natural, and, for many purposes, still the most effective thing. They collocated with formula (3) at a finite number of points P_i on L, and they assumed that both ω and $\partial\omega/\partial n$ could be approximated by piecewise constant functions over a finite number N of chosen subdivisions of L. Then, choosing $i = 1,2,\ldots N$ and integrating (sometimes numerically) the logarithm functions still left under the integrals over each interval, they formed a square system of algebraic equations in the unknown piecewise constant values of nonprescribed ω and $\partial\omega/\partial n$. Whatever resemblance to finite elements the intervals over which the functions were assumed piecewise constant may have had is a matter of 'after the fact' labeling. Such was not the 1963 thinking at all. Of course, the BEM, for it to become the BEM as we now know it, needed to and did profit greatly by borrowing concepts from the finite element method. But that comes later in the story.

To return to my own thinking as a student in 1963, my task, as I saw it then, was to try to extend the above ideas (as contained in the cited papers) for potential theory to elasticity theory viewed as a type of *vector* potential theory. Specifically, what I hoped ultimately to do was derive a formula, comparable to formula (3), applicable to plane boundary value problems of classical *elasticity* (without body force), which can now be written

$$\underset{\sim}{c}^T(p)\,\underset{\sim}{u}(p) = \int_L [\underset{\sim}{U}^T(p,q)\underset{\sim}{t}(q) - \underset{\sim}{T}^T(p,q)\underset{\sim}{u}(q)]dL(q), \qquad (4)$$

wherein $\underset{\sim}{u}$ is the elastic displacement vector of an elastostatic body occupying the region D, and $\underset{\sim}{t}$ is the (equilibrated) traction vector on L. The quantities $\underset{\sim}{c}$, $\underset{\sim}{U}$, and $\underset{\sim}{T}$ are known tensor functions, the same functions for all problems, the superscript T indicates tensor transpose, and the products $\underset{\sim}{c}^T\underset{\sim}{u}$, $\underset{\sim}{U}^T\underset{\sim}{t}$, etc. indicate inner products yielding vectors (cf. (24)). If $\underset{\sim}{u}$ is thought of as the vector counterpart of ω, one may even regard the governing Navier-Cauchy equation of elastostatics

$$(\lambda+\nu)\nabla(\nabla\cdot\underset{\sim}{u}) + \mu\nabla^2\underset{\sim}{u} = 0, \qquad (5)$$

which u must satisfy throughout D (λ and μ are elastic constants), as a kind of vector Laplace equation. The traction $\underset{\sim}{t}$ is thought of as the vector counterpart of $\partial\omega/\partial n$, with $\underset{\sim}{U}$ formally replacing log r and T formally replacing $\partial \log r/\partial n$ in Betti's reciprocal theorem (cf. (25)). This theorem plays the (vector) role of Green's reciprocal identity for elasticity.

With these identifications, the strategy for elasticity boundary value problems via formula (4) is virtually identical to that already described in connection with formula (3) for potential theory. Specifically, with p on L, specify part of the pair $\{\underset{\sim}{u},\underset{\sim}{t}\}$ on L corresponding to a well posed (displacement, traction, mixed, mixed-mixed, etc.) problem, and solve Eq. (4) for that part of the pair not specified. Once this is done, put p in D and using Eq. (4) generate $\underset{\sim}{u}(p)$ by simple quadrature. Stresses, both on L and at points in D, may be easily obtained by now standard procedures (22) once $\{\underset{\sim}{u},t\}$ on L are fully known.

It should be noted that $\underset{\sim}{U}$, $\underset{\sim}{T}$, and the very existence of such a formula as (4) with p in D, if not on L, was the product of the work of Kelvin, Betti, Lauricella, and Somigliana, and others as well (cf. (25)). My contribution was merely to see the possibility of direct extension of the 'Jaswon-and-associates' ideas of 1963 as outlined above, and to work out the details of requiring p on L. Of course, the process had to be implemented numerically too.

I saw what I wanted to do, but I was worried. Why hadn't Jaswon already done what I intended to do since he was clearly interested in elasticity also? He approached the elasticity problem (17) in two dimensions from an entirely different viewpoint using concepts of the (biharmonic) stress function instead of the 'vector-extension idea' I had in mind. Also, since dealing 'directly' with boundary values of a harmonic function ω and $\partial\omega/\partial n$ seemed so straightforward and advantageous according to Jaswon, why was there so much attention given in the classical literature to the ingredients of an 'indirect' approach[3] even for potential theory? Further, the elasticity counterparts of such 'indirect potentials', as discussed principally by Betti (3) and Lauricella (23), received some formal attention in the literature too, but no one, including Jaswon, seemed to even consider the strategy I was proposing. This bothered me greatly. Finally, there seemed to be cause for concern about the stability of a numerical attack upon the integral equations for (scalar) potential theory, whether of the direct or indirect type. Was I hoping for too much to expect numerical success with the vector problem?

In retrospect, to go forward in the face of such questions, without trying to answer them first, seems a bit reckless and a little naive. But I was younger then, very lucky, and the beneficiary of some good timing. In any case, I had committed myself to a direct form of an integral equation for plane elastostatics problems when I first met Maurice Jaswon in person on May 4, 1964, on his visit to Urbana.

In the shadow of Talbot Laboratory's 3 million pound testing machine (which seemed to make him very uncomfortable for some reason) Jaswon expressed his approval and surprise at my thinking. The answer to why he hadn't already done it himself was simple - he hadn't thought of it! However, regarding the prospects for numerical success, he was less than optimistic, and he informed me of his own numerical difficulties, but not in much detail. It seemed apparent that I could expect only more such difficulties with a vector version of everything. Jaswon was not put off, though, and he suggested that I not be either, about the literature's indirect strategy versus 'our' direct one, which, by the way, did not have those names then.

The Indirect BEM.--It all comes out that there are good, but not critical, reasons for the classical attention to indirect versus direct integral methods based on analytical properties of the equations arising with the respective methods. The issues in outline form are as follows. The direct approach favored by Jaswon and subsequently by me and many others, first of all is a little more complicated to formulate than the indirect approach used historically, and it seemed to be a poorer risk numerically. With the indirect approach, only one integral appears in the associated boundary formula, not two, and prescribed data go into the problem freely, not as part of an integrand. Also, with the indirect BEM it is usually possible to get integral equations of the so-called second kind for which there seems to be a richer

---

[3]An indirect integral approach involves only one of the two integrals involved in formula (3), not both, (see (2) and subsequent discussion here), with density functions, say $\lambda(q)$ and $\mu(q)$, replacing the more physical $\omega(q)$ and $\partial\omega(q)/\partial n$. These density functions are not related to $\omega(q)$ in an obvious way.

background theory and general level of understanding (cf. (26)). With the direct approach, on the other hand, integral equations of the first or mixed kind often arise. However, unknowns in the integral boundary formula are always meaningful and are usually physical quantities that can be prescribed in a well posed boundary value problem, whereas the density functions in the indirect approach are often obscure and ill-behaved at domain corners. With the direct approach, one and the same formula is applicable to all types of boundary data prescription; knowns and unknowns appear in a kind of reciprocal way in the boundary integral equation. I remember being delighted when I realized that this very formula is a constraint between surface displacement and surface traction pertaining to one and the same elastic field in a body occupying D. From this formula traction can be determined from displacement and vice versa without explicit reference to or knowledge of the field itself throughout D. Such unity and reciprocity is not available with the indirect formulas. The direct formula then was the answer to a question that occurred to me a few years earlier: if displacements are prescribed on the surface L of D, must we solve for the stress field throughout D in order to obtain the tractions on L causing those (prescribed) displacements on L, or is there a 'shorter' route to this? I think it was an at least subconscious awareness of this question that made me so interested in the direct BEM in the first place.

In any event, on the practical side, there is valuable flexibility for general purpose computer codes based on this idea, not to mention the inherent conceptual simplicity of the direct BEM. Also, the matter of good numerical conditioning of the entire approximate solution process based on the direct BEM is an issue for which I shall be eternally grateful. It appears that the spectral characteristics of the integral equations and the associated numerical conditioning of the algebraic equations, for certain kinds of prescribed data using the direct BEM, are inferior to that which would be the case with the indirect method. However, the mesh sizes actually needed and used in practical discretizations for calculation are (usually) such that the mentioned differences prove to be of relatively little consequence. The work of Ulrich Heise (e.g. (15)) is, to my knowledge, the most illustrative and definitive on this issue.

UW, Seattle; Some Numerical Work - Tom Cruse

In summer 1964, the understanding of such matters and, indeed, all of my computational experience with the BEM was to come. I was awarded my degree at Illinois without having obtained any numbers from my equations, and I moved to the University of Washington, Seattle in autumn 1964. With the help of C. C. Chang, an M.S. student there, I eventually solved some plane elasticity problems numerically, using the direct formulation. The formulation with numerical examples finally appeared in print as (34) in early 1967.

Most of the numerical difficulties which Jaswon anticipated implementing the direct BEM for elastostatics never materialized for several reasons. First, as indicated earlier, any reasonable numerical procedure based on the direct formulation proves to be more effective and stable than anyone would have had the right to expect in advance. Second, I regarded the computer at that time as a rather hostile beast, and Mr. Chang and I asked it to do only the most elementary and predictable things. Anything remotely sophisticated or unpredictable associated with the inevitable singularities we did analytically. Also, Mr. Chang was very careful, diligent, and very optimistic about everything. Ultimately, proceeding independently of some of Jaswon's methods, we were lucky and apparently did not make a number of what may now be called, 'strategic numerical errors'. On the other hand, on viewing our computational results, I believed much too strongly that our numerical success was a result of performing analytically as many 'element integrations' as we did. This belief, I think, slowed my subsequent confidence in and acceptance of the use of Gaussian quadrature with BEM. Such use is now, of course, the mainstay of all modern BEM procedures, but I was no help whatsoever in the initiation of this important development.

In early 1966 I met Tom Cruse who was in the late stages of his graduate work for the Ph.D. at Washington and looking for a thesis topic. He was attracted by the 'integral ideas' and Tom's subsequent first work (7,9) is now well known. Therein he clearly indicated an effective mechanism for treating time-dependent problems via the BEM. This groundbreaking work was and still is important and was done virtually on his own. I had gone permanently to Kentucky by the fall of 1966 before Tom had really gotten a good start in his research. We worked together on the same campus for less than a year.

I remember admiring not only Tom's intellect and drive but also his courage in choosing to work on the BEM at Washington, when the two fashionable things to be doing then were modern continuum mechanics and finite elements. Those were lonely days indeed, but Tom's instincts have always been sound. His next choices, namely to do the first 3-D work in elasticity (8) and then to show that the BEM was an excellent choice for crack problems (e.g. (11)) and stress concentrations of all sorts, did more to attract much needed attention and to establish respectability and additional promise for the BEM than anything else. His subsequent work on inelasticity with Swedlow (10) was seminal as well. But this, like his elastodynamics work and, indeed, like a lot of good work everywhere, sat awhile before being really appreciated. To this day, no single individual is more responsible for the importance, good name, and visibility of the BEM than Tom Cruse.

Tom had given the integral equation method a proper name in 1969. He called it the boundary integral equation (BIE) method, and I believe he was also the first to use the terminology 'direct BIE method' for the approach taken by Jaswon and subsequently emphasized by all of us. It is curious, and to my mind regrettable, that while Maurice, Tom, and I were all so very interested in the same things in the middle to late sixties, no two of us were physically in the same place for very long. The three of us were in the same room only once, I believe, at a meeting in Paris in 1978. Anyway, the term BIE is still in use although the terminology BEM is inexorably taking over.

To Kentucky - Dave Shippy

At Kentucky meanwhile, Dave Shippy and I started working together in about 1967, and we made steady if not overly rapid progress with two-dimensional applications in heat conduction (37), anisotropic elasticity (36), isotropic inclusion problems (35), and viscoelasticity

(38), etc. Jaswon, whose appointment at Kentucky had originally made me aware of opportunities there, had exited for England not long after I moved from Seattle. He has been in England ever since. As for Dave Shippy, he and I are still doing BEM research together at Kentucky.

Association with Dave for nineteen years now has been one of the real joys in my professional life. No one that I know, including myself, understands, appreciates, or has greater affection for and facility with the BEM than Dave Shippy. I am convinced that he knows more about the analytical and computational aspects of the Cauchy Principal Value (cf. (30)) than Cauchy himself. He has been a master applied mathematical modeler since his book (29) appeared in 1963, and his relationship with the hardware and software of computing is like that of a composer/conductor with a musical score and orchestra. Dave's individual contributions to BEM are harder to isolate than some of the works already mentioned. But, like insight, or inspiration, or like friendship itself, which are all hard to bring into focus, Dave's independent contributions have been numerous and positively essential to the BEM. Fortunately for me, but not always so for him, I think, our respective contributions are pretty hard to separate at this point. In any case, we both agree that we have accomplished more, and more enjoyably, working together, than ever would have been possible working separately.

The French Connection - Lachat and Watson

While the BIE was receiving plenty of attention by the early 1970's, the next massive dose of interest in the method, especially for solid mechanics, came as a result of the important work of Lachat and Watson (21), (22) in about 1975. They showed in (22) that the systematic functional representation procedures, using the so-called shape functions, so long in use in finite element analysis, could be systematically incorporated into the BEM. I was shocked and chagrined on two counts when I saw this paper. First, I was still so unfamiliar in principal with the finite element method in 1975 that I was surprised at the systematic role which shape functions played even there, let alone the role they could play in BEM. Secondly, as alluded to earlier, I was erroneously under the impression that Gaussian quadrature, needed of course if anything like shape functions were introduced (and as now routinely used in most BEM codes), would not provide enough accuracy for the BEM matrix elements to lead to acceptable solution accuracy. I couldn't have been more mistaken.

To better appreciate the significance of the shape function contribution, recall that the classical numerical approach, as employed by Jaswon, involved the assumption that the 'boundary-only' functions, e.g., $\omega(q)$ and $\partial\omega(q)/\partial n$ in formula (3) were assumed to be adequately representable by functions piecewise constant over intervals (now called elements) of the boundary L. This was the basic strategy for a long time, extended to three dimensions, as well, over (usually flat triangular) surface patches or elements. Some ad hoc attempts to use a linear or (rarely) a quadratic variation of functions for certain purposes were made earlier, but nothing systematic like the Lachat and Watson effort seems to have been done prior to 1975. Therein, of course, we have a comprehensive representation of boundary functions (e.g. ω and $\partial\omega/\partial n$) as varying linearly, quadratically, cubically, etc.

over boundary elements which themselves, through use of the same shape functions, are represented as flat, locally quadratic, or cubic surfaces through preselected nodal points. The associated mapping operation maps all elements and integrations thereover to a standard space of squares or equilateral triangles. Integration may thus be systematically performed using Gaussian quadrature in the standard space, and, wonder of wonders (to me!), the whole thing works beautifully. It was just what was needed to provide the major improvement in computational flexibility and efficiency that the BEM needed and was waiting for. It was a milestone contribution.

The conference[4] organized by Tom Cruse and Jean-Claude Lachat in Versailles, France, in May 1977, was the first real meeting of people from virtually everywhere interested in BEM research, and it was the occasion for lots of good fellowship. I enjoyed the hospitality of Jean-Claude in Chamonix, and the conversations with him and John Watson and many others at the conference were most enjoyable and memorable. On that same trip, the hospitality of Jaswon and Symm in England, Zienkiewicz in Wales, and Heise and Rieder in Germany was especially warm and generous. Maurice and George were very pleased, and understandably so, at the size and beauty of the garden that had grown from the seeds they planted in 1963. There was a general feeling that the BIE 'had arrived', but as of then, 1977, it was still called the BIE!

The Larger BEM

With such power and respectability though, as afforded by the infusion of shape functions, and with now recognizable kinship with the finite element method, the boundary integral method, 'wearing the right clothes' and with 'such important work' to do, needed a 'proper' name. How about the *boundary* element method, BEM? Perfect! This suggests that it has been a member of 'the family' all along, and maybe that it is really a finite element method at heart but 'somehow' you just need them on the boundary!

Banerjee and Butterfield (2) claim credit for first use of the term 'boundary elements' or the acronym BEM. Brebbia (4) and the authors of (2) indicate that the term originated in about 1976 in Southampton. Southampton and the folks from there are more than welcome to whatever distinction is involved here. Frankly, although it was an inevitable choice and in many ways a good one, I'm not too fond of the term BEM for reasons to be made explicit below.

What has been described above is of course not the whole BEM story by any means, nor even the whole story from, say, 1962-1968. Indeed, in that six year period, the papers by Friedman and Shaw (14), Shaw (39), Rieder (33), Kupradze (20), Banaugh and Goldsmith (1), Chen and Schweihert (5), Hess (16), Massonnet (27), Oliveira (31), and Doyle (12), are just a sample of work done in that period which may now be regarded as important, or in some cases, seminal BEM for a variety of disciplines (cf. (2,41)). I became aware of almost all of this work much after it was published, unfortunately, and too late for it to have been a help in my early research. Also, to my utter astonishment, in the 1970's I found a paper by William Prager (32) published in 1928

---

[4] First International Symposium on Innovative Numerical Analysis in Applied Engineering Science.

wherein he not only has what is now clearly a bonafide BEM formulation for a fluid flow problem, but he also did a numerical computation using a hand 'drum' calculator! As for work subsequent to 1968, I won't even try to list here any of the BEM papers, not already mentioned, for reasons already discussed.

One interesting aspect of having been in on the 'ground floor', so to speak, of a development like the direct BEM as Maurice, Tom, and a number of others of us have been, is that many years later we are the beneficiaries of lots of hindsight which appears to have been foresight. It is tempting to believe that we perhaps knew at the beginning, as some people now think is the case, that the BEM would eventually be good for problems involving cracks, axisymmetric piles, stresses in veins of coal and diamond, analysis of Rayleigh surface waves, bending of plates, flow in porous soils, soil-structure interaction, and on and on. Well it just isn't so, certainly not with me. I am as surprised as can be at what has developed. It is great to be associated with all this - but anticipate it? In all honesty, never!

It now appears that some form of the BEM can be brought to bear on almost any problem which can be posed in terms of linear differential equations. This includes many problems in structural mechanics. Even non-linear problems can often be attacked with profit in an incremental, piecewise fashion, using iteration, (i.e. (2,28)). One can identify a linear operator (governing differential equation, like Laplace's or the Navier-Cauchy equation), a reciprocal relation (like Green's or Betti's), and a fundamental solution (like log r or $\tilde{U}$) and construct a direct, (or indirect) BEM rather readily. Then some kind of numerical procedure, e.g., like the classical procedure of Jaswon, or the more modern procedures of Lachat and Watson, could form the basis of an effective 'boundary element' computer code. Everyone seems to now appreciate the reduction in dimension feature and the accuracy/computational and discretization effort that exists in reality or potentially with the BEM. The method seems to work to best advantage on 'bulky bodies', i.e., those with low surface to volume ratio, wherein there may be high stress or other field gradients, or perhaps even singularities, e.g., in the vicinity of cracks. The regions should be at least piecewise homogeneous, and the BEM should be inherently the method of choice for regions of indefinite extent. By contrast, very complex nonlinear behavior, especially for finite, slender, shell-like bodies, which may also be continuously inhomogeneous, would be problem characteristics least suited to advantageous BEM application. The papers to follow in this volume should clarify these issues and provide specific examples.

A 'Rose' by any Name

I would like to close by expressing my view that the BEM is not just another (often better) numerical method, among many others, now so popular in the age of the modern digital computer. The BEM, regardless of what is emphasized about it today or what name it is given, as I tried to show, is based on an important mathematical point of view with a rich and classical heritage having nothing to do, necessarily, with numerical computation.

Indeed, the fundamental theorem of integral calculus states that if we add (integrate) the slopes (rate of change df/dx) of a function

$f(x)$ from a point x_A to another x_B, we get the difference in the function values at those (boundary) points, i.e., $f(x_B) - f(x_A)$. Green's theorem in the plane and Gauss' divergence theorem in three dimensions are really just generalizations of that idea, and these theorems are clearly the result of an overall attempt to relate vector (and tensor) functions of spatial variables to their boundary values[5]. One would expect then that since the very idea of solution to a boundary value problem, physical or otherwise, involves such a relation, some form of these theorems would be a good starting place, at least for linear problems. Indeed, with virtually nothing else but these theorems and the concept of a fundamental solution to the original differential equation, some limits, and a particular point of view, one is led to the direct boundary integral formula (e.g. Eqs. (3) and (4)). *This formula can be regarded as the single relation which contains information about the governing differential equation, the region over which it must hold, and boundary data which could be prescribed; and with the field point inside the domain, this same formula provides the desired solution to the original boundary value problem by simple quadrature.* No other method that I know of, especially any numerical one, is built on such a fundamental analytical base, reducing the very dimension of the problem, by one. This method focuses all activity on the boundary before any numerical work is done. The key to all of this, of course, is the fundamental (singular) solution which has no counterpart in the (non-boundary) finite element method. With the BEM as described here, one has gone far analytically, perhaps as far as one can go in general, before any approximations are introduced. Such a tactic, whenever possible, represents an imaginative and optimum view of the roles that can be played by classical analysis and digital computation by machine. It is at least a refreshing alternative to methods which approximate the domain or differential operators at the very outset.

Regardless of such issues, the boundary element method or BEM is an inevitable name for the described technique when all is said and done, since the integrations must be done, in general, in a nonexact fashion, and boundary elements as identified are perhaps the best way to accomplish the computational objective. However, they are not the only way, and elements are in no way an *essential* part of the process as is the case with finite element methods. I applaud the goal of engendering a feeling of familiarity over the widest possible applications-oriented audience. But I believe that the BEM terminology and too early attention to numerical approximation theory often disguises or discourages appreciation of the richness and simplicity of the ideas in the previous paragraph. This is unfortunate in my judgment. After all 'boundary integrals' and not 'boundary elements' are the distinctive feature of the BEM, i.e., integrals rather than elements are what the method is all about.

Regardless of what this 'rose' is called or what is currently thought 'really important about it', researchers all over the world are recognizing the place of the BEM in computational analysis. Some of their views are expressed on the following pages.

---

[5]This observation was brought home to me first by Professor R. G. Langebartel at Illinois in 1964.

Acknowledgement

I am grateful for the moral and material support extended over the years for my research by my family, the Department of Engineering Mechanics at the University of Kentucky, our Graduate School, the Air Force Office of Scientific Research and the National Science Foundation. It is a pleasure to also acknowledge my friendship with most of the authors already cited and the contributions of all of the graduate students at Kentucky who have worked on the BEM. I would like further to acknowledge friendship and profitable BEM associations with Graeme Fairweather, Andy Seybert, George Blandford, M. Maiti, Dave Clements, Jan Sladek, Dimitri Beskos, George Manolis, Marijan Dravinski, Ken Kline, Ray Wilson, M. Mayr, Subrata Mukherjee, Jim Liggett, Tom Rudolphi, Gordon Holze, and from way back Art Robinson and Morris Stern. Special thanks are due to Y. H. Pao whose work we seem to continually refer to. All of these people have made first rate contributions to BEM research. Each of them and the others cited probably have a tale to tell, like that told above, about their respective experiences with the BEM; it would be good to hear them all sometime. Finally, although it is impossible to adequately acknowledge the influence of the late Marvin Stippes on all of the BEM, reverence is the most appropriate word I can attach to his memory.

Appendix I - References

1. Banaugh, R. P., and Goldsmith, W., "Diffraction of Steady Acoustic Waves by Surfaces of Arbitrary Shape," *J. Acoust. Soc. Am.*, Vol. 35, No. 10, 1963, pp. 1590-1601.

2. Banerjee, P. K., and Butterfield, R., *Boundary Element Methods in Engineering Science*, McGraw-Hill, London, 1981.

3. Betti, E., "Teoria dell Elasticita," *Il Nuovo Ciemento*, t. 7-10, 1872.

4. Brebbia, C. A., Telles, J. C. F., and Wrobel, L. C., *Boundary Element Techniques*, Springer-Verlag Berlin, Heidelberg, 1984.

5. Chen, L. H., and Schweikert, J., "Sound Radiation from an Arbitrary Body," *J. Acoust. Soc. Am.*, Vol. 35, 1963, pp. 1626-1632.

6. Crouch, S. L., and Starfield, A. M., *Boundary Element Methods in Solid Mechanics*, George Allen & Unwin, London, 1983.

7. Cruse, T. A., "A Direct Formulation and Numerical Solution of the General Transient Elastodynamic Problem. II," *J. Math. Anal. and Appl.*, Vol. 22, No. 2, May 1968, pp. 341-355.

8. Cruse, T. A., "Numerical Solutions in Three Dimensional Elastostatics," *Int. J. Solids Struct.*, Vol. 5, 1969, pp. 1259-1274.

9. Cruse, T. A., and Rizzo, F. J., "A Direct Formulation and Numerical Solution of the General Transient Elastodynamic Problem. I," *J. Math. Anal. and Appl.*, Vol. 22, No. 1, Apr. 1968, pp. 244-259.

10. Cruse, T. A., and Swedlow, J. L., "Formulation of Boundary Integral Equations for Three-Dimensional Elasto-Plastic Flow," *Int. J. Solids Struct.*, Vol. 7, pp. 1673-1683, 1971.

11. Cruse, T. A., and Vanburen, W., "Three-Dimensional Elastic Stress Analysis of a Fracture Specimen with an Edge Crack," *Int. J. Fracture Mechanics*, Vol. 7, No. 1, March 1971, pp. 1-15.

12. Doyle, J. M., "Integration of the Laplace Transformed Equations of Classical Elastokinetics," *J. Math. Anal. Appl.*, Vol. 13, 1966.

13. Fredholm, I., "Solution d'un problème fondamental de la theorie de l'elasticité," *Arch. Mat. Astronom. Fysik*, Vol. 2, 1905.

14. Friedman, M. B., and Shaw, R. P., "Diffraction of a Plane Shock Wave by an Arbitrary Rigid Cylindrical Obstacle," *J. Appl. Mech.*, Vol. 29, No. 1, 1962, pp. 40-46.

15. Heise, U., "Systematic Compilation of Integral Equations of the Rizzo Type and of Kupradze's Functional Equations for Boundary Value Problems of Plane Elastostatics," *J. Elasticity*, Vol. 10, No. 1, Jan. 1980, pp. 23-56.

16. Hess, J. L., and Smith, A. M. D., "Calculations of Potential Flow About Arbitrary Bodies," in *Progress in Aeronautical Sciences*, Vol. 8, pp. 1-138, Pergamon Press, New York, 1966.

17. Jaswon, M. A., "Integral Equation Methods in Potential Theory. I," *Proc. Royal Soc., A*, Vol. 275, 1963, pp. 23-32.

18. Jaswon, M. A., and Ponter, A. R., "An Integral Equation Solution of the Torsion Problem," *Proc. Royal Soc., A*, Vol. 273, 1963, pp. 237-246.

19. Kellogg, O. D., *Foundation of Potential Theory*, Dover, New York, 1953.

20. Kupradze, V. D., *Potential Methods in the Theory of Elasticity*, translated from Russian by Israel Program for Scientific Translation, Jerusalem, 1963.

21. Lachat, J. C., "Further Developments of the Boundary Integral Techniques for Elastostatics," *Ph.D. thesis*, Southampton University, 1975.

22. Lachat, J. C., and Watson, J. O., "Effective Numerical Treatment of Boundary Integral Equations: a Formulation for Three Dimensional Elastostatics," *Int. J. Num. Meth. Engng.*, Vol. 10, 1976, pp. 991-1005.

23. Lauricella, G., "Sur l'intégration de l'equation relative à l'equilibre des plaques élastiques encastreés," *Acta. Math.*, Vol. 32, 1909.

24. Leigh, D. C., *Nonlinear Continuum Mechanics*, McGraw-Hill, New York, 1968.

25. Love, A. E. H., *A Treatise on the Mathematical Theory of Elasticity*, Dover, New York, 1944.

26. Lovitt, W. V., *Linear Integral Equations*, Dover, New York, 1950.

27. Massonnet, C. E., "Numerical Use of Integral Procedures," in O. C. Zienkiewicz and G. S. Holister (eds), *Stress Analysis*, Chap. 10, Wiley, London, 1965.

28. Mukherjee, S., "Time-dependent Inelastic Deformation of Metals by Boundary Element Methods," in P. K. Banerjee and R. P. Shaw (eds), *Developments in Boundary Element Methods-2*, Applied Science, London, 1982.

29. Murphy, G., Shippy, D. J., and Luo, H. L., *Engineering Analogies*, Iowa State University Press, Ames, Iowa, 1963.

30. Muskhelishvili, N. I., *Singular Integral Equations*, P. Noodrhoff Ltd., Groningen, Holland, 1953.

31. Oliveira, E. R. A., "Plane Stress Analysis by a General Integral Method," *J. ASCE, Engng. Mech. Div.*, Feb. 1968, pp. 79-85.

32. Prager, W., "Die Druckverteilung an körpern in ebener Potentialströmung," *Physik Zeitschr.*, 29, 1928.

33. Rieder, G., "Iterationsverfahren und Operatorgleichungen in der Elastizitätstheorie," *Abh. Braunschweig. Wiss. Ges.*, Vol. 14, 1962, pp. 109-343.

34. Rizzo, F. J., "An Integral Equation Approach to Boundary Value Problems of Classical Elastostatics," *Quart. Appl. Math.*, Vol. 25, 1967, pp. 83-95.

35. Rizzo, F. J., and Shippy, D. J., "A Formulation and Solution Procedure for the General Non-homogeneous Elastic Inclusion Problem," *Int. J. Solids Struct.*, Vol. 4, 1968, pp. 1161-1179.

36. Rizzo, F. J., and Shippy, D. J., "A Method for Stress Determination in Plane Anisotropic Elastic Bodies," *J. Compos. Mater.*, Vol. 4, Jan. 1970, pp. 36-61.

37. Rizzo, F. J., and Shippy, D. J., "A Method of Solution for Certain Problems of Transient Heat Conduction," *AAIA Journal*, Vol. 8, No. 11, Nov. 1970, pp. 2004-2009.

38. Rizzo, F. J., and Shippy, D. J., "An Application of the Correspondence Principle of Linear Viscoelasticity Theory," *SIAM J. Appl. Math.*, Vol. 21, No. 2, Sept. 1971, pp. 321-330.

39. Shaw, R. P., "Diffraction of Acoustic Pulses by Obstacles of Arbitrary Shape with a Robin Boundary Condition - Part A," *J. Acoust. Soc. Am.*, Vol. 41, No. 4, 1966, pp. 855-859.

40. Symm, G. T., "Integral Equation Method in Potential Theory. II," *Proc. Royal Soc., A*, Vol. 275, 1963, pp. 33-46.

41. Tanaka, M., "Some Recent Advances in Boundary Element Methods," *Appl. Mech. Rev.*, Vol. 36, No. 5, May 1983, pp. 627-634.

Torsion of Elastic Bars

Dimitris L. Karabalis*, A. M. ASCE

Introduction

Pure torsion problems result in one of the simplest three-dimensional stress states and undoubtly this is one of the main reasons for the considerable amount of work that has been carried out in this area. One could probably find one more reason in the fact that the formulation of pure torsion problems leads eventually to the solution of Laplace's equation and thus to the various well known mathematical aspects of the powerful Potential Theory. However, analytical solutions have been reported only for a few cases of bars whose cross section has a simple geometrical shape (17,18).

The more general problems of cylinders of arbitrary cross section or solids of revolution can be tackled effectively only by modern numerical methods such as the Finite Difference Method (FDM), the Finite Element Method (FEM), and the Boundary Element Method (BEM). Eventhough all three of the above methods, and a variety of other specialized techniques, are very powerful in treating boundary value problems, the BEM seems to be the best suited one for the class of problems encountered in this work. This is due, of course, to the particular character of the method which introduces a number of advantages over other numerical methods, such as:

i) There is no need for use of conformal mapping.

ii) Any type of well posed boundary conditions, i.e., uniform or mixed, can be handled with equal ease.

iii) Multiply connected regions need no special treatment.

iv) Internal stresses and displacements at any point are computed only if needed but they are not required for the solution of the problem.

v) A discretization of *only* the boundary of the domain of interest is necessary.

The last point is probably the most important one, since it reduces the dimensions of the problem by one and thus, it succeeds in minimizing the required amount of discretization when compared to the FDM or the FEM both of which require volume discretization schemes. However, a well known disadvantage of the BEM is that the resulting matrices are full, in contrast to the sparse and usually banded matrices that are encountered in FDM or FEM analyses.

The purpose of this chapter is to provide an introduction to the fundamental analytical and numerical aspects of the BEM in a simple and easy to understand manner. Thus, the use of the BEM for the solution of the Saint-Venant torsion problem will be presented from the applied point of view. Questions pertaining to the mathematical point of view, such as the existence, the uniqueness, or the convergence of the solutions given, are not discussed in this chapter. The interested reader can find detailed discussions on these subjects in the various classical works on Potential Theory, e.g., Kellogg (23), Sternberg and Smith (49), MacMillan (33), Kupradze (30) as well as the more recent BEM oriented works of Jaswon and Symm (20,21) and Symm (52).

---

* Assistant Professor, Department of Civil Engineering, University of South Carolina, Columbia, SC 29208.

A general discussion, along with an extensive list of references, on the various applications of the Potential Theory to engineering-related problems and its numerical treatment by the BEM has been presented by Beskos (9). Other general references of information on the BEM and its applications are the texts by Brebbia (10), Banerjee and Butterfield (6), and Brebbia, Telles and Wrobel (11).

1. Formulation of the Torsion Problem

Detailed formulations of the Saint-Venant torsion problem are presented in many classical treatises on the theory of elasticity, e.g., Sokolnikoff (48) and Timoshenko and Goodier (53). Therefore, only some results of the various possible formulations are stated below for easy reference.

Consider a homogeneous isotropic cylindrical bar of arbitrary cross section D, bounded by an outer contour S_o and inner contours S_i , $i = 1,2,...,n$. The bar is subjected to no body forces and is free from external forces on its lateral surface. One end of the bar is fixed while the other is twisted by a couple whose resultant moment is parallel to the axis z of the bar, as shown in Fig. 1. If Ω is the angle of twist per unit length of the bar, the resulting displacement field can be described as

$$u = -\Omega yz , \qquad v = \Omega xz , \qquad w = \Omega \, \phi(x,y) , \tag{1.1}$$

where $\phi(x,y)$ is the so-called warping function expressing Saint-Venant's assumption that, in the general case of cylindrical bars with arbitrary cross sections, plane sections do not remain plane but are warped and that each section along the z-axis is warped by the same amount.

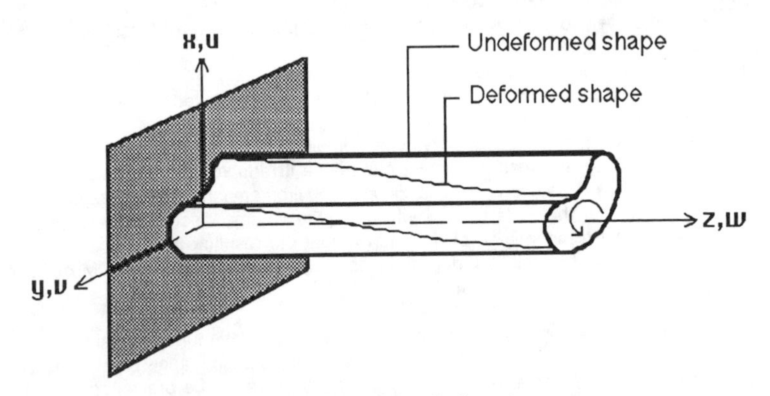

Figure 1. Geometry and coordinate system.

Subsequently, the torsion problem is usually formulated in any of the following three ways:

(a) The warping function must satisfy the Laplace's equation

$$\nabla^2\phi = 0 \qquad \text{in } D , \tag{1.2}$$

with the boundary condition

$$\partial/\partial n\,(\phi) = ly - mx \qquad \text{on } S_i \ (i = 0,1,\ldots,n), \tag{1.3}$$

where l and m are the direction cosines of the outward normal n on the boundary S_i.

(b) The conjugate harmonic function to $\phi(x,y)$ is denoted by $\psi(x,y)$ and is defined by the Cauchy - Riemann equations

$$\partial/\partial x\,(\phi) = \partial/\partial y\,(\psi)\,, \qquad \partial/\partial x\,(\psi) = -\partial/\partial y\,(\phi)\,. \tag{1.4}$$

The conjugate of the warping function must also satisfy the Laplace's equation

$$\nabla^2\psi = 0 \qquad \text{in } D, \tag{1.5}$$

and the boundary conditions

$$\psi = (x^2 + y^2)/2 + k_i \qquad \text{on } S_i \ (i = 0,1,\ldots,n), \tag{1.6}$$

where k_i is a constant on the boundary S_i. The values of one of these constants can be specified arbitrarily, say $k_o = 0$, but the remaining constants must be determined from the requirement that the warping function $\phi(x,y)$ is singled-valued within D.

(c) Prandtl's stress function $F(x,y)$ is defined by

$$F(x,y) = \psi(x,y) - (x^2 + y^2)/2\,, \tag{1.7}$$

and must satisfy the Poisson's equation

$$\nabla^2 F = -2 \qquad \text{in } D, \tag{1.8}$$

and the boundary conditions

$$F = k_i \qquad \text{on } S_i \ (i = 0,1,\ldots,n)\,, \tag{1.9}$$

where k_i is a constant on the boundary S_i which must also fulfill the requirements stated in case (b).

All of the above formulations constitute classical problems in Potential Theory. Formulation (a) can be easily recognized as the Neumann's problem for the Laplace's equation, while formulations (b) and (c) represent the Dirichlet's problem for the Laplace's equation and the Poisson's equation, respectively.

2. Integral Representations

In this section the basic steps towards a BEM solution of the three formulations of the torsion problem are presented. In general, a BEM formulation is based on an *integral representation* of the problem and requires knowledge of the *fundamental solution* of its governing equation. A BEM formulation can also be classified into any of the following two

general categories:
i) Direct, if the problem is formulated in terms of physically meaningful quantities, e.g., displacements, stresses, etc., and
ii) Indirect, if the problem is formulated in terms of density functions of no physical meaning.
A brief presentation of both mathematical formulations is provided with the torsion problem in mind. A detailed review of the integral formulations of the elastic torsion problem along with the existing connections among them has been reported by Christiansen (13).

For the sake of simplicity, only bars with a simply connected cross section D bounded by a contour S are considered in this work, and $k_o = 0$ in Eqs. 1.6 and 1.9. It is apparent, in view of the formulations presented in the previous section, that this assumption simplifies the solution of the problem when formulations (b) or (c) are used but no distinction between simply and multiply connected cross sections is necessary in connection with formulation (a). However, the interested reader can find more details on boundary integral based formulations and solutions of the torsion problem of bars with multiply connected cross sections in the works of Caulk (12), Danson and Kuich (14), Danson and Brebbia (15), Jaswon and Ponter (19), Katsikadelis and Sapountzakis (22), Lo and Niedenfuhr (32), Ponter (43), Sauer and Mehlhorn (47), etc., and in the references made in these works. In particular, Refs. (22) and (43) adress the problem of inhomogeneous (composite) cylindrical bars.

2.1 Fundamental Solution

The fundamental solution of Laplace's equation is defined as the two point function G(p,q) that satisfies the equation

$$\nabla^2 G(p,q) = \delta(p,q) , \tag{2.1}$$

where p and q are two points in the infinitely extended plane domain and $\delta(p,q)$ is the Dirac delta function characterized by the properties

$$\delta(p,q) = 0 \qquad \text{for } p \neq q \tag{2.2}$$

$$\int_D f(p)\, \delta(p,q)\, dD(p) = f(q) \tag{2.3}$$

for the function f(p) defined in the domain D. The solution of Eq. 2.1 has the form

$$G(p,q) = \ln r_{pq} \tag{2.4}$$

where r_{pq} is the distance between the points p and q. It is obvious from Eq. 2.4 that as p approaches q the function G becomes singular.

2.2 Direct Formulation (Green's Theorem)

The direct boundary integral formulation is based on a reciprocal integral identity, i.e. Green's theorem. Given two functions f(x,y) and g(x,y) with continuous first and second derivatives in the two- dimensional region D, Green's theorem states that

$$\iint_D \left(f\, \nabla^2 g - g\, \nabla^2 f \right) dx\, dy = \int_S \left(f\, \partial/\partial n(g) - g\, \partial/\partial n(f) \right) ds , \tag{2.5}$$

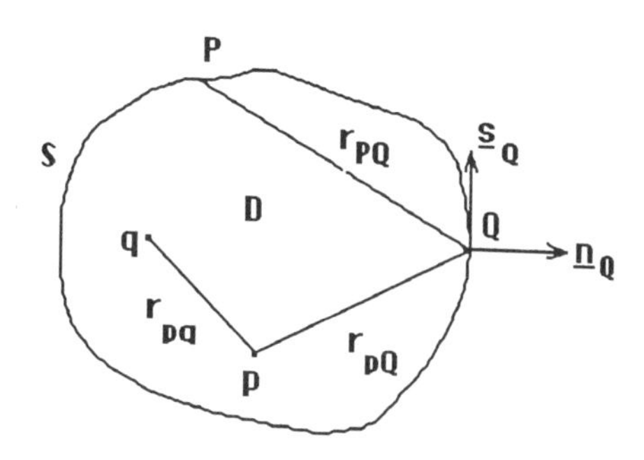

Figure 2. Definitions and geometrical quantities appearing in boundary integral formulations.

where, as shown in Fig. 2, $\underline{n}$ is the outward unit normal vector on the boundary S which in this work is considered as positive. If one chooses to substitute the function g with the fundamental solution G of Eq. 2.4, then Eq. 2.5 yields, in view of the Eq. 2.1,

$$-\iint_D G(p,q)\ \nabla^2 f(q)\ dD(q) + f(p) =$$

$$\int_S f(Q)\ \partial/\partial n_Q[G(p,Q)]\ dS(Q) - \int_S G(p,Q)\ \partial/\partial n_Q[f(Q)]\ dS(Q) \qquad (2.6)$$

where, as shown in Fig.2, Q represents a point on the boundary S, p and q are interior points, and $\partial/\partial n_Q$ denotes the normal derivative at point Q. In what follows, boundary points are always designated by capital letters, while lower case letters are used for interior points. If point p moves to the boundary S the contour integrals in the right-hand side of Eq. 2.6 become singular when $P \equiv Q$. Indeed when $r_{pQ} \to 0$, $G = 0(\ln r)$ and $\partial/\partial n_Q[G(p,Q)] = 0(1/r)$, where the symbol 0( ) means "order of". However, these singularities can be treated analytically through a standard limiting process that excludes the singular point P from the integration. Finally, for a point P on the boundary S Eq. 2.6 yields

$$\beta(P)\ f(P) = \iint_D G(P,q)\ \nabla^2 f(q)\ dD(q) +$$

$$+ \int_S \Big[f(Q)\ \partial/\partial n_Q[G(P,Q)] - G(P,Q)\ \partial/\partial n_Q[f(Q)] \Big]\ dS(Q) \qquad (2.7)$$

Eq. 2.7 is general in the sense that no assumption has been made with regard to the geometry of the boundary S in the neighborhood of point P. However, depending on the location of point P the function $\beta(P)$ can take the following values:

$$\beta(P) = \begin{cases} 0, & \text{if } P \notin D \cup S \\ 2\pi, & \text{if } P \in D \text{ and } P \notin S \\ \theta(P), & \text{if } P \in S, \end{cases} \qquad (2.8)$$

where $\theta(P)$ is the angle in radians made by the two tangents drawn on each side of point P, as shown in Fig. 3. In the most frequently met case of a smooth boundary $\theta(P) = \pi$.

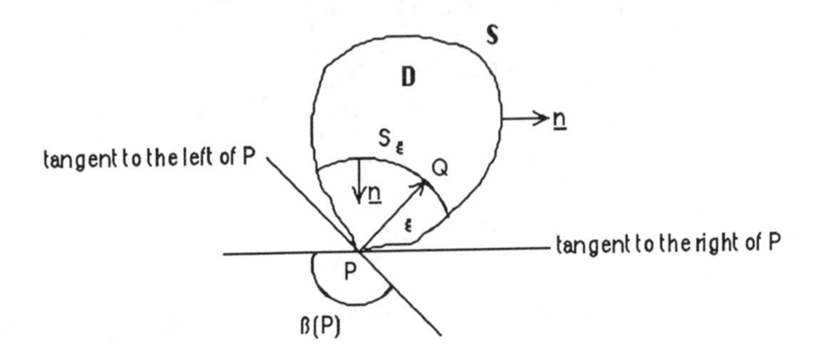

Figure 3. Integration contour excluding the singular point P.

Eq. 2.7 is essentially a boundary integral equation (BIE) since the volume integral it contains vanishes for harmonic functions f, e.g. the functions ϕ and ψ of Eqs. 1.2 and 1.5, respectively, or it assumes a known value, e.g. if the function f is substituted by the function F of Eq. 1.8. Eq. 2.7 gives the values of the function f at any interior or boundary point in terms of f and its normal derivative $\partial/\partial n(f)$ on the boundary S, by properly choosing the value of $\beta(P)$ from Eq. 2.8. Therefore , it presents the unique advantage that it can accomodate with equal ease any type of boundary conditions. For example, if f is prescribed on the boundary (Dirichlet's problem), Eq. 2.7 becomes an integral equation for the normal derivative $\partial/\partial n(f)$ on the boundary. Similarly, if the boundary condition $\partial/\partial n(f)$ is given (Neumann's problem), it turns into an integral equation for the complimentary boundary condition f. If mixed boundary conditions are given, i.e., f is prescribed over parts of the boundary and $\partial/\partial n(f)$ on the remaining parts, then Eq. 2.7 becomes an integral equation for those parts of f and $\partial/\partial n(f)$ that are not known.

2.3 Indirect Formulation (Single- or Double-Layer Potentials)

The function f(p) can, alternatively, be expressed as the superposition of unit singular solutions G(p,q) and source densities of amplitude h(Q) specified on the boundary S, i.e.,

$$f(p) = \int_S h(Q)\, G(p,Q)\, dS(Q)\,. \tag{2.9}$$

In classical potential theory the function f(p) in Eq. 2.9 is usually known as the single-layer potential of the density function h(Q). Eventhough, Eq. 2.9 is written for points p in the interior it also holds for points P on the boundary. Taking the normal derivative of both sides of Eq. 2.9 at points P on the boundary, in a manner similar to that outlined for Eq. 2.7, one has

$$\partial/\partial n_P\big[f(P)\big] = \int_S h(Q)\, \partial/\partial n_P[G(P,Q)]\, dS(Q) - \beta(P)\, h(P) \tag{2.10}$$

Use of Eq. 2.10 in conjunction with the appropriate boundary conditions, i.e. Eq. 1.3, results in a BIE for $h(Q)$. Once $h(Q)$ has been computed, Eq. 2.9 can provide $f(p)$ throughout the domain D or the boundary S.

Similarly, the double-layer potential of the density function $h(Q)$ is defined as

$$f(p) = \int_S h(Q)\, \partial/\partial n_Q[G(p,Q)]\, dS(Q) \tag{2.11}$$

which for points P on the boundary becomes

$$f(P) = \int_S h(Q)\, \partial/\partial n_Q\, [G(P,Q)]\, dS(Q) \;-\; \beta(P)\, h(P) \tag{2.12}$$

Thus, use of Eq. 2.12 in conjunction with the appropriate boundary conditions, i.e. Eq. 1.6, results in a BIE for $h(Q)$. Once $h(Q)$ has been computed, Eq. 2.11 can provide $f(p)$ throughout the domain D or the boundary S.

Treatment of Poisson type equations like Eq. 1.8 is also possible by indirect means as is described, e.g., in Banerjee and Batterfield (6).

3. BEM Formulation and Solution

The boundary integral equations developed in the previous section are used here towards a BEM formulation and numerical solution of the torsion problem as it is expressed in its three forms in section 1.

3.1. Torsion Formulation (a)

i) Direct formulation

The direct boundary integral formulation of the torsion problem, i.e., in terms of a physically meaningful quantity, proceeds on the basis of Eq. 2.7. Substitution of the warping function $\phi(x,y)$ in place of the arbitrary function $f(x,y)$, and in view of Eqs. 1.2, 1.3, and 2.4, yields

$$\int_S \phi(Q)\, \partial/\partial n_Q(\ln r_{PQ})\, dS(Q) \;-\; \beta(P)\, \phi(P) =$$

$$= \int_S \ln r_{PQ}\, [ly - mx]_Q\, dS(Q)\,. \tag{3.1}$$

Eq. 3.1 is a singular Fredholm equation of the second kind a solution of which can obviously produce only the distribution of the warping function ϕ over the boundary S. An analytic solution of Eq. 3.1 is, generally speaking, out of the question. Therefore, a numerical procedure is sought in an effort to replace the integral equation by a system of simultaneous linear equations.

To this end, the boundary S of the region D is divided into an M number of intervals, the i-th interval being of length $2h_i$ and associated with a central point Q_i, midway between the two end points $Q_{i\pm1|2}$, as shown in Fig. 4. A decision of essential consequences on the efficiency and accuracy of the method must be made at this point with regard to the assumed variation of the functions of interest within each boundary interval (9). The choice of preference in the pertinent literature seems to be the uniform distribution of all boundary functions within each interval. This assumption is justified, in part, since i) it simplifies the required integrations and therefore improves the efficiency of the entire computational process, and ii) various tests have shown that the resulting accuracy is well within the acceptable limits of error if reasonably fine discretization meshes are used, see for example Refs. (19,22,28,43,51]. However, various higher order approximations have also been used, e.g., linear variation (26,32,55), or quadratic and higher order variations (4,40,47). A brief description of a BEM formulation based on higher order elements is presented in section 4. In addition to the above approximation of the boundary

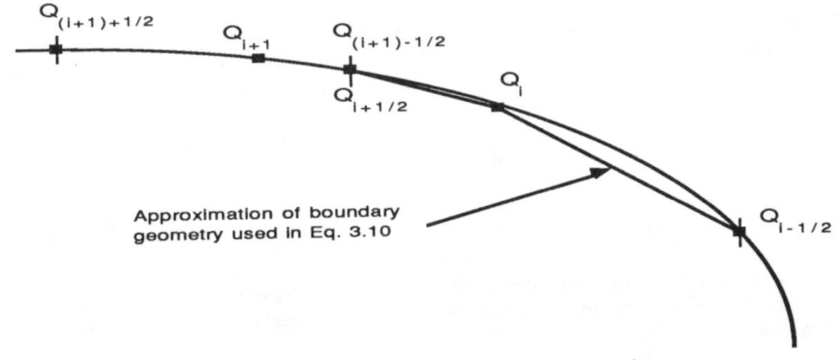

Figure 4. Boundary discretization and nomenclature.

quantities, it is in various occasions convenient and sometimes necessary, as it is shown later, to approximate the physical boundary S itself, e.g., Refs. (4,19,32,40), etc. A rule of thumb, however, in choosing some interpolation curve for the boundary quantities of interest is that higher order approximations should be used only if the physical boundary S is approximated at least in the same quality. An interesting discussion on the subject of the selection of a BEM discretization as well as on the development of a p-adaptive boundary element algorithm, suitable to the torsion problem, can be found in the work of Alarcón and Reverter (2).

As an introduction to the BEM approach a simple uniform distribution of all boundary quantities is assumed in this section over each boundary interval in which the contour S has been discretized. Then, Eq. 3.1 can be written in a summation form as

$$\sum_{i=1}^{M} (A_{ji} - \delta_{ji} B_i)\, \phi_i = \sum_{i=1}^{M} B_{ji} C_i , \qquad j = 1,2,\ldots,M \tag{3.2}$$

where

$$A_{ji} = \int_{(i)} \partial/\partial n_i (\ln r_{ij})\, dS(Q) ,$$

$$B_{ji} = \int_{(i)} \ln r_{ij}\, dS(Q)\,,$$
$$C_i = \left[\, ly - mx \right]_{Q=Q_i}\,, \tag{3.3}$$
$$\beta_i \;=\; \beta\,(P \equiv P_i)\,,$$

δ_{ij} is the Kronecker' s delta, r_{ij} denotes the distance from the midpoint of the j-th interval to a point of the i-th interval, and the subscript (i) indicates integration over the length of the interval i. In performing the integrations indicated by Eqs. 3.3_a and 3.3_b, two distinct possibilities exist: i) $i \neq j$, where the integrations can be performed in a straightforward manner, e.g. Simpson's rule, Gaussian quadrature, and ii) $i \equiv j$, where the integrals are singular and should be evaluated with special care.

To evaluate Eq. 3.3_a for $i \neq j$, it suffices to write (19), in view of the relatively slow variation of $\partial/\partial n_i(\ln r_{ij})$ at "large" distances r_{ij},

$$\int_{(i)} \partial/\partial n_i(\ln r_{ij})\, dS(Q) \;=\; 2h_i \left[\, \partial/\partial n(\ln r_{ij})\, \right]_{Q=Q_i} \tag{3.4}$$

where

$$\left[\partial/\partial n(\ln r_{ij})\right]_{Q=Q_i} \;=\; \left[(x_{Q_i} - x_{Q_j})\, l_{Q_i} \;+\; (y_{Q_i} - y_{Q_j})\, m_{Q_i}\right] \,/\, (r_{ij})^2\,, \tag{3.5}$$

with l_{Q_i}, m_{Q_i} being the direction cosines of the outward normal at point Q_i, and similarly for x_{Q_i}, y_{Q_i}, etc. If $i \equiv j$, one can utilize the well known relationship (21)

$$\int_S \partial/\partial n_Q(\ln r_{PQ})\, dS(Q) \;=\; \pi \tag{3.6}$$

to write

$$\int_{(i)} \partial/\partial n_Q(\ln r_{PQ})\, dS(Q) \;=$$

$$= \pi - \left\{ \sum_{k=1}^{i-1} 2h_k \left[\partial/\partial n(\ln r_{ij})\right]_{Q=Q_k} + \sum_{k=i+1}^{M} 2h_k \left[\partial/\partial n(\ln r_{ij})\right]_{Q=Q_k} \right\} \tag{3.7}$$

Some other more direct ways of evaluating these types of integrals can be demonstrated with regard to the integration indicated by Eq. 3.3_b. Thus, if $i \neq j$ a direct application of Simpson's rule of integration yields

$$\int_{(i)} \ln r_{ij}\; dS(Q) \;=\; (h_i/3)\left\{ [\ln r_{ij}]_{Q=Q_{i-1/2}} \;+\; 4\, [\ln r_{ij}]_{Q=Q_i} \;+\; [\ln r_{ij}]_{Q=Q_{i+1/2}} \right\} \tag{3.8}$$

For the case $i \equiv j$, either the trapezoidal rule of integration can be used with only two points located at the ends of the interval i, or one can use the relationship

$$\int \ln r \, dr = r \ln r - r . \tag{3.9}$$

If the latter of the two possibilities is chosen, the curved boundary interval (i) can be replaced by two straight lines, as shown in Fig. 4, and then the right-hand side of Eq. 3.3_b is approximated as

$$\int_{(i)} \ln r_{ij} \, dS(Q) = = [r_{ij}]_{Q=Q_{i-1/2}} \left\{ [\ln r_{ij}]_{Q=Q_{i-1/2}} - 1 \right\} + [r_{ij}]_{Q=Q_{i+1/2}} \left\{ [\ln r_{ij}]_{Q=Q_{i+1/2}} - 1 \right\} \tag{3.10}$$

Following these explanations on the evaluation of the various quantities appearing in Eq. 3.2, a system of M linear algebraic equations can be formed for the M unknown values of the warping function over the entire set of boundary intervals considered for the problem at hand, i.e.,

$$[H] \{\phi\} = \{E\}, \tag{3.11}$$

where

$$H_{ji} = A_{ji} - \delta_{ji} \beta_i$$
$$E_j = B_{ji} C_i . \tag{3.12}$$

The variation of the function ϕ over the entire region D (interior points) can be found, if it is of any interest, by a direct substitution of the boundary solution into Eq. 2.7 which for interior points p takes the form

$$2\pi \, \phi(p) = \int_S \left[\phi(Q) \, \partial/\partial n_Q(\ln r_{pQ}) - \ln r_{pQ} \, \partial/\partial n_Q(ly - mx) \right] dS(Q) , \tag{3.13}$$

and after discretization can be written in a matrix form as

$$2\pi \{\phi(p)\} = [A] \{\phi(Q)\} - \{E\}, \tag{3.14}$$

where the $\{\phi(p)\}$ and $\{\phi(Q)\}$ are used to indicate values at interior cells and boundary elements, respectively, and the elements of the matrices $[A]$ and $\{E\}$ are given by Eqs. 3.3_a and 3.3_b, respectively, for points j in the interior of the domain D.

At this point, one should notice that the direct formulation of the torsion problem remains the same for simply- or multiply-connected regions D.

ii) Indirect formulation

Alternatively, one can use a single-layer potential for the solution of the problem under consideration. To this end, the boundary condition 1.3 for the warping function ϕ is substituted into Eq. 2.10, in the place of the arbitrary function f, to yield

$$\int_S h(Q)\, \partial/\partial n_P(\ln r_{PQ})\, dS(Q) \;-\; \beta(P)\; h(P) \;=\; [ly - mx]_P . \tag{3.15}$$

Eq. 3.15 is a singular Fredholm equation of the second kind for the unknown boundary density function h(Q). A numerical solution to Eq. 3.15 can be obtained along the same lines described previously for the direct formulation. Thus, the integral on the left-hand side can be converted into a summation of the form

$$\sum_{i=1}^{M} (A_{ji} - \delta_{ji}\, \beta_i)\, h_i \;=\; C_j , \qquad j = 1,2,...,M \tag{3.16}$$

where the coefficients A_{ji} and C_j can be obtained from Eq. 3.3_a and 3.3_c, through the same numerical procedures described in the previous section. Then, Eq. 3.16 can also be viewed as forming a system of M linear algebraic equations for the M unknown values of the density function at prescribed nodal points (centers of elements) on the boundary S. Following the solution of this system of equations the warping function ϕ can be obtained everywhere on the region D by means of Eq. 2.9 through the usual the discretization procedure.

3.2. Torsion Formulation (b)

The Dirichlet's problem for Laplace´s equation, as it is formulated by Eqs. 1.5 and 1.6, admits any one of the general solutions described in section 2. The application of the direct formulation (Green´s theorem) to the solution of this problem becomes a mere recapitulation of the procedure already described in detail in section 3.1 for the torsion formulation (a). Thus, two alternative solutions based on a single- or a double-layer potential are presented in this section.

If a single-layer potential is used, the conjugate of the warping function ψ appearing in Eq. 1.5 can be substituted in the place of the arbitrary function f in Eq. 2.9. Then, on the basis of the observation that Eq. 2.9 holds the same for boundary points P as well, and in view of the boundary condition 1.6, it becomes

$$\psi(P) = \int_S h(Q)\, \ln r_{PQ}\; dS(Q) = (x_P{}^2 + y_P{}^2)/2 \tag{3.17}$$

Eq. 3.17 is a singular Fredholm equation of the first kind for the unknown boundary density function h(Q). Following the same boundary discretization techique as before, the integral on the right-hand side can be transformed into the summation form

$$\sum_{i=1}^{M} B_{ji}\, h_i \;=\; F_j , \qquad j = 1,..., M \tag{3.18}$$

where B_{ji} is given by Eq. 3.3_b and

$$F_j = \left[(x^2 + y^2)/2\right]_{P=P_j} . \tag{3.19}$$

Apparently, Eq. 3.18 can be cast in a matrix form similar to that of Eq. 3.11. However, the reader should be aware of the fact that using a single-layer potential for this Dirichlet problem does involve some mathematical difficulties related to the existence of a solution to the Fredholm equation of the first kind (39).

As an alternative one might choose to use a double-layer potential, in which case the solution of Eq. 1.5 can be expressed, in view of Eq. 2.11, for interior points p as

$$\psi(p) = \int_S h(Q)\, \partial/\partial n_Q(\ln r_{pQ})\; dS(Q) . \tag{3.20}$$

As point p moves to the boundary the double-layer potential suffers the jump indicated by formula 2.12. Thus, in view also of the boundary condition 1.6, Eq. 3.20 becomes

$$\psi(P) = \int_S h(Q)\, \partial/\partial n_Q(\ln r_{PQ})\; dS(Q) \; - \beta(P)\, h(P) \; = \; (x_P^2 + y_P^2)/2 \tag{3.21}$$

which is a singular Fredholm equation of the second kind for the unknown boundary density function h(P). A numerical solution of this BIE follows exactly the same basic steps described in the previous sections. Thus, in summation form it can be written as

$$\sum_{i=1}^{M} (A_{ji} - \delta_{ji}\, \beta_i)\, h_i \; = F_j \, , \qquad j = 1,\ldots,M \tag{3.22}$$

where A_{ij} and β_i are given, respectively, by Eqs. 3.3_a and 3.3_d, and F_j by Eq. 3.19, and in matrix form as

$$[H]\,\{h\} = \{F\} . \tag{3.23}$$

The solution in terms of the conjugate of the warping function ψ for interior or boundary points can be obtained through the usual discretization of Eqs. 3.20 or 3.21, respectively.

3.3. Torsion Formulation (c)

Torsion formulation (c), as it is expressed by Eqs. 1.8 and 1.9, is described by an inhomogeneous governing equation (Poisson's equation) for the stress function F and thus appears essentially different than the two previous formulations. This is so because the existence of the inhomogeneous part of the governing equation creates the need for a particular solution and thus it does not allow for a solution in terms of a single- or a double-layer potential. However, a straightforward application of the direct formulation (Green's theorem) provides an easy solution to this problem. Thus, after substituting in Eq. 2.7 the arbitrary function f(x,y) by the stress function F(x,y) and taking into consideration the boundary condition 1.9 one has, for interior points p,

$$2\pi\, F(p) = -2 \iint_D \ln r_{pq}\; dD(q) \; - \int_S \ln r_{pQ}\; \partial/\partial n_Q[F(Q)]\; dS(Q) . \tag{3.24}$$

The complimentary unknown boundary condition $\partial/\partial n_Q[F(Q)]$ which appears in Eq. 3.24 can be computed from the same equation by letting the point p approach the boundary S, in which case one has for $k_0 = 0$

$$2 \iint_D \ln r_{Pq} \, dD(q) = - \int_S \ln r_{PQ} \, \partial/\partial n_Q[F(Q)] \, dS(Q) , \tag{3.25}$$

i.e., a singular Fredholm equation of the first kind. As it has been shown in the previous sections, the integrals appearing in Eq. 3.25 can be replaced by summations, and the resulting system of linear algebraic equations can be solved for the values of the unknown function at specific boundary nodal points. It should be noted, however, that the required surface integration over the entire cross section D of the cylindrical bar, which is a result of the inhomogenity of the governing equation, could cause some inconvinience in the related computations. However, it has been shown by Mendelson (37,38) that this formulation is very well suited for the solution of the elastoplastic problem. Certain improvements on the above boundary integral formulation of the elastic torsion problem have been reported, e.g., Fukui (16) and Melnikov (36). Since the solution to the problem posed by formulation (c), as it is expressed by Eqs. 3.24 and 3.25, does not involve artificial density functions but distributions of the "physically more meaningful" stress function F and its derivative, it is usually called a "semidirect" boundary integral formulation.

4. Higher Order Elements

In the previous section constant boundary elements are used, in the sense that all boundary quantities are assumed to be uniform over each element and the geometry of the boundary is approximated by straight line segments. The simplicity of these elements in formulating the described numerical technique and the subsequent minimization of programming effort is probably their most important attribute. It is also proven by various investigations, some of which are mentioned in the Numerical Results and Discussion section,

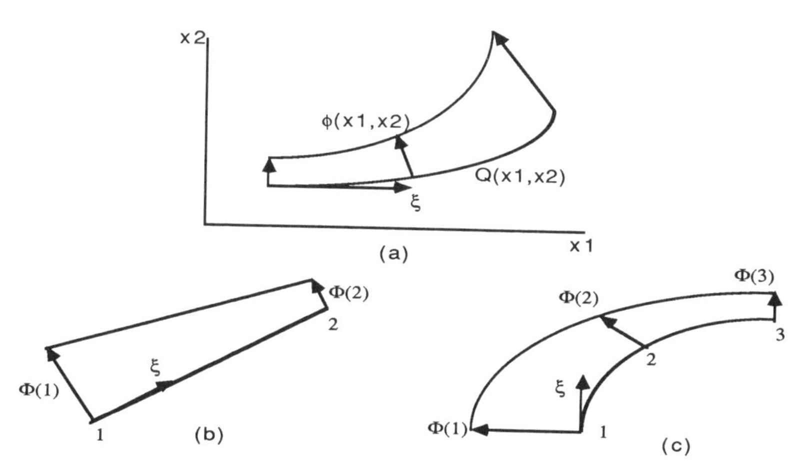

Figure 5. Higher order elements: (a) nomenclature, (b) linear element, (c) quadratic element.

that constant elements can provide "accurate" results for a wide variety of problems. However, in an effort to better describe the geometry of the boundary and the variation of the functional behavior within each element, higher order (linear, quadratic, etc.) variations of these quantities can be considered within each boundary element.

Consider the case of a line boundary element as the one shown in Fig. 5a. The Cartesian co-ordinates x_i (i=1,2) of an arbitrary point on the element can be expressed, in terms of the nodal co-ordinates X_i and a set of chosen shape functions of the intrinsic co-ordinate ξ, as

$$x_i = \sum_{Q=1}^{E} N_Q(\xi)\, X_i(Q)\,, \tag{4.1}$$

with E being the number of nodal values necessary to describe the element according to the variation chosen. For a linear variation, two nodal values X_i are required, see Fig. 5b, and the corresponding shape functions are

$$N_1 = (1-\xi)/2\,, \qquad N_2 = (1+\xi)/2 \tag{4.2}$$

Similarly, three nodal values are necessary for a quadratic element, as shown in Fig. 5c, and the related shape functions are given by

$$N_1 = -\,\xi\,(1-\xi)/2, \qquad N_2 = \xi\,(1+\xi)/2, \qquad N_3 = (1-\xi)\,(1+\xi). \tag{4.3}$$

Higher order elements can also be developed in the same fashion, e.g., Brebbia, Telles and Wrobel (11), and Lachat and Watson (31).

The variation within a boundary element of the several functions appearing in the previous formulations can similarly be expressed in terms of the values of those functions at specified nodal points. As an example, the warping function ϕ appearing in Eq.3.1 can be written in the form

$$\phi_i = \sum_{Q=1}^{E} M_Q(\xi)\, \Phi_i\,(Q)\,, \tag{4.4}$$

where $\Phi_i\,(Q)$ are the values of the warping function at an E number of nodal points Q related to the chosen variation of ϕ within each boundary element. The shape functions M_Q of Eq. 4.4 are not, in general, the same shape functions as those used in Eq. 4.1. However, it has become common practice to select the same order of variation for the geometry of the boundary and the various related functions, since it simplifies the assemblage and manipulation of the resulting matrix equations. Thus, the so-called isoparametric element emerges. Mixed elements, using different order shape functions for different quantities, have also been used, e.g. Martin et al (34).

After these definitions, the solution of the problem proceeds in the same basic lines as before. For the direct formulation (a) of the torsion problem, for example, Eq. 4.4 is substituted into Eq. 3.1 to yield

$$\sum_{m=1}^{M}\Big\{\sum_{Q=1}^{E_m}\Big[\int_{S_m} \partial/\partial n_R[\ln r_{PR}(\xi)]\, N^m{}_Q(\xi)\, J(\xi)\, dS[R(\xi)]\ \ \Phi(Q)\Big]\Big\} - \beta(P)\,\Phi(P)$$

$$= \sum_{m=1}^{M}\Big\{\int_{S_m} \ln r_{PR}(\underline{x})\ \ [ly - mx]_R\, dS[R(\underline{x})]\Big\}, \tag{4.5}$$

where $J(\xi)$ is the Jacobian transforming the cartesian co-ordinate system $\underline{x}$ into the intrinsic co-ordinate system ξ used for the element, S_m is the area of the boundary element m, E_m is the number of nodal points within the element m, and $N^m{}_Q$ is the shape function corresponding to the nodal point Q of the element m. It should be noticed at this point that the locations P and Q, used previously to represent the centers of the uniform elements, now correspond to nodal points, R is a dammy variable point on the element m, and M is still the number of boundary elements. Writing Eq. 4.5 for the complete set of nodal points P on the boundary S results to a system of linear algebraic equations for the unknown nodal values of the warping function Φ. In matrix form this can be expressed as

$$[H']\,\{\Phi\} = \{E'\}. \tag{4.6}$$

The vector $\{E'\}$ on the right hand-side of Eq. 4.6 can be computed in exactly the same way as for Eq. 3.11, i.e., by using Eqs. $3.3_{b,c}$. However, a more accurate approach, consistent with the higher order element formulation, would be to employ the same shape functions utilized for the left hand-side of Eq. 4.5, in which case the various elements of the vector $\{E'\}$ are

$$E'_P = \sum_{m=1}^{M}\Big\{\int_{S_m} \ln r_{PR}(\xi)\, N^m{}_Q(\xi)\, J(\xi)\, dS[R(\xi)]\Big\}\ \ \partial/\partial n_Q\ [\Phi(Q)], \tag{4.7}$$

where

$$\partial/\partial n_Q\ \ [\Phi(Q)] = [ly - mx]_Q\quad, \tag{4.8}$$

is the known boundary condition accessed at nodal points Q. The elements of the matrix $[H']$ on the left hand-side of Eq. 3.31 represent the influence of each nodal point Q on the nodal point P. Depending upon the location of the nodal point Q, the H'_{PQ} element of the matrix $[H']$ can be computed as follows:

i) If the nodal point Q belongs only to the boundary element m, i.e. it is an internal point of the element m, see Fig. 6_a,

$$H'_{PQ} = \int_{S_m} \partial/\partial n_R[\ln r_{PR}(\xi)]\, N^m{}_Q(\xi)\, J(\xi)\, dS[R(\xi)]\ \ -\ \delta_{PQ}\,\beta_P\,. \tag{4.9}$$

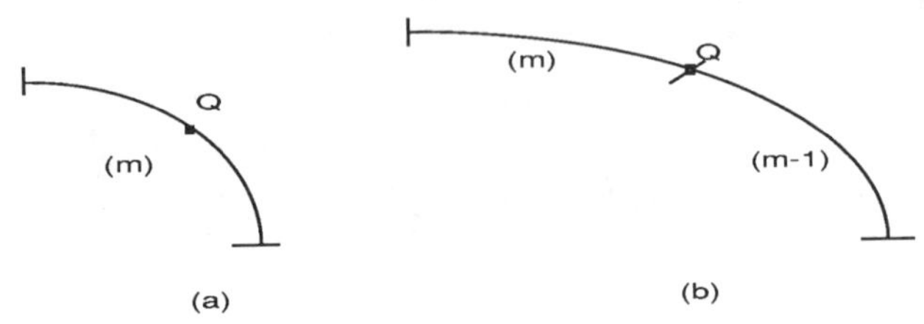

Figure 6. (a) Interior node Q, (b) Common node Q to two adjacent elements.

ii) If the nodal point Q is common to elements m-1 and m, see Fig. 6b,

$$H'_{PQ} = \sum_{i=m-1}^{m} \left\{ \int_{S_i} \partial/\partial n_R[\ln r_{PR}(\xi)]\, N^i_Q(\xi)\, J(\xi)\, dS[R(\xi)] \right\} - \delta_{PQ}\, \beta_P . \qquad (4.10)$$

The integrations indicated in Eqs. 4.7-10 are ususally performed numerically with the help of Gaussian quadrature formulae. The diagonal terms H'_{ii}, where a logarithmic singularity occurs, can also be computed numerically using special quadrature rules for integrals exibiting such a singularity. For a thorough discussion on the Gaussian quadrature scheme of integration the interested reader is refered to the work of Stroud and Secrest (50). However, for the simple case of linear shape functions some analytical formulae have also been reported and found useful particularly for the singular case (11).

Higher order elements, as those briefly described above, appear computationally more intense than the simple uniform elements of the previous sections. This is true, of course, with regard to the numerical integration part of the BEM programme. However, higher order elements have been proven to produce more accurate results than uniform elements for a smaller number of nodal points. Therefore, the resulting system of equations is easier to assemble and solve, particularly for problems requiring a very fine discretization. Detailed discussions on several aspects pertaining to programming for the BEM can be found, for example, in the work of Alarcón and his co-workers (1-3), Brebbia (10), Brebbia, Telles and Wrobel (11), and in Chapter 11 of this book.

5. Numerical Results and Discussion

A great number of torsion problems have been solved by the numerical techniques developed in the previous sections of this work. Some of the most representative examples can be found in the work of Jaswon and Ponter (19) who used the direct approach in conjunction with formulation (a) and uniform elements. A few of the results reported in this reference are reproduced in what follows.

It was shown in the previous sections that the various quantities, i.e., the warping function ϕ, the conjugate of the warping function ψ, and the stress function F, appearing in a number of formulations of the torsion problem of cylindrical bars, can be evaluated numerically on the boundary of the cross section under consideration. This is feasible by means of a discretized boundary integral equation and thus the method is applicable to arbitrary boundary geometries. Once the unknown boundary quantities have been computed, the complete set of boundary conditions corresponding to the problem under consideration can be used to calculate various other quantities of interest throughout the cross section, e.g., the torsional rigidity which in terms of the warping function can be expressed as (48)

$$G = \iint_D r^2\, dx\, dy + \int_S \phi\, \partial/\partial n(\phi)\, ds \ , \tag{5.1}$$

or the shear stress which becomes maximum at the boundary and is given by

$$\sigma = \left| \partial/\partial s(\phi) - \partial/\partial n(r^2/2) \right| . \tag{5.2}$$

In the case of Eq. (5.2) one can calculate the unknown tangential derivative $\partial/\partial s(\phi)$ either by means of a numerical differentiation or, more accurately, by differentiating under the integral sign in Eq. 3.1 and then computing directly the stresses following the numerical integration procedure developed in section 3. Furthermore, assuming that the complete set of boundary conditions is known, an appropriate straightforward application of the general formulae 2.6, 2.7, 2.9 or 2.11, depending of course on the formulation used for the solution of the problem,

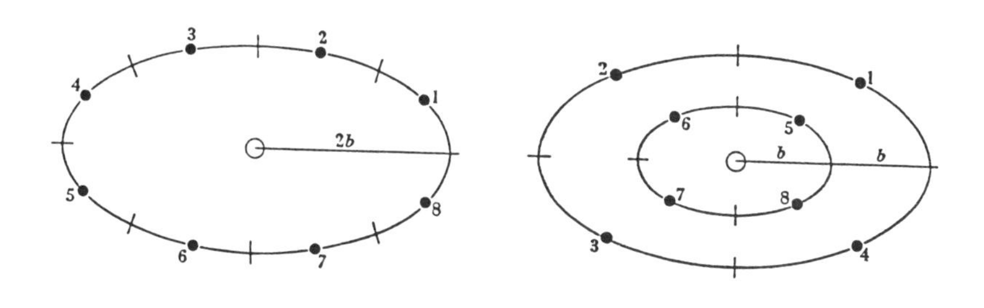

Figure 7. Initial discretization for an ellipse and a hollow ellipse: σ_{max} occurs at the ends of the minor axis of the outer ellipse.

Ellipse

n	G/b^4	σ_{max}/b
8	4·985	2·134
16	5·087	1·645
32	5·033	1·603
anal.	5·026	1·600

Hollow ellipse

n	G/b^4	σ_{max}/b
8	2·595	1·553
16	4·160	1·557
32	4·717	1·585
64	4·715	1·602
anal.	4·712	1·600

Table 1. Computed data for the ellipse and the hollow ellipse of Figure 7: n = number of nodal points, b = minor semi-axis of outer ellipse [after Jaswon and Ponter (19)].

can provide the distribution of either the warping function, or the conjugate of the warping function, or the stress function throughout the cross section D. This information, in turn, opens the way for calculations of displacements, see Eq. 1.1, and stresses over the entire domain of interest.

Starting with the relatively simple case of curved boundaries the results obtained in Ref. (19) for the torsional rigidity G and maximum shear stress σ_{max} of an ellipse and a hollow ellipse, as they are shown in Fig. 7, are reproduced in Table 1. In either case the summetry of the boundary geometry allows for a substantial reduction of the number of independent equations that must be solved. It is also apparent, in view of the results shown, that even "coarse" discretization schemes and uniform elements produce an acceptable error which decreases rapidly as the numerical solution converges to the "exact" analytical solution for finer discretization schemes.

In some problems an additional difficulty is encountered due to the discontinuity of the boundary geometry at one or more "corner" points. At a first look the problem can be circumvented by avoiding to locate any nodal points at a point of boundary discontinuity. This is

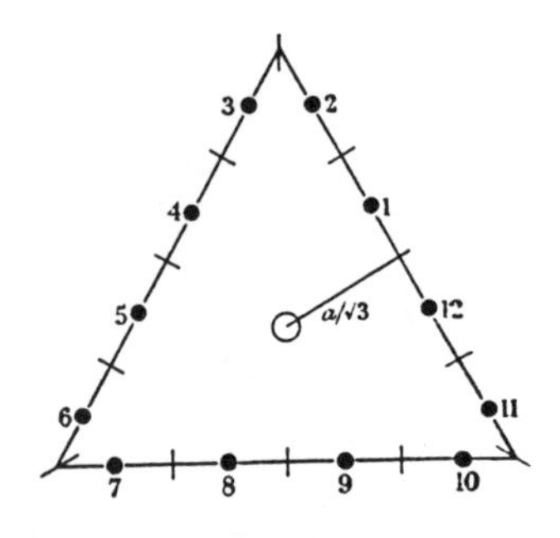

Figure 8. Initial discretization of an equilateral triangle: σ_{max} occurs at the center of each side.

n	G/a^4	$\sigma_{max.}/a$
12	0·3377	0·867
24	0·3461	0·866
48	0·3466	0·866
anal.	0·3464	0·866

Table 2. Computed data for the triangle of Figure 8: n = number of nodal points, α = length of half-side. [after Jaswon and Ponter (19)].

the case, for example, in the discretizations shown in Figures 8 and 9 for an equilateral triangle and a notched circle, respectively. In either case the results given in the corresponding Tables 2 and 3 for a number of discretization schemes show a rapid convergence to the analytically obtained solution. Eventhough an appropriate use of Eq. 2.8 seems to allow the location of nodal points at points of boundary discontinuity, such an approach often results in inaccurate results and/or severe singularities (16). Moreover, such a use of Eq. 2.8 would disrupt the automatic generation of the required system of equations and would substantially increase the user's involvement in the numerical solution of the problem. However, a number of alternative methods of dealing with the problem in its general form have been reported, e.g., Refs. (6,7,10,16). In particular, some applications of such specialized techniques to the torsion problem can be found in the works of Athanasiadis (4), Athanasiadis and Mitakidis (5), Kermanidis (27), and Windisch (54). Furthermore, Patterson and Sheikh (42) have reported on the use of discontinuous higher order boundary elements at points exibiting geometrical or boundary condition dicontinuities.

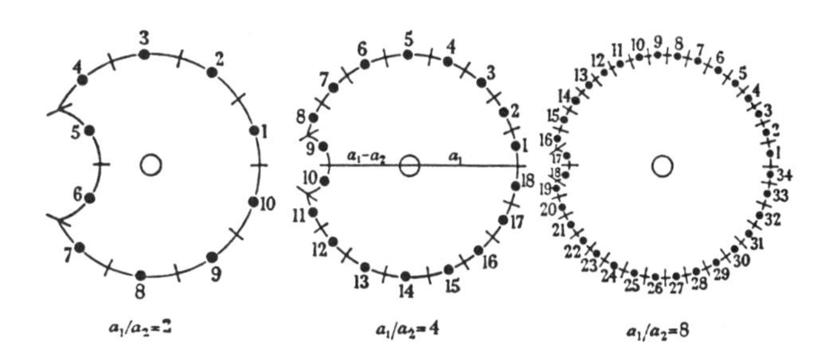

Figure 9. Notched circles: σ_{max} occurs at the center of the notch.

$a_1/a_2 = 2$

n_1	n_2	G/a_1^4	$\sigma_{max.}/a_1$
8	2	1·125	1·373
16	4	1·092	1·474
32	8	1·078	1·488
48	12	1·075	1·495
anal.		1·072	1·500

$a_1/a_2 = 4$

n_1	n_2	G/a_1^4	$\sigma_{max.}/a_1$
16	2	1·424	1·646
32	4	1·418	1·728
48	6	1·416	1·744
anal.		1·413	1·750

$a_1/a_2 = 8$

n_1	n_2	G/a_1^4	$\sigma_{max.}/a_1$
32	2	1·529	1·780
64	4	1·528	1·865
anal.		1·527	1·875

$a_1/a_2 = 16$

n_1	n_2	G/a_1^4	$\sigma_{max.}/a_1$
64	2	1·560	1·910
anal.		1·559	1·938

Table 3. Computed data for the notched circles of Figure 9: α_1 = radius of major circle, α_2=radius of minor circle, n_1 = number of nodal points on major arc, n_2 = number of nodal points on minor arc [after Jaswon and Ponter (19)].

Comparison studies on the accuracy and efficiency of the various methods reported in this work have been conducted by Athanasiadis and Mitakidis (5), and Kobayashi and Nishimura (29). In Ref. (29) the direct solution based on Green's formula and the indirect single-layer potential solution of Neumann's problem [formulation (a)] are compared and the use of the single-layer potential method is recommended on the basis of accuracy alone. The indirect method is also recommended in Ref. (5) over the direct method on the basis of the required computational effort. However, one should point out that the use of the direct boundary integral formulation presents the advantage that the computed boundary data can be used directly in the calculation of stresses or displacements, see for example Eqs. 5.1 and 5.2. This is not the case, of course, with the indirect methods where only boundary distributions of potential functions are computed and additional calulations are necessary in order to arrive at physically meaningful quantities.

The single-layer potential method has been applied by a number of authors, e.g., Fukui (16) and Kishida (28) in connection with formulation (a), and Melnikov (36) and Lo and Niedenfuhr (32) for formulation (c). The double-layer potential method has also been applied by Wong and Aguirre-Ramirez (55) and Sauer and Melhorn (47) in conjunction with formulation (b). All three formulations along with their numerical treatment are briefly discussed by Mendelson (37). A variety of other boundary integral formulations applied to the torsion problem of prismatic bars has also been reported, e.g., Kermanidis (26,27), Patterson and Sheikh (41), Rieder (44), Symm (51), and Windisch (54).

The more general problem of axially symmetrical bars of variable cross-section can be also treated effectively by the BEM. The development of the BEM solution of this problem falls out of the scope of this introductory article, but the interested reader can find detailed discussions in the works of, e.g., Battenbo and Baines (8), Kermanidis (24,25), Mayr and Neureiter (35), Miyamoto (40), and Rizzo, Gupta and Wu (46).

Acknowledgement

The author wishes to express his appreciation to the editor of this book Professor D. E. Beskos for his valuable comments during the writting of this article.

References

1. Alarcón, E., Martin, A. and Paris, F., "Boundary Elements in Potential and Elasticity Theory," Computers and Structures, Vol. 10, 1979, pp. 351-362.

2. Alarcón, E. and Reverter, A., "p-Adaptive Boundary Elements," International Journal for Numerical Methods in Engineering, Vol. 23, 1986, pp. 801-829.

3. Alarcón, E., Reverter, A. and Molina, J., "Hierarchical Boundary Elements," Computers and Structures, Vol. 20, 1985, pp. 151-156.

4. Athanasiadis, G., "Torsion Prismatischer Stäbe nach der Singularitätenmethode," Ingenieur - Archiv, Vol. 49, 1980, pp.89-96.

5. Athanasiadis, G. and Mitakidis, A., "Die Untersuchung Einiger Integralgleichungen des St.-Venantschen Torsionsproblems," Ingenieur - Archiv, Vol. 53, 1983, pp.303-316.

6. Banerjee, P. K. and Butterfield, R., "Boundary Element Methods in Engineering Science," McGraw-Hill, London, 1981.

7. Barone, M. R. and Robinson, A. R., "Determination of Elastic Stresses at Notches and Corners by Integral Equations," International Journal of Solids and Structures, Vol. 8, 1972, pp. 1319-1338.

8. Battenbo, H. and Baines, B. H., "A Boundary Integral Solution to Torsion of Cylinders and Solids of Revolution," International Journal for Numerical Methods in Engineering, Vol. 9, 1975, pp. 461-476.

9. Beskos, D. E., "Potential Theory," ch. 2, in: D. E. Beskos, Ed., "Boundary Element Methods in Mechanics," North-Holland, Amsterdam, 1987.

10. Brebbia, C. A., "The Boundary Element Method for Engineers," Pentech Press,

London, 1978.

11. Brebbia, C. A., Telles, J. C. F. and Wrobel, L. C., "Boundary Element Techniques," Springer-Verlag, Berlin, 1984.

12. Caulk, D. A., "Analysis of Elastic Torsion in a Bar with Circular Holes by a Special Boundary Integral Method," Journal of Applied Mechanics, Vol. 50, 1983, pp. 101-108.

13. Christiansen, S., "A Review of Some Integral Equations for Solving the Saint-Venant Torsion Problem," Journal of Elasticity, Vol. 8, No. 1, 1978, pp.1-20.

14. Danson, D. J. and Kuich, G., "Using BEASY to Solve Torsion Problems," pp. 821-834, in: C. A. Brebbia, T. Futagami and M Tanaka, Eds., "Boundary Elements," Springer-Verlag, Berlin, 1983.

15. Danson, D. J. and Brebbia, C. A., "Further Engineering Applications of BEASY," pp. 857-880, in: C. A. Brebbia, T. Futagami and M Tanaka, Eds., "Boundary Elements," Springer-Verlag, Berlin, 1983.

16. Fukui, T., "On Corner Solutions by the Indirect B.I.E.M.," pp. 929-939, in: C. A. Brebbia, T. Futagami and M Tanaka, Eds., "Boundary Elements," Springer-Verlag, Berlin, 1983.

17. Higgins, T. J., "A Comprehensive Review of St.-Venant's Torsion Problem," American Journal of Physics, Vol. 10, 1942, pp. 248-259.

18. Higgins, T. J., "The Approximate Mathematical Methods of Applied Physics as Exemplified by Application to St.-Venant's Torsion Problem," Journal of Applied Physics, Vol. 14, 1943, pp. 469-480.

19. Jaswon, M. A. and Ponter, A. R., "An Integral Equation Solution of the Torsion Problem," Proceedings, Royal Society of London, Series A, Vol. 273, 1963, pp. 237-246.

20. Jaswon, M. A., "Integral Equation Methods in Potential Theory: I," Proceedings, Royal Society of London, Series A, Vol. 275, 1963, pp. 23-32.

21. Jaswon, M. A. and Symm, G. T., "Integral Equation Methods in Potential Theory and Elastostatics," Academic Press, London, 1977.

22. Katsikadelis, J. T. and Sapountzakis, E. J., "Torsion of Composite Bars by Boundary Element Method," Journal of Engineering Mechanics, Proceedings ASCE, Vol. 111, EM9, 1985, pp.1197-1210.

23. Kellogg, O. D., "Foundations of Potential Theory," Springer-Verlag, 1967 (reprint from the first edition: Ungar, 1929).

24. Kermanidis, T., "Eine Integralgleichungsmethode zur Lösung des Torsionsproblems des Umdrehungskörpers," Acta Mechanica, Vol. 16, 1973, pp.175-181.

25. Kermanidis, T., "A Numerical Solution for Axially Symmetrical Elasticity Problems," International Journal of Solids and Structures, Vol. 11, 1975, pp. 493-500.

26. Kermanidis, T., "Kupradze's Functional Equation for the Torsion Problem of Prismatic Bars - Part 1," Computer Methods in Applied Mechanics and Engineering, Vol. 7, 1976, pp. 39-46.

27. Kermanidis, T., "Kupradze's Functional Equation for the Torsion Problem of Prismatic Bars - Part 2," Computer Methods in Applied Mechanics and Engineering, Vol. 7, 1976, pp. 249-259.

28. Kishida, M., "On Fictitious-Boundary Method in Boundary-Integral Methods," Theoretical and Applied Mechanics, Vol. 28, University of Tokyo Press, Tokyo, 1980, pp. 139-151.

29. Kobayashi, S. and Nishimura, N., "Some Considerations on the Improvement of Integral Equation Method," Transactions of JSCE, Vol. 11, 1979, pp. 98-99.

30. Kupradze, V. D., "Potential Methods in the Theory of Elasticity," Israel Programme for Scientific Translations, 1965.

31. Lachat, J. C. and Watson, J. O., "Effective Numerical Treatment of Boundary Integral Equations: A Formulation for Three-Dimensional Elastostatics," International Journal for Numerical Methods in Engineering, Vol. 10, 1976, pp. 991-1005.

32. Lo, C. C. and Niedenfuhr, F. W., "Singular Integral Equation Solution for Torsion," Journal of the Engineering Mechanics Division, Proceedings ASCE, Vol. 96, EM4, 1970, pp. 535-542.

33. MacMillan, W. D., "The Theory of the Potential," Dover, New York, 1958.

34. Martin, A., Rodriguez, I. and Alarcón, E., "Mixed Elements in the Boundary Theory," pp. 34-42, in: C. A. Brebbia, Ed., New Developments in Boundary Element Methods, Computational Mechanics Centre Publication, Southampton, 1980.

35. Mayr, M. and Neureiter, W., "Ein numerisches Verfahren zur Lösung des Axialsymmetrischen Torsionsproblems," Ingenieur - Archiv, Vol. 46, 1977, pp.137-142.

36. Melnikov, Y. A., "Some Applications of the Green's Function Method in Mechanics," International Journal of Solids and Structures, Vol. 13, 1977, pp.1045-1058.

37. Mendelson, A., "Boundary-Integral Methods in Elasticity and Plasticity," NASA TN D-7418 (1973).

38. Mendelson, A., "Solution of Elastoplastic Torsion Problem by Boundary Integral Method," NASA TN D-7872 (1975).

39. Mikhlin, S. G., "Integral Equations and their Applications," Second Edition, Pergamon Press, Oxford, 1964.

40. Miyamoto, Y., "Analysis of Axisymmetrical Torsion Problems by the Boundary Element Method," pp. 355-365, in: C. A. Brebbia, T. Futagami and M Tanaka, Eds., "Boundary Elements," Springer-Verlag, Berlin, 1983.

41. Patterson, C. and Sheikh, M. A., "A Modified Trefftz Method for Three Dimensional Elasticity," pp. 427-437, in: C. A. Brebbia, T. Futagami and M Tanaka, Eds., "Boundary Elements," Springer-Verlag, Berlin, 1983.

42. Patterson, C. and Sheikh, M. A., "Interelement Continuity in the Boundary Element Method," Chapter 6, in: C. A. Brebbia, Topics in Boundary Element Research, Volume 1, Springer-Verlag, Berlin, 1984.

43. Ponter, A. R. S., "An Integral Equation Solution of the Inhomogeneous Torsion Problem," SIAM Journal on Applied Mathematics, Vol. 14, 1966, pp. 819-830.

44. Rieder, G., "Eine Variante zur Integralgleichung von Windisch für das Torsionsproblem," Zeitschrift für Angewandte Mathematik und Mechanik, Vol. 49, 1969, pp. 351-358.

45. Rizzo, F. J., "An Integral Equation Approach to Boundary Value Problems of Classical Elastostatics," Quarterly of Applied Mathematics, Vol. 25, 1967, pp.83-95.

46. Rizzo, F. J., Gupta, A. K. and Wu, Y., "A Boundary Integral Equation Method for Torsion of Variable Diameter Circular Shafts and Related Problems," pp. 373-380, in: R. P. Shaw et. al., Eds., "Innovative Numerical Analysis for the Engineering Sciences," University of Virginia, Charlottesville, 1980.

47. Sauer, E. and Mehlhorn, G., "Application of the Boundary Integral Equation Method to Shear and Torsion Problems in Elastic Prismatic Members," pp. 656-665, in: R. P. Shaw et. al., Eds., "Innovative Numerical Analysis for the Engineering Sciences," University of Virginia, Charlottesville, 1980.

48. Sokolnikoff, I. S., "Mathematical Theory of Elasticity," , Second Edition, McGraw-Hill, New York, 1956.

49. Sternberg, W. J. and Smith, T. L., "Theory of Potential and Spherical Harmonics," Toronto Press, 1946.

50. Stroud, A. H. and Secrest, D., "Gaussian Quadrature Formulas," Prentice-Hall, Englewood Cliffs, NJ, 1966.

51. Symm, G. T., "An Integral Equation Method in Conformal Mapping," Numerische Mathematik, Vol. 9, 1966, pp. 250-258.

52. Symm, G. T.,"Integral Equation Methods in Potential Theory: II," Proceedings, Royal Society of London, Series A, Vol. 275, 1963, pp. 33-46.

53. Timoshenko, S. and Goodier, J. N., "Theory of Elasticity," Second Edition, McGraw-Hill, New York, 1951.

54. Windisch, E., "Eine Numerische Methode zur Lösung des Torsionsproblems," Acta Mechanica, Vol. 4, 1966, pp. 191-199.

55. Wong, J. P. and Aguirre-Ramirez, G., "A Finite Element-Integral Equation Solution to Torsion Problem," Computers and Structures, Vol. 9, 1978, pp. 53-55.

Static Analysis of Beams, Plates and Shells

Morris Stern*

Introduction

In this chapter we consider boundary element formulations for the solution of equilibrium problems involving flexure of beams, plates and shells. Although boundary element methods do not generally produce new and useful computational algorithms in one-dimensional problems, the small deflection analysis of linear elastic beams provides a simple and familiar basis for introducing the ideas which underpin the method as applied to plates and shallow shells. It is therefore not surprising that the treatment of beams in the boundary element literature is basically of a tutorial nature; two excellent examples are the early article by Butterfield (5) and the treatment of beams in the book (2). The section in this chapter on linear elastic beams is in the same vein.

The boundary element literature on plate bending is considerably richer. Probably the first significant modern entry is due to Massonnet in 1965 (17) where he observed that for computational purposes an integral equation formulation suitable for clamped plates is suggested by analogy with the problem of determining an Airy stress function for a particular elastrostatic state. In 1968 Jaswon and Maiti (11) proposed a boundary integral treatment for uniformly loaded clamped or simply supported plates in terms of two source distribution densities generating harmonic potentials which are then related to the plate displacement. Other early examples of such indirect formulations are found in Maiti and Chakrabarty (16) and Hansen (8). These methods usually work quite well for the specific class of problems for which they are intended, but attempts to use them more generally often lead to very poor numerical behavior. Other forms of indirect methods have also been proposed, for example Altiero and Sikarskie (1) suggest embedding the original problem in a larger plate for which the Green's function is known, and determining a loading outside or on the original plate boundary to yield the original boundary conditions. Wu and Altiero (29) distribute discrete concentrated loads outside the boundary and are able to extend the method to include anisotropic material behavior.

A direct approach in which the primary variables of the formulation are the boundary variables of deflection, normal slope, bending moment and shear would appear more suited to a general setting. The earliest direct formulation (for smooth boundaries) is due to Forbes and Robinson (7) in 1969. The introduction of corners in the plate boundary was done independently by Bézine (3) and Stern (21), apparently without knowledge of the prior work by Forbes. Tottenham (26) also proposed several

* Professor of Aerospace Enginering and Engineering Mechanics, The University of Texas, Austin, TX 78712

direct and indirect formulations. In addition to the already quoted primary sources, treatments of the direct formulation of boundary element equations for bending of thin plates are found in the excellent article on elastostatics by Hartmann (9) and in the books by Brebbia, Telles and Wrobel (4) and Banerjee and Butterfield (2).

More recently, Stern and Lin (23) have reconsidered the question of what constitutes suitable fundamental solutions in the derivation of boundary integral representations for thin plates, and as a result have proposed a new treatment which reduces the severity of the singularity in the representations. It is this development we will follow in the exposition later.

Extensions of the boundary element method to anisotropic material behavior can be found in (12) and the previously cited (29). A novel approach to some orthotropic plate problems (transforming the domain coordinates so that with respect to the transformed coordinates the material behavior appears isotropic) has been proposed by Irschik (10). In a different direction Morjaria and Mukherjee (18) extended the boundary element analysis to include inelastic effects in plate bending, and a very recent paper by Moshaiov and Vorus (19) treats this same problem from a somewhat different point of view. Considerations related to singularities which might occur at reentrant corners or at the tips of through-cracks (stress intensity factors) are dealt with in (15) and (22). There are also extensions to Reissner's plate theory (27) and von-Karman plate theory (13), (30). Finally, boundary element formulations for plates on elastic foundations have been put forth by several authors; the most recent papers containing some numerical results are (6) and (14).

The application of boundary element methods to thin shells has not met with much success, in general. However, for the case of shallow shells some progress has been made. The earliest formulation with substantial generality is again due to Forbes and Robinson (7) and it is their aproach which will be followed in the final section of this chapter. It is a direct method which does not lead to a pure boundary integral representations but requires that unknown variables be determined over the entire shell domain. A semi-direct method in which the membrane resultants are related to an Airy stress function has been exploited by Newton and Tottenham (20); the method is also described in (26), and applies only to special cases (which includes cylindrical shells and spherical caps). More recently Tosaka and Miyake (25) have proposed a similar formulation which appears more general.

Linear Elastic Beams

Consider the small deflection analysis of a linear elastic beam subject to distributed bending loads and supported at its ends $x = 0$ and $x = L$ as in Figure 1. We denote the deflection $w(x)$ and write $\theta(x) = w'(x)$ ($\equiv dw/dx$) for the slope. Then the constitutive equation relating the bending moment M to the (approximate) curvature is

$$M(x) = EI\theta'(x) = EIw''(x) \tag{1}$$

where EI is the bending stiffness of the beam. The equilibrium equations defining the bending moment M and shear resultant S in terms of the loading are

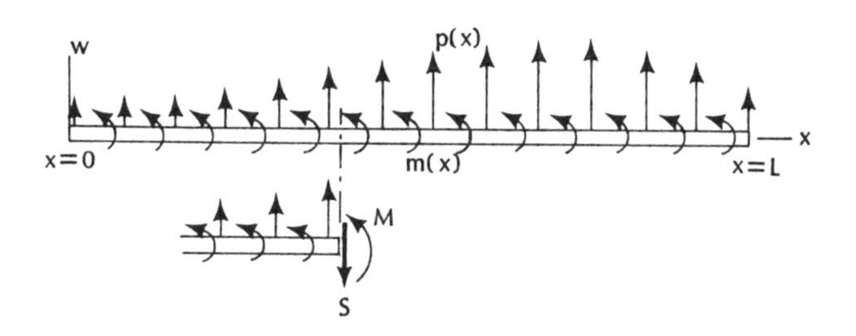

Figure 1. Beam Notation

$$M'(x) = S(x) - m(x) \tag{2}$$

$$S'(x) = p(x) \tag{3}$$

where p is a distributed transverse load and m is a distributed couple. Eliminating M and S produces the usual beam deflection equation

$$[EIw'']'' = q(x) \equiv p(x) - m'(x) \tag{4}$$

where q may be regarded as an "effective" transverse load.

A typical well posed boundary value problem requires that for given load q the deflection satisfy equation (4) in the "region" R (i.e. the interval $0 < x < L$) while at each point of the boundary C (which in this case consists only of the two points $x = 0$ and $x = L$) two independent relations arising from the nature of the beam's support and connecting the variables w, θ, M and S (the boundary conditions) be satisfied. Now suppose $w^*(x)$ is any smooth function on R which we regard as the deflection of the beam for some particular support conditions and loading q^* defined by

$$q^* = [EIw^{*\prime\prime}]'' \tag{5}$$

and denote by θ^*, M^* and S^* the corresponding slope, moment and shear. Then if we integrate the reciprocal work difference twice by parts, the result is

$$\int_R [w(x)q^*(x) - w^*(x)q(x)]\,dx \quad (\equiv \int_0^L \{w[EIw^{*\prime\prime}]'' - w^*[EIw'']''\}\,dx)$$
$$= w_L S_L^* - \theta_L M_L^* + M_L\theta_L^* - S_L w_L^* - w_o S_o^* + \theta_o M_o^* - M_o\theta_o^* + S_o w_o^* \tag{6}$$

where we use the subscripts o, L to denote values at the boundary points $x = 0$ and

$x = L$.

As a particular choice for the auxiliary function w^* we introduce the idea of a *fundamental solution* of the homogeneous form of equation (4) associated with the point ξ in the interior R; $0 < \xi < L$. This is a smooth function on R except at $x = \xi$ where it has a particular type of singularity. The reciprocal work identity (6) can therefore be applied only in a deleted region $R_\varepsilon(\xi)$ which, for some $\varepsilon > 0$, is that part of R where $|x - \xi| \geq \varepsilon$. Then on $R_\varepsilon(\xi)$ there are two additional boundary points (at $x = \xi - \varepsilon$ and at $x = \xi + \varepsilon$) so that, with $q^* = 0$, the reciprocal work equation (6) becomes

$$\begin{aligned} -\int_{R_\varepsilon(\xi)} & w^*(x;\xi)q(x)\,dx \\ &= w_L S_L^* - \theta_L M_L^* + M_L\theta_L^* - S_L w_L^* - w_o S_o^* + \theta_o M_o^* - M_o\theta_o^* + S_o w_o^* \\ &\quad - [wS^* - \theta M^* + M\theta^* - Sw^*]\Big|_{x=\xi-\varepsilon}^{x=\xi+\varepsilon} \end{aligned} \tag{7}$$

The nature of the singularity in what we shall call a fundamental solution is such that it "extracts" from equation (7) the value of w, θ, M, or S at $x = \xi$ in the limit as $\varepsilon \to 0$. For example, a fundamental solution to extract deflection will have the properties

$$\text{i)}\quad [\![w^*]\!] \equiv \lim_{\varepsilon\to 0} w\Big|_{x=\xi-\varepsilon}^{x=\xi+\varepsilon} = 0 \tag{8}$$

$$\text{ii)}\quad [\![\theta^*]\!] = 0 \tag{9}$$

$$\text{iii)}\quad [\![M^*]\!] = 0 \tag{10}$$

$$\text{iv)}\quad [\![S^*]\!] = c \neq 0 \tag{11}$$

with

$$\text{v)}\quad [EIw^{*\prime\prime}]'' = 0 \quad \text{for} \quad x \neq \xi \tag{12}$$

No generality is lost if we take $c = 1$ in equation (11) since the equations are linear and we can always divide w^* by c. With these properties for $w^*(x;\xi)$ equation (7) produces the representation

$$\begin{aligned} w\big|_{x=\xi} &= w_L S_L^* - \theta_L M_L^* + M_L\theta_L^* - S_L w_L^* - w_o S_o^* + \theta_o M_o^* - M_o\theta_o^* + S_o w_o^* \\ &\quad + \lim_{\varepsilon\to 0}\int_{R_\varepsilon(\xi)} w^*(x;\xi)q(x)\,dx \end{aligned} \tag{13}$$

In the particular case where the beam stiffness EI is constant, the general solution of equation (12) is a cubic polynomial and thus, to within an arbitrary smooth cubic, the fundamental solution characterized by equations (8) through (12) is

$$w^*(x;\xi) = \frac{1}{12EI}\,[\operatorname{sgn}(x-\xi)](x-\xi)^3 \tag{14}$$

with

$$\theta^*(x;\xi) = \frac{1}{4EI}\,[\operatorname{sgn}(x-\xi)](x-\xi)^2 \tag{15}$$

$$M^*(x;\xi) = \frac{1}{2}[\operatorname{sgn}(x-\xi)](x-\xi) \tag{16}$$

$$S^*(x;\xi) = \frac{1}{2}[\operatorname{sgn}(x-\xi)] \tag{17}$$

Furthermore, so long as ξ is in the interior of R, equation (13) may be differentiated with respect to ξ to produce representations for slope, moment and shear at $x = \xi$. (Identical formulas would result if we sought directly fundamental solutions to extract slope, moment or shear at $x = \xi$.)

So far we have required that EI and q be smooth functions on R, but it is easy to extend equation (13) to include concentrated force and moment loading. For example, suppose we have a concentrated force P_a applied at $x = a$. We regard this as the limiting case ($\varepsilon \to 0$) of a smooth distribution of transverse load $q = p_\varepsilon(x)$ which vanishes for $|x - a| \geq \varepsilon$ but whose resultant is the concentrated load. Then in this case

$$\lim_{\varepsilon \to 0} \int_R w^*(x;\xi)q(x)\,dx = P_a w^*(a;\xi) \tag{18}$$

In the same manner, if a concentrated couple M_b is applied at $x = b$, then for loading which consists of a distributed couple $m_\varepsilon(x)$ which vanishes for $|x - b| \geq \varepsilon$ but with resultant M_b, we find, upon integrating by parts,

$$-\int_R w^*(x;\xi)m'_\varepsilon(x)\,dx = -w^*(x;\xi)m_\varepsilon(x)\Big|_{x=0}^{x=L} + \int_R \theta^*(x;\xi)m_\varepsilon(x)\,dx \tag{19}$$

and thus in the limiting case

$$\int_R w^*(x;\xi)q(x)\,dx = M_b\theta^*(b;\xi) \tag{20}$$

Since equation (13) is linear we merely add terms of the form (18) and (20) to the integral of the distributed load for each concentrated force or couple in the interior of R.

Thus, once the fundamental solution is known, equation (13) furnishes a complete representation of the solution of the original boundary value problem in terms of the values of deflection, slope, moment and shear at each point of the boundary. Recall that boundary conditions furnish two relations at each boundary point so that the solution of the original problem has been reduced to determining the unknown boundary values, and for this we need two additional independent relations for each boundary point. The basic idea of the boundary integral equation method is to obtain equations analogous to equation (13) for this purpose.

A fundamental solution associated with a boundary point is, as before, a solution of the homogeneous equation except at the point of interest where it is singular, but now the region R_ε is a connected interval. If we denote the boundary point x_p (so that either $x_p = 0$ or $x_p = L$) then equations (8) to (12) defining a fundamental solution $w_f^*(x;x_p)$ to extract deflection at x_p are replaced with

$$\text{i) } \lim_{x \to x_p} w_f^*(x;x_p) = 0 \qquad x \text{ in } R \tag{21}$$

$$\text{ii) } \lim_{x \to x_p} \theta_f^*(x;x_p) = 0 \qquad x \text{ in } R \tag{22}$$

$$\text{iii) } \lim_{x \to x_p} M_f^*(x;x_p) = 0 \qquad x \text{ in } R \tag{23}$$

$$\text{iv) } \lim_{x \to x_p} S_f^*(x;x_p) = 1 \qquad x \text{ in } R \tag{24}$$

$$\text{v) } [EIw_f^{*\prime\prime}]'' = 0 \qquad x \text{ in } R \tag{25}$$

The representation corresponding to equation (15) is

$$\begin{aligned} & w_L S_{fL}^* - \theta_L M_{fL}^* + M_L \theta_{fL}^* - S_L w_{fL}^* \\ & \quad - w_o S_{fo}^* + \theta_o M_{fo}^* - M_o \theta_{fo}^* + S_o w_{fo}^* + \int_R w_f^*(x;x_p) q(x)\, dx = 0 \end{aligned} \tag{26}$$

where it is understood that the indicated evaluations are to be limiting values from inside R. Thus equation (26), written for each boundary point, involves only the boundary values of the variables (and, of course, the prescribed load).

We need another relation at each boundary point. For this purpose we introduce a fundamental solution to extract the slope, which then has the properties

$$\text{i) } \lim_{x \to x_p} w_m^*(x;x_p) = 0 \qquad x \text{ in } R \tag{27}$$

$$\text{ii) } \lim_{x \to x_p} \theta_m^*(x;x_p) = 0 \qquad x \text{ in } R \tag{28}$$

$$\text{iii) } \lim_{x \to x_p} M_m^*(x;x_p) = 1 \qquad x \text{ in } R \tag{29}$$

$$\text{iv) } \lim_{x \to x_p} S_m^*(x;x_p) = 0 \qquad x \text{ in } R \tag{30}$$

$$\text{v) } [EIw_m^{*\prime\prime}]'' = 0 \qquad x \text{ in } R \tag{31}$$

and leads to a representation of the same general form as equation (26) but with different coefficients:

$$w_L S^*_{mL} - \theta_L M^*_{mL} + M_L \theta^*_{mL} - S_L w^*_{mL}$$
$$- w_o S^*_{mo} + \theta_o M^*_{mo} - M_o \theta^*_{mo} + S_o w^*_{mo} + \int_R w^*_m(x;x_p) q(x)\, dx = 0 \qquad (32)$$

Then equations (26) and (32), together with the prescribed boundary conditions at each boundary point, are sufficient to determine all the boundary values needed in the representation (13).

For the particular case of constant beam stiffness, fundamental solutions are readily found as

$$w^*_f(x;x_p) = \frac{1}{6EI}(x - x_p)^3 \qquad (33)$$

$$w^*_m(x;x_p) = \frac{1}{2EI}(x - x_p)^2 \qquad (34)$$

Then if we write equations (26) and (32) at $x = 0$ and $x = L$, the resulting four equations in the eight boundary variables can be put in matrix form

$$\mathbf{KW} = \mathbf{Q} \qquad (35)$$

where

$$\mathbf{K} = \begin{bmatrix} -1 & 0 & 0 & 0 & 1 & -L & \frac{L^2}{2EI} & \frac{-L^3}{6EI} \\ 0 & 1 & 0 & 0 & 0 & -1 & \frac{L}{EI} & \frac{-L^2}{2EI} \\ -1 & -L & \frac{-L^2}{2EI} & \frac{-L^3}{6EI} & 1 & 0 & 0 & 0 \\ 0 & 1 & \frac{L}{EI} & \frac{L^2}{2EI} & 0 & -1 & 0 & 0 \end{bmatrix} \qquad (36)$$

$$\mathbf{W} = \begin{bmatrix} w_o \\ \theta_o \\ m_o \\ S_o \\ w_L \\ \theta_L \\ M_L \\ S_L \end{bmatrix} \quad (37), \qquad \mathbf{Q} = \frac{1}{6EI} \begin{bmatrix} \int_0^L x^3 q(x)\, dx \\ \int_0^L 3x^2 q(x)\, dx \\ \int_0^L (x - L)^3 q(x)\, dx \\ \int_0^L 3(x - L)^2 q(x)\, dx \end{bmatrix} \qquad (38)$$

With some combination and rearrangement it is easy to see that equation (35) is equivalent to the consistent load stiffness equations for a uniform beam which are usually obtained in a more direct manner by standard methods of structural analysis. However, the ideas used in developing these relations may be applied in other less transparent cases. For example, by developing suitable fundamental solutions associated with the equation

$$[EIw'']'' + kw = q \tag{39}$$

one can treat beams on elastic foundations. A thorough discussion of the boundary element formulation of this problem has been given by Butterfield (5). Our main purpose in exposing the details of the development for beams is to provide background for the development of boundary integral equation methods for plates and shallow shells.

Thin Elastic Plates

The direct development of boundary integral equations for elastic plates follows the same general outline as the development just concluded for elastic beams. Of course the algebraic details are more complex and there are some obvious differences due to the difference in the dimensionality of the two problems, but the essence of the method is the same. A discussion of some details, and analysis of some of the questions which arise in the development, may be found in references (3), (7), (9), (21), (22), (23), and (26). The last also contains some discussion of boundary element formulations which are different from the direct method to be discussed here.

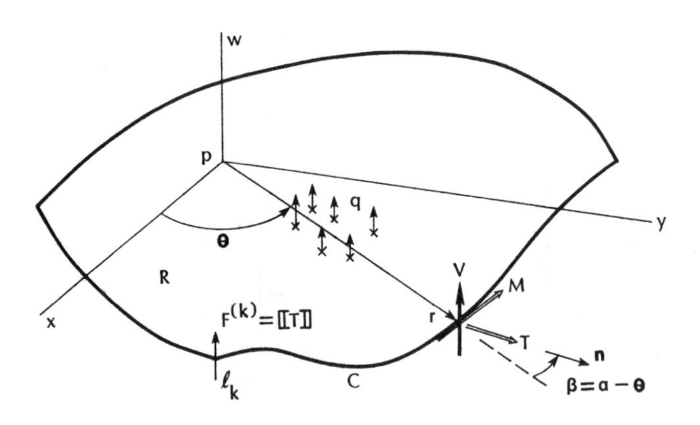

Figure 2. Plate Notation

The notation we will use is illustrated in Figure 2. The plate middle surface occupies a region R with boundary C which we suppose is smooth except for a finite number of corners denoted $l_1, l_2, \ldots, l_K$. The plate deflection is denoted w and the transverse load is q. For a thin plate of linearly elastic isotropic material (modulus of elasticity E, Poisson's ratio ν, thickness h) small deflection theory (see (9), chapter 4) requires

$$D\,\Delta\Delta w = q \tag{40}$$

where $D = Eh^3/12(1-\nu^2)$ is the (constant) flexural stiffness of the plate, and

$$\Delta = \frac{\partial^2}{\partial x^2} + \frac{\partial^2}{\partial y^2} = \frac{\partial^2}{\partial r^2} + \frac{1}{r}\frac{\partial}{\partial r} + \frac{1}{r^2}\frac{\partial^2}{\partial \theta^2} \tag{41}$$

is the plane Laplacian differential operator referred to either Cartesian or polar coordinates. The normal slope N, bending moment M, twisting moment T, and equivalent shear resultant V on the boundary are related to the deflection through

$$N = \frac{dw}{dn} = \cos\alpha\frac{\partial w}{\partial x} + \sin\alpha\frac{\partial w}{\partial y} = \cos\beta\frac{\partial w}{\partial r} + \frac{\sin\beta}{r}\frac{\partial w}{\partial \theta} \tag{42}$$

$$M = -\frac{D}{2}\{(1-\nu)\Delta_1 w + (1+\nu)\Delta w\} \tag{43}$$

$$T = \frac{D(1-\nu)}{2}\Delta_2 w \tag{44}$$

$$V = -\frac{d}{dn}(D\Delta w) - \frac{d}{ds}T \tag{45}$$

where the boundary differential operators Δ_1 and Δ_2 are defined as

$$\begin{aligned}\Delta_1 &= \cos 2\alpha\,[\frac{\partial^2}{\partial x^2} - \frac{\partial^2}{\partial y^2}] + 2\sin 2\alpha\,\frac{\partial^2}{\partial x\partial y}\\ &= \cos 2\beta\,[\frac{\partial^2}{\partial r^2} - \frac{1}{r}\frac{\partial}{\partial r} - \frac{1}{r^2}\frac{\partial^2}{\partial \theta^2}] + 2\sin 2\beta\,\frac{\partial}{\partial r}[\frac{1}{r}\frac{\partial}{\partial \theta}]\end{aligned} \tag{46}$$

$$\begin{aligned}\Delta_2 &= \sin 2\alpha\,[\frac{\partial^2}{\partial x^2} - \frac{\partial^2}{\partial y^2}] - 2\cos 2\alpha\,\frac{\partial^2}{\partial x\partial y}\\ &= \sin 2\beta\,[\frac{\partial^2}{\partial r^2} - \frac{1}{r}\frac{\partial}{\partial r} - \frac{1}{r^2}\frac{\partial^2}{\partial \theta^2}] - 2\cos 2\beta\,\frac{\partial}{\partial r}[\frac{1}{r}\frac{\partial}{\partial \theta}]\end{aligned} \tag{47}$$

with α and β orienting the outward normal from the x- and r- directions respectively. Also, because T depends explicitly on the direction of the boundary normal the tangential derivative in equation (45) has an additional term:

$$\frac{dT}{ds} = -\sin\alpha\,\frac{\partial T}{\partial x} + \cos\alpha\,\frac{\partial T}{\partial y} + \frac{1}{\rho}\frac{\partial T}{\partial \alpha}$$
$$= -\sin\beta\,\frac{\partial T}{\partial r} + \frac{\cos\beta}{r}\frac{\partial T}{\partial \theta} + \left(\frac{1}{\rho} - \frac{1}{r}\right)\frac{\partial T}{\partial \beta} \qquad (48)$$

where $1/\rho$ is the curvature of the boundary, positive if the center of curvature is on the inward normal.

Now suppose w^* is another smooth function on R which we regard as a candidate plate deflection corresponding to the load

$$q^* = D\,\Delta\Delta w^* \qquad (49)$$

We form the reciprocal work integral and again integrate by parts twice to produce the reciprocal work identity

$$\int_R (w^* q - w q^*)\,da \quad \left(\equiv \int_R [w^*(D\,\Delta\Delta w) - w(D\,\Delta\Delta w^*)]\,da\right)$$
$$= \int_C [V^* w - M^* N + N^* M - w^* V]\,ds + \sum_{k=1}^{K} [F^{*(k)} w^{(k)} - F^{(k)} w^{*(k)}] \qquad (50)$$

where at each corner there is the possibility of a concentrated force reaction which we denote $F^{(k)}$ at l_k, and we also write $w^{(k)}$ for the deflection there. The corner forces arise in the course of integrating the tangential derivative of the twisting moment between corners and may thus be evaluated as

$$F^{(k)} = [\![T]\!]_{l_k} \equiv [T^+ - T^-]|_{l_k} \qquad (51)$$

With the origin p of the polar coordinate system in Figure 2 at an interior point of the region R, we identify the auxiliary function w^* in equation (50) with a fundamental solution to extract the displacement at p. Such a solution is

$$w^* = \frac{1}{8\pi D} r^2 \ln r \qquad (52)$$

$$N^* = \frac{r}{8\pi D}(1 + 2\ln r)\cos\beta \qquad (53)$$

$$M^* = -\frac{1+\nu}{4\pi}(1 + \ln r) - \frac{1-\nu}{8\pi}\cos 2\beta \qquad (54)$$

$$V^* = -\frac{\cos\beta}{4\pi r}[2 + (1-\nu)\cos 2\beta] + \frac{1-\nu}{4\pi R}\cos 2\beta \qquad (55)$$

$$F^{*(k)} = \frac{(1-\nu)}{8\pi}[\![\sin 2\beta]\!]_{l_k} \qquad (56)$$

and produces the representation

$$w|_{p \text{ in } R} = \int_C [-V^*w + M^*N - N^*M + w^*V]ds - \sum_{k=1}^{K} [F^{*(k)}w^{(k)} - F^{(k)}w^{*(k)}] + \int_R w^* q\, da \tag{57}$$

Once the boundary values of w, N, M, V and the corner forces $F^{(k)}$ are known, equation (57) furnishes a complete representation of the solution in the interior R. Furthermore, equation (57) can be differentiated with respect to the coordinates of p to produce representations of other quantities of interest. For example, differentiating with respect to the Cartesian coordinates of p while holding the boundary point fixed may be accomplished with the aid of the formulas

$$\frac{\partial}{\partial x} = -\cos\theta \frac{\partial}{\partial r} + \frac{\sin\theta}{r}\left(\frac{\partial}{\partial \theta} - \frac{\partial}{\partial \beta}\right) \tag{58}$$

$$\frac{\partial}{\partial y} = -\sin\theta \frac{\partial}{\partial r} - \frac{\cos\theta}{r}\left(\frac{\partial}{\partial \theta} - \frac{\partial}{\partial \beta}\right) \tag{59}$$

Then the Cartesian moment resultants at p (as defined in Figure 3) are given by

$$M_x = -\frac{D}{2}\{(1+\nu)\Delta^* w + (1-\nu)\Delta_1^* w\} \tag{60}$$

$$M_y = -\frac{D}{2}\{(1+\nu)\Delta^* w - (1-\nu)\Delta_1^* w\} \tag{61}$$

$$M_{xy} = \frac{D(1-\nu)}{2}\Delta_2^* w \tag{62}$$

with w given by equation (57) where the differentiation is carried out before integrating and the differential operators have the form

$$\Delta^* = \frac{\partial^2}{\partial r^2} + \frac{1}{r}\frac{\partial}{\partial r} + \frac{1}{r^2}\frac{\partial^2}{\partial \beta^2} \tag{63}$$

$$\Delta_1^* = \cos 2\theta \left[\frac{\partial^2}{\partial r^2} - \frac{1}{r}\frac{\partial}{\partial r} - \frac{1}{r^2}\frac{\partial^2}{\partial \beta^2}\right] + 2\sin 2\theta \frac{\partial}{\partial r}\left(\frac{1}{r}\frac{\partial}{\partial \beta}\right) \tag{64}$$

$$\Delta_2^* = \sin 2\theta \left[\frac{\partial^2}{\partial r^2} - \frac{1}{r}\frac{\partial}{\partial r} - \frac{1}{r^2}\frac{\partial^2}{\partial \beta^2}\right] - 2\cos 2\theta \frac{\partial}{\partial r}\left(\frac{1}{r}\frac{\partial}{\partial \beta}\right) \tag{65}$$

It remains to determine the boundary values needed in equation (57). Generally, for a well posed problem two distinct boundary conditions involving the boundary values will be prescribed on each segment of the boundary; for example, where the plate is clamped we should have $w = 0, N = 0$, while if the plate were simply supported then w and M would vanish. A free boundary would be characterized by $M = 0, V = 0$. In any event, we suppose that along each segment of boundary there is

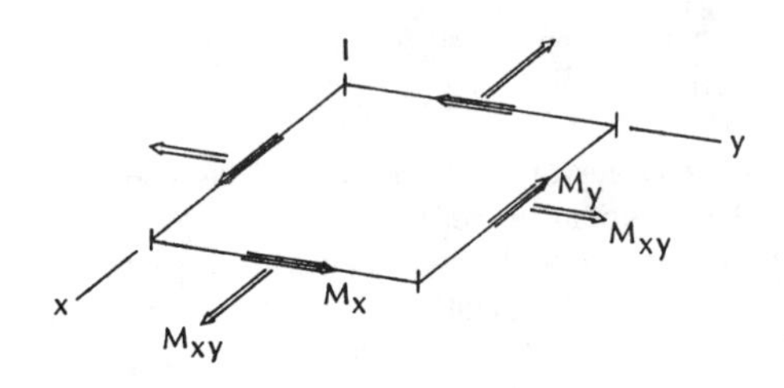

Figure 3. Moment Sign Conventions

prescribed two independent (linear homogeneous) boundary conditions involving the boundary variables. Thus two additional relations are needed to determine the boundary variables, and for this purpose we develop two independent boundary integral equations.

Put the origin point p on the boundary as in Figure 4a and orient the polar coordinate system so that $\theta = 0$ is along the inward normal at p. Fundamental solutions to extract deflection and normal slope at a boundary point are respectively

$$w_f^* = r^2 \ln r \tag{66}$$

$$N_f^* = r(1 + 2\ln r)\cos\beta \tag{67}$$

$$M_f^* = -D\{2(1 + \nu)(1 + \ln r) + (1 - \nu)\cos 2\beta\} \tag{68}$$

$$V_f^* = 2D\{\frac{-\cos\beta}{r}[2 + (1 - \nu)\cos 2\beta] + (1 - \nu)\frac{\cos 2\beta}{R}\} \tag{69}$$

$$F_f^*(k) = D(1 - \nu)[\![\sin 2\beta]\!]_{l_k} \tag{70}$$

and

$$w_m^* = 2r\ln r\cos\theta - (1 + \nu)r\theta\sin\theta \tag{71}$$

$$N_m^* = 2\ln r\cos(\theta + \beta) - (1 + \nu)\theta\sin(\theta + \beta) \tag{72}$$

$$M_m^* = \frac{D(1 - \nu)}{r}\{2\sin 2\beta\sin\theta - (1 + \nu)(1 + \cos 2\beta)\cos\theta\} \tag{73}$$

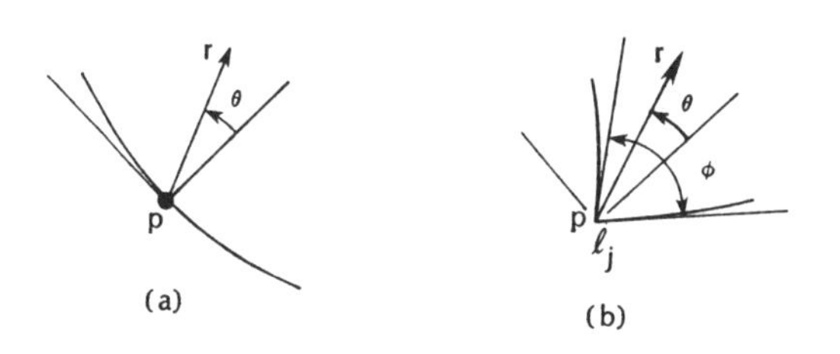

Figure 4. a) Regular boundary point, b) Corner point.

$$V_m^* = \frac{D(1-\nu)}{r^2}\{2(1-\nu\cos 2\beta)\cos\beta\cos\theta + 2(1+\cos 2\beta)\sin\beta\sin\theta + [(3-\nu)\cos\beta\sin\theta + (1+\nu)\sin\beta\cos\theta]\sin 2\beta\} + \frac{2D(1-\nu)}{rR}\{(1+\nu)\cos 2\beta\cos\theta - 2\sin 2\beta\sin\theta\} \tag{74}$$

$$F_m^{*(k)} = \frac{D(1-\nu)}{r}\{(1+\nu)\cos\theta\,[\![\sin 2\beta]\!] + 2\sin\theta\,[\![\cos 2\beta]\!]\}|_{l_k} \tag{75}$$

These lead to representations in which the free term is multiplied by a constant that is not unity. The constants can be evaluated analytically as in (23), but by noting that the representation must hold for *all* solutions of well posed boundary value problems, in particular for the two special cases of uniform deflection $w|_p$, and for uniform rotation to slope $N|_p$ with deflection $w|_p$ at p, the representations can be put in the form

$$\int_C \{V_f^*(w - w|_p) - M_f^*N + N_f^*M - w_f^*V\}\,ds + \sum_{k=1}^{K}\{F_f^{*(k)}(w^{(k)} - w|_p) - w_f^*F^{(k)}\} = \int_R w_f^* q\,da \tag{76}$$

$$\int_C \{V_m^*[w - w|_p + N|_p r\cos\theta] - M_m^*[N + N|_p\cos(\theta+\beta)] + N_m^*M - w_m^*V\}\,ds + \sum_{k=1}^{K}\{F_m^{*(k)}[w^{(k)} - w|_p + N|_p(r\cos\theta)|_{l_k}] - w_m^*F^{(k)}\} = \int_R w_m^* q\,da \tag{77}$$

So far we have avoided corners on the boundary where the normal slope, bending moment and equivalent shear may have discontinuities due to the abrupt change in the orientation of the normal, and a concentrated force reaction (given by the discontinuity in twisting moment) can also exist. There are therefore eight boundary variable values associated with a corner; two limiting values, one from each side, of normal slope, bending moment, and equivalent shear; a single value for deflection, and a concentrated force. We suppose that boundary conditions, and if needed, asymptotic smoothness requirements deduced from an analysis similar to that in (28), will yield five independent (linear homogeneous) boundary condition relations involving the eight variables. Some of the more common cases are given in the following table.

Corner Support		Boundary Condition Relations
-side	+side	
clamped	clamped	$w = 0,\ N^- = N^+ = 0,\ M^- = M^+ = 0$
ss	ss	$w = 0,\ M^- = M^+ = 0,\ N^- = N^+ = 0$
free	free	$M^- = M^+ = 0,\ V^- = V^+ = 0,\ F = 0$
clamped	ss	$w = 0,\ N^- = 0,\ M^+ = 0,\ N^+ = 0,\ M^- = 0$
clamped	free	$w = 0,\ N^- = 0,\ M^+ = 0,\ V^+ = 0,\ N^+ = 0$
ss	free	$w = 0,\ M^- = M^+ = 0,\ V^+ = 0,\ N^+ + N^- \cos\phi = 0$

We therefore need three boundary integral equations at each corner. Equation (76) can be used as it stands, and with the notation indicated in Figure 4b, equation (77) is also available if we interpret the notation $N|_p$ (with $p = l_j$ a corner point) as the weighted average of the limiting slopes from either side:

$$N\big|_{p=l_j} = \frac{N^+\big|_{l_j} + N^-\big|_{l_j}}{2\sin\phi} \tag{78}$$

The third equation at the corner is obtained from the fundamental solution

$$w^*_{m'} = 2r\ln r sin\theta + (1+\nu)\theta\cos\theta \tag{79}$$

where $N^*_{m'}$, $M^*_{m'}$, $V^*_{m'}$ and $F^{*(k)}_{m'}$ may be written directly from equations (72) thru (75) by replacing all occurances of $\cos\theta$ and $\sin\theta$ with $\sin\theta$ and $-\cos\theta$ respectively. The resulting representation is

$$\int_C \{ V^*_{m'}[w - w|_{p=l_j} + N'|_{p=l_j} r\sin\theta] - M^*_{m'}[N + N|_{p=l_j}\sin(\theta+\beta)] + N^*_{m'}M - w^*_{m'}V\}\,ds$$
$$+ \sum_{k=1}^{K}\{F^{*(k)}_{m'}[w^{(k)} - w|_{p=l_j} + N'|_{p=l_j}(r\sin\theta)|_{l_k}] - w^{*(k)}_{m'}F^{(k)}\} = \int_R w^*_{m'}q\,da \tag{80}$$

where

$$N'\big|_{p=l_j} = \frac{N^+\big|_{l_j} - N^-\big|_{l_j}}{2\cos\phi} \tag{81}$$

It should be remarked that the singular part of the auxiliary solution used to obtain the boundary integral equations is not unique. Indeed, most developments, for example (3), (7) and (21), use a fundamental solution equivalent to $w^* = r \ln r \cos\theta$ rather than w_m^* given in equation (71). The former has the property that it, and its derivatives, are single valued for $r \neq 0$. However, as one approaches $r = 0$ on the boundary, the bending moment and equivalent shear become so singular that special care must be exercised in the numerical calculations involving the boundary integral in equation (77). On the other hand, if we use w_m^* as the fundamental solution, then the bending moment and equivalent shear are much better behaved as $r \to 0$ on the boundary. The price to be paid is that w_m^* is multivalued and therefore one must be careful to place the branch cut so that it is not crossed in completing the boundary integration.

Numerical Considerations

Given the plate geometry and stiffness, and the nature of the boundary supports, we should like to determine the deflection and other quantities such as moment resultants throughout the plate region. The boundary integral equations (76) and (77), and the additional corner equation (80) where needed, together with boundary conditions, determine the boundary variables in terms of which the representation (57) "solves" the problem. To accomplish this solution numerically, we start by discretizing the boundary into segments which we might call boundary elements. On each such element the boundary variables are interpolated in terms of nodal values; for example linear interpolation between endpoint nodes (so that the number of nodes is the same as the number of elements), or quadratic interpolation using an additional interior node on each element. Note that at each regular node there are nodal values for four variables while at a corner node eight variables are defined. Boundary conditions furnish two equations relating nodal values at each regular node and five at corner nodes. The remaining equations needed to effect a solution are the boundary integral equations written for the origin p at each node. Integration over each boundary element is completed using a convenient quadrature formula to calculate the (integrated) coefficients of the nodal variables on that segment. Then the resulting linear equations are solved for the remaining unknown nodal variables.

To get an idea of the nature and quality of the results from such a formulation we present some example calculations using quadratic interpolation for the variables on each boundary element, and regular six point Gauss quadrature to evaluate integrals on each element. The first example is a uniformly loaded simply supported equilateral triangular plate. The discretized boundary integral equations are solved for the normal slope and equivalent shear (and the corner force) using 1, 2 and 4 elements on a side of the plate. Shear results are shown in Figure 5a. We have also calcuated the bending moment along the central line using the boundary data computed above, and these results are shown in Figure 5b. In this example the results converge rapidly to fairly accurate values with relatively crude meshes. Similar results for a uniformly loaded simply supported square plate (for which case the corner force does not vanish) are shown in Figure 6. Finally, the bending moment distribution on the boundary of a clamped circular plate with an off-center load is calcualted using several meshes and the results plotted in Figure 7. Again, excellent results can be obtained even with relatively crude meshes.

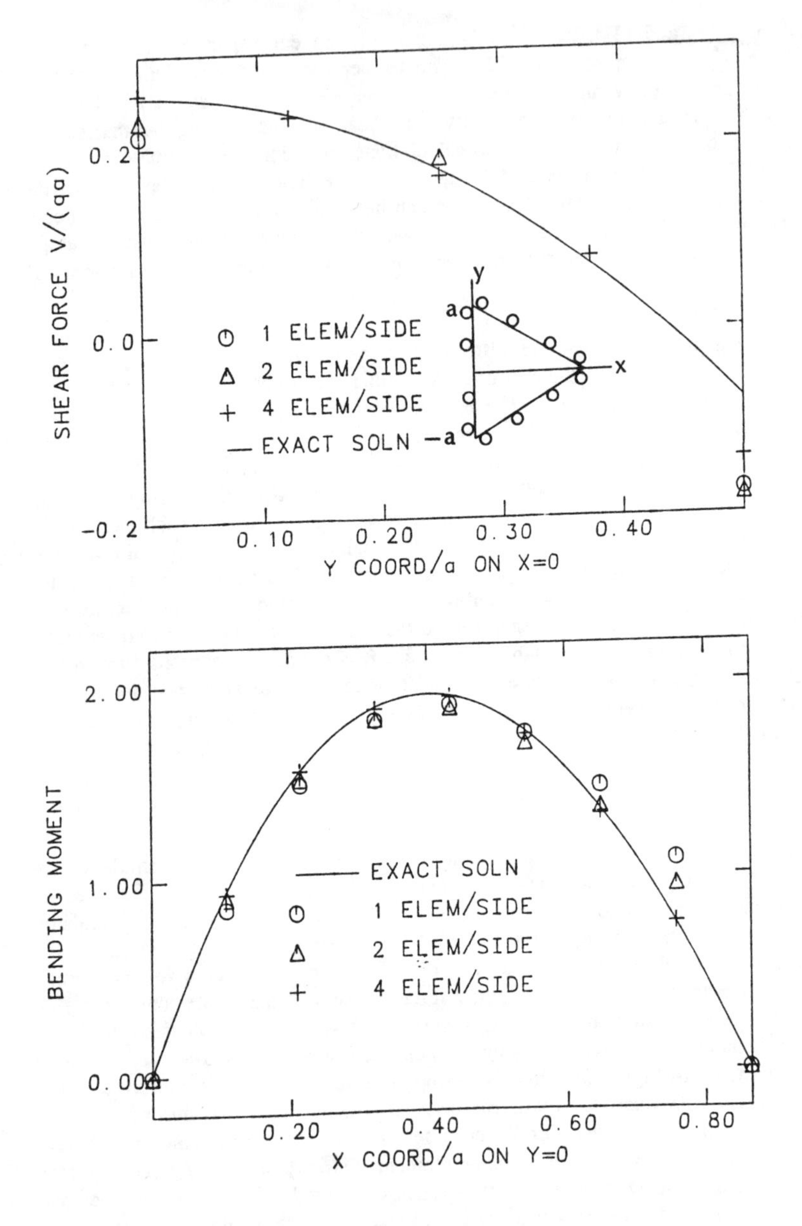

Figure 5. Simply Supported Triangular Plate.
a) Equivalent shear on boundary. b) Bending moment on central line.

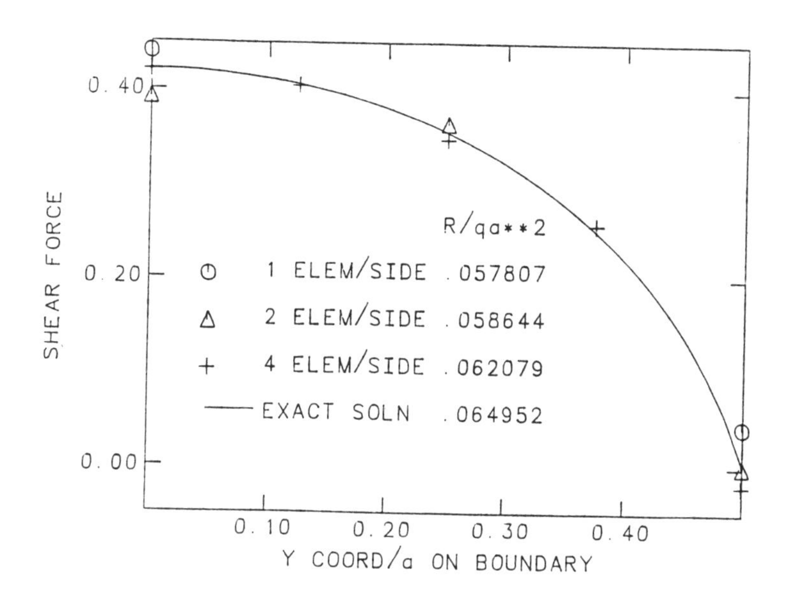

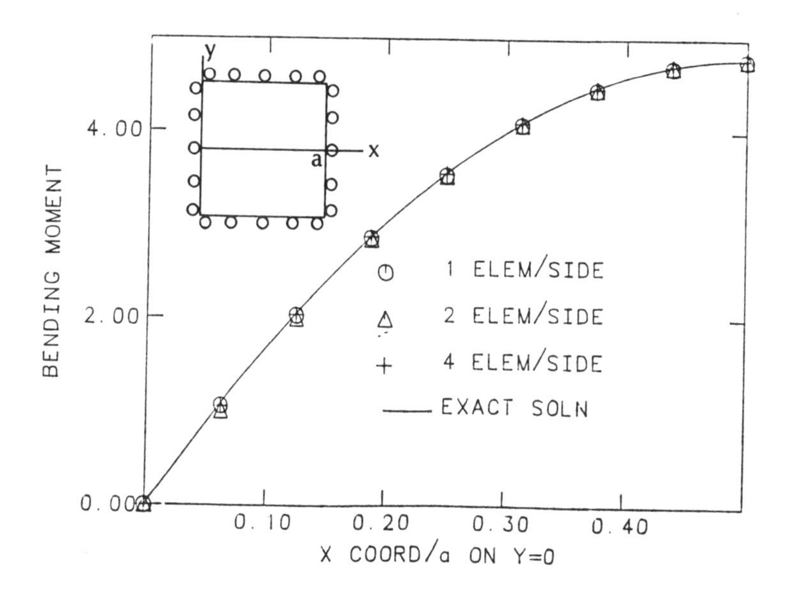

Figure 6. Simply Supported Square Plate.
a) Equivalent shear on boundary. b) Bending moment on central line.

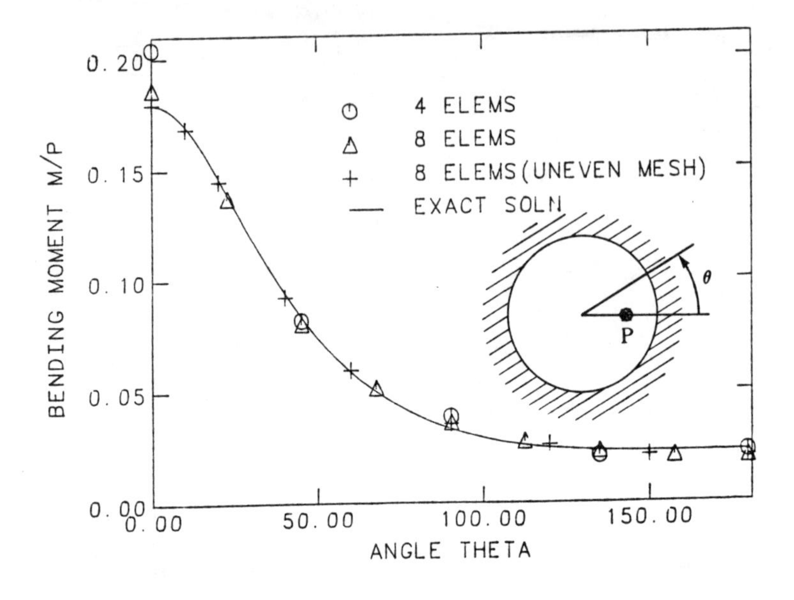

Figure 7. Boundary Bending Moment for Clamped Circular Plate.

Shallow Shells

The development of boundary element methods for thin shells has met with only limited success. Indeed, except for some rather specialized cases, the only developments with any general applicability appear to be in the area of shallow shell analysis. The ideas we will outline here are based on one of the earliest, and possibly most general, approaches to the problem due to Forbes and Robinson (7) in 1969.

In addition to the usual assumptions of linear thin shell theory we add approximations following from the requirement that the shell be "nearly flat", that is, the difference in slope between any two points in the shell middle surface is small, and the curvatures throughout the shell vary (if at all) so slowly that their rates may be neglected. The result which follows is that the equations for the in-plane membrane displacement (corresponding to isotropic plane stress elasticity) and the equation for the transverse deflection (corresponding to thin plate flexure as treated in the preceeding section) are weakly coupled through the shell curvature.

For definiteness, suppose the middle surface of the shell (thickness h) is defined by $z = f(x,y)$ where z measures distance from the tangent plane at the origin and x,y are cartesian coordinates in this plane. The projection of the shell middle surface is the region R with smooth boundary C. For shallow shells the curvatures are approximated by

$$k_x = \frac{\partial^2 f}{\partial x^2}\,, \qquad k_y = \frac{\partial^2 f}{\partial y^2}\,, \qquad k_{xy} = \frac{\partial^2 f}{\partial x \partial y} \tag{82}$$

If we denote the in-plane membrane displacement components by u, v and the normal displacement component by w, then the equilibrium equations for the membrane forces can be put in the form of a plane stress problem:

$$\Delta u + \frac{1+\nu}{2}\frac{\partial}{\partial y}\left[\frac{\partial v}{\partial x} - \frac{\partial u}{\partial y}\right] = \frac{P_x}{B} \tag{83}$$

$$\Delta v + \frac{1+\nu}{2}\frac{\partial}{\partial x}\left[\frac{\partial u}{\partial y} - \frac{\partial v}{\partial x}\right] = \frac{P_y}{B} \tag{84}$$

while the bending equation looks like a plate flexure problem:

$$D\,\Delta\Delta w = Q \tag{85}$$

The coupling shows up in that the generalized loads depend on the displacement field through the curvatures:

$$P_x = -p_x + B\left\{(k_x + \nu k_y)\frac{\partial w}{\partial x} + (1-\nu)k_{xy}\frac{\partial w}{\partial y}\right\} \tag{86}$$

$$P_y = -p_y + B\left\{(k_y + \nu k_x)\frac{\partial w}{\partial x} + (1-\nu)k_{xy}\frac{\partial w}{\partial x}\right\} \tag{87}$$

$$Q = q - B\Bigg\{[(k_x + k_y)^2 - 2(1-\nu)(k_x k_y - k_{xy}^2)]w - (k_x + \nu k_y)\frac{\partial u}{\partial x} - (k_y + \nu k_x)\frac{\partial v}{\partial y} - (1-\nu)k_{xy}\left[\frac{\partial v}{\partial x} + \frac{\partial u}{\partial y}\right]\Bigg\} \tag{88}$$

where p_x, p_y, and q are the components of the applied load (per unit area of middle surface) and $B = Eh/(1-\nu^2) = 12D/h^2$.

Fundamental solutions to extract u, v, w (and $N = dw/dn$ on the boundary of the shell) for the coupled system of equations associated with a general shallow shell have not been explicitly given. Nor is it clear that such solutions would be useful for computation since evaluation is likely to be extremely complicated and hence computationaly expensive. What Forbes and Robinson proposed was to use the fundamental solutions associated with the plane stress problem and the plate flexure problem in the reciprocal work relations to derive boundary integral representations for the solution of shallow shell boundary value problems. This has the advantage of involving only a few expressions with functions which are well known and easily calculated (i.e., trigonometric functions and logarithms); however, the price to be paid is that the boundary integral representations also contain integrals involving the (unknown) displacement components in the interior R as well as on the boundary C .

Without furnishing details of the developoment of boundary integral equations for plane stress, we note that appropriate fundamental solutions are

$$u_{1x}^{*} = \frac{1+\nu)}{4\pi(1-\nu)B}\left[\cos^2\theta - \frac{3-\nu}{1+\nu}\ln r\right] \tag{89}$$

$$u_{1y}^{*} = \frac{1+\nu}{4\pi(1-\nu)B}\sin\theta\cos\theta \tag{90}$$

$$u_{2x}^{*} = \frac{1+\nu}{4\pi(1-\nu)B}\sin\theta\cos\theta \tag{91}$$

$$u_{2y}^{*} = \frac{1+\nu}{4\pi(1-\nu)B}\left[\sin^2\theta - \frac{3-\nu}{1+\nu}\ln r\right] \tag{92}$$

with associated plane stress traction components

$$t_{1x}^{*} = -\frac{1}{4\pi r}[2+(1-\nu)\cos 2\theta]\cos\beta \tag{93}$$

$$t_{1y}^{*} = -\frac{1}{4\pi r}[(1-\nu)\sin\beta + (1+\nu)\sin 2\theta\cos\beta] \tag{94}$$

$$t_{2x}^{*} = -\frac{1}{4\pi r}[(1-\nu)\sin\beta + (1+\nu)\sin 2\theta\cos\beta] \tag{95}$$

$$t_{2y}^{*} = -\frac{1}{4\pi r}[2-(1+\nu)\cos 2\theta]\cos\beta \tag{96}$$

where r, θ, α, and β have the same meaning as in Figure 2. Upon writing **u**, **t**, and **p** for the vector displacement, traction and applied membrane load, the boundary integral representation formulas take the form

$$c\,u|_p + \int_C (\mathbf{t}_1^*\cdot\mathbf{u} - \mathbf{t}\cdot\mathbf{u}_1^*)\,ds = -\int_R \mathbf{u}_1^*\cdot\mathbf{P}\,da \tag{97}$$

$$c\,v|_p + \int_C (\mathbf{t}_2^*\cdot\mathbf{u} - \mathbf{t}\cdot\mathbf{u}_2^*)\,ds = -\int_R \mathbf{u}_2^*\cdot\mathbf{P}\,da \tag{98}$$

where $c = 1$ if p is in the interior R and $c = 1/2$ if p is on the (smooth) boundary C. To these we append the corresponding plate flexure representation formulas, equation (57) for p in the interior and equations (76) and (77) if p is on the boundary, with q replaced by the generalized load Q in the integrals over R.

A well posed boundary value problem for a shallow shell will contain four independent boundary conditions: two involving the variables of the membrane problem (displacement and traction components) and two involving the flexure variables. It should be noted that the boundary tractions are not the tractions of the plane stress problem but are coupled to the transverse displacement through the curvature. The physical traction on the boundary is identified with $\mathbf{t}_C = \mathbf{t} + \mathbf{s}$ where

$$s_x = -B\{(k_x + \nu k_y)\cos\alpha + (1-\nu)k_{xy}\sin\alpha\}w \tag{99}$$

$$s_y = -B\{(k_y + \nu k_x)\sin\alpha + (1-\nu)k_{xy}\cos\alpha\}w \tag{100}$$

There are two distinct sets of variables in the problem: the eight boundary variables (two displacement and two traction components associated with the plane stress-like problem, and the deflection, normal slope, bending moment and equivalent shear associated with the plate flexure-like problem) which are involved in the boundary integrals, and the three displacement field components in the interior which are needed to evaluate the generalized forces **P** and Q. The solution method suggested by Forbes and Robinson is direct in that both sets of variables are approximated on their respective domains in terms of a finite number of unknown values and sufficient discretized integral equations with p on the boundary and in the interior are written to determine the unknown values. In any but the simplest problems this could lead to a large densely populated system of algebraic equations to solve .

An alternative iterative method more in keeping with the spirit of a boundary element formulation is to treat the boundary variables as primary unknowns which, once determined, yield the displacement components at any point p in the interior through the boundary integral representation formulas (which may be formally differentiated with respect to the location of p to yield the derivatives needed to evaluate the generalized forces **P** and Q). The idea then is to adopt an initial guess for the generalized forces, say $P_x^{(0)} = p_x$, $P_y^{(0)} = p_y$ and $Q^{(0)} = q$. Using essentially the same method as described in the section on Numerical Considerations, solve the four boundary integral equations - (76), (77), (97) and (98) with the point p on the boundary - together with the four boundary conditions, to get an initial approximation to the boundary variables. With these boundary variables and the same generalized forces, calculate new values for **P** and Q in the interior and repeat the process. Since the plane stress problem and the plate flexure problem are coupled only through the generalized loads (and possibly the traction boundary condition) we can further subdivide the process by first solving the plate flexure equations, then using these results in the plane stress problem, and alternating.

Before closing this section we remark that Tottenham (26), and Tosaka and Miyake (25), have proposed shallow shell boundary integral formulations which are based on a somewhat different approach. In these formulations the membrane problem is stated in terms of an Airy stress function ϕ which leads to coupled equations of the form

$$\Delta\Delta\phi + \alpha\Delta_k w = \bar{p} \tag{101}$$

$$\Delta\Delta w + \beta\Delta_k\phi = \bar{q} \tag{102}$$

where $\bar{p}$ and $\bar{q}$ are determined from the prescribed loading, the constants α and β depend on material stiffness and the shell geometry, and the differential operator Δ_k is defined as

$$\Delta_k = k_x \frac{\partial^2}{\partial x^2} + k_y \frac{\partial^2}{\partial y^2} - 2k_{xy} \frac{\partial^2}{\partial x \partial y} \tag{103}$$

Numerical results from boundary element calculations for shallow shells are scarce, and of those results which have been reported almost all deal only with spherical caps or shallow cylindrical shells. While this hardly constitutes a fair test of the methods, these authors do report very satisfactory results, at least in this limited class of applications.

Finally, it might be noted that the flexure problem for a thin plate supported on an elastic (Winkler) foundation is governed by the equation

$$P\,\Delta\Delta w + kw = q \tag{104}$$

where k is the spring constant of the foundation. This problem is contained in the shallow shell equations if we ignore the membrane equations (set $\mathbf{u} \equiv 0$ and replace the curvature term in the flexure equation with the foundation stiffness,

$$B\,[(k_x + k_y)^2 - 2(1-\nu)k_x k_y - k_{xy}^2] = k \tag{105}$$

This problem has been treated directly by Katsikadelis and Armenakas (14) and by Costa and Brebbia (6). The fundamental solution required involves Kelvin functions.

References

1. Altiero, N. J. and Sikarskie, D. L., "A Boundary Integral Method Applied to Plates of Arbitrary Plan Form", *Computers and Structures*, **9** (1978) 163-168.
2. Banerjee, P. K. and Butterfield, R., *Boundary Element Methods in Engineering Science*, McGraw Hill, (1981).
3. Bézine, G., "Boundary Integral Formulation for Plate Flexure with Arbitrary Boundary Conditions", *Mechanics Research Communications*, **5** (1978) 197-206.
4. Brebbia, C. A., Telles, J. C. F. and Wrobel, L. C., *Boundary Element Techniques; Theory and Applications in Engineering*, Springer-Verlag (1984).
5. Butterfield, R., "New Concepts Illustrated by Old Problems", *Developments in Boundary Element Methods - 1*, eds. P. K. Banerjee and R. Butterfield, Applied Science Publishers, Ltd. (1979) 1-20.
6. Costa, J. A. and Brebbia, C. A., "The Boundary Element Method Applied to Plates on Elastic Foundations", *Engineering Analysis*, **4** (1985) 174-183.
7. Forbes, D. J. and Robinson, A. R., *Numerical Analysis of Elastic Plates and Shallow Shells by an Integral Equation Method*, Structural Research Series Report 345, The University of Illinois (1969).
8. Hanson, E. B., "Numerical Solution of Integro-Differential and Singular Integral Equations for Plate Bending Problems", *Journal of Elasticity*, **6** (1976) 39-56.
9. Hartmann, F., "Elastostatics", *Progress in Boundary Element Methods-1*, ed. C.A. Brebbia, Pentech Press (1981) 84-167.

10. Irschik, H., "A Boundary Integral Equation Method for Bending of Orthotropic Plates", *International Journal of Solids and Structures*, **20** (1984) 245-255.
11. Jaswon, M. A. and Maiti., M., "An Integral Equation Formulation of Plate Bending Problems", *Journal of Engineering Mathematics*, **2** (1968) 83-93.
12. Kamiya, N. and Sawaki, Y., "A General Boundary Element Method for Bending Analysis of Orthotropic Elastic Plates", *Res Mechanica* **5** (1982) 329-334.
13. Kamiya, N., and Sawaki, I., "Finite Deflection of Plates", *Topics in Boundary Element Research-1*, ed. C. A. Brebbia, Springer Verlag (1984) 204-224.
14. Katsikadelis, J. and Armenakas, A., "Plates on Elastic Foundation by BIE Method", *Journal of Engineering Mechanics*, **110** (1984) 1086-1105.
15. Kim, J. W. "On the Computation of Stress Intensity Factors in Elastic Plate Flexure via Boundary Integral Equations", *Boundary Element Methods in Engineering*, ed. C. A. Brebbia, Springer Verlag (1982) 540-554.
16. Maiti, M. and Chakrabarty, S. K., "Integral Equation Solutions for Simply Supported Polygonal Plates", *International Journal of Engineering Science*, **12** (1974) 793-806.
17. Massonnet, C. E., "Numerical Use of Integral Procedures", *Stress Analysis*, eds,O. C. Zienkiewicz and G. S. Holister, Wiley (1965) 198-235.
18. Morjaria, M. and Mukherjee, S., "Inelastic Analysis of Transverse Deflection of Plates by the Boundary Element Method", *Journal of Applied Mechanics*, **47** (1980) 291-296.
19. Moshaiov, A. and Vorus, W. S., "Elasto-Plastic Plate Bending Analysis by a Boundary Element Method with Initial Plastic Moments", *International Journal of Solids and Structures*, **22** (1986) 1213-1230.
20. Newton, D. A. and Tottenham, H., "Boundary Value Problems in Thin Shells of Arbitrary Plan Form", *Journal of Engineering Mathematics*, **2** (1968) 211-224.
21. Stern, M., "A General Boundary Integral Formulation for the Numerical Solution of Plate Bending Problems", *International Journal of Solids and Structures*, **15** (1979) 769-782.
22. Stern, M., "Boundary Integral Equation for Bending of Thin Plates", *Progress in Boundary Element Methods - 2*, ed. C. A. Brebbia, Pentech Press (1983) 158-181.
23. Stern, M. and Lin, T. L., "Thin Elastic Plates in Bending", *Developments in Boundary Element Methods -4*, eds. P. K. Banerjee and J. O. Watson, Applied Science Publishers Ltd. (1986) 91-119.
24. Timoshenko, S. P. and Woinowsky-Kreiger, S., *Theory of Plates and Shells*, Second ed., McGraw Hill (1959).
25. Tosaka, N. and Miyake, S., "A Boundary Integral Formulation for Elastic Shallow Shell Bending Problems", *Boundary Elements - V*, eds. C. A. Brebbia et. al., Springer Verlag (1983) 527-538.
26. Tottenham, H., "The Boundary Element Method for Plates and Shells", *Developments in Boundary Element Methods - 1*, eds. P. K. Banerjee and R. Butterfield, Applied Science Publishers (1979) 173-205.

27. Weeën, F.V., "Application of the Boundary Integral Equation Method to Reissner's Pla Model",International Journal for Numerical Methods in Engineering,**18** (1982) 1-10.

28. Williams, M. L., "Surface Stress Singularities Resulting from Various Boundary Conditions in Angular Corners of Plates Under Bending", *Proceedings First U.S. National Congress of Applied Mechanics* (1951) 325-329.

29. Wu, B. C. and Altiero, N. J., "A New Numerical Method for the Analysis of Anisotropic Thin-Plate Bending Problems", *Computer Methods in Applied Mechanics and Engineering*, **25** (1981) 343-353.

30. Ye, T. Q. and Liu, Y., "Finite Deflection Analysis of Elastic Plate by the Boundary Element Method", *Applied Mathematical Modelling*, **9** (1985) 183-188.

Elastostatics

by

José Domínguez*

Introduction

Linear elastic isotropic behavior is commonly assumed for stress analysis of solids in civil and mechanical engineering. In addition to that, most numerical methods were developed for elastostatic problems and later extended to other fields of engineering. Consequently, this chapter, dedicated to the study of the Boundary Element Method (BEM) in elastostatics, is rather general and, to a large extent, treats basic aspects which are also valid for more specific problems studied in other chapters.

The existence of solution to the integral equation in terms of which the boundary value problem of potential theory can be formulated was demonstrated in 1903 by Fredholm (17). Discretization of the boundary to solve the integral equation for potential problems was first used sixty years later, once digital computers were available, by Jaswon (21) and Symm (40). The formulation of the direct BEM for elastostatics in its present form was first presented by Rizzo in 1967 (34). For the first time Somigliana's integral representation (cf. 28) was taken to the boundary to obtain a boundary integral equation that was solved discretizing the boundary into elements in a way similar to that used a few years before for the potential problems in the aforementioned papers.

In the next four sections the general formulation of the BEM will be presented. A short summary of the basic equations of elastostatics is included first for completeness and in order to introduce some notation. Next, the fundamental solutions, the direct BEM formulation, and formulae for stresses on the boundary are presented. The basic equations of the indirect BEM are then obtained from the direct BEM formulation.

* Professor, School of Industrial Engineering, University of Seville, 41012 Seville, Spain.

The following section treats the problem of traction discontinuities at corners in the most simple way. This procedure is also used, where zoned media are studied, for corner points belonging to internal boundaries.

In the next two sections body forces are studied. First, the general case is treated and then the particular case of thermal loads.

The application of the BEM to foundations and underground structures for which an infinite or semi-infinite medium has to be modeled is taken up next, and is followed by the formulation for axisymmetric problems for both axisymmetric and non-axisymmetric loading.

Some computational aspects of the implementation of the method are also reviewed and the chapter is closed with the treatment of the combination of the BEM with other numerical methods used in elastostatics.

Basic Equations of Elastostatics

The equilibrium equations, that must be satisfied at every point, are

$$\sigma_{ij,j} + b_i = 0 \tag{1}$$

where, the indices i and j range from 1 to 2 and from 1 to 3 in two and three-dimensional problems, respectively, σ_{ij} are components of the stress tensor and b_i stands for the components of the body forces. Space derivatives are indicated by a comma.

The stress-strain relations are

$$\sigma_{ij} = \lambda\, \delta_{ij}\, \varepsilon_{kk} + 2\mu\, \varepsilon_{ij} \tag{2}$$

where ε_{ij} are components of the strain tensor, λ and μ are the Lamé's constants and δ_{ij} is the Kronecker Delta. A summation is implied when an index is repeated in any particular term.

The strain tensor in terms of displacements is

$$\varepsilon_{ij} = \frac{1}{2}\,(u_{i,j} + u_{j,i}) \tag{3}$$

and the tractions on a point of the surface Γ of the region Ω under study are given by

$$t_i = \sigma_{ij} \eta_j \tag{4}$$

where η_j is the j component of the unit normal to Γ at the point considered.

Fundamental Solutions

The formulation of the BEM requires knowledge of the solution of a basic elastic problem corresponding to an infinite domain loaded with a concentrated unit point load. This solution due to Kelvin is known as "fundamental solution".

When Eqs. 2 and 3 are substituted in Eq. 1, Navier's equation

$$\frac{1}{1-2\nu} u_{k,ki} + u_{i,kk} + \frac{1}{\mu} b_i = 0 \tag{5}$$

is obtained, where ν is the Poisson's ratio.

Kelvin's solution is obtained from Eq. 5 when a unit concentrated load is applied at a point $\underline{y}$ of the complete space following the direction given by the unit vector e_i, i.e.,

$$b_i = \delta(\underline{x}-\underline{y})\, e_i \tag{6}$$

Taking each load as independent, the displacement at a point $\underline{x}$, when loads are applied along any direction, may be written as

$$u_j = U_{ij}(\underline{x},\underline{y})\, e_i \tag{7}$$

where $U_{ij}(\underline{x},\underline{y})$ represents the displacement at $\underline{x}$ in the j direction when a unit load is applied at $\underline{y}$ following the i direction. This displacement is given by

$$U_{ij}(x,y) = \frac{1}{16\pi\mu\,(1-\nu)r}\left[(3-4\nu)\delta_{ij} + r_{,i}\, r_{,j}\right] \tag{8}$$

for three-dimensional problems and

$$U_{ij}(\underline{x},\underline{y}) = \frac{1}{8\pi\,\mu(1-\nu)}\left[(3-4\nu)\ln\frac{1}{r}\,\delta_{ij} + r_{,i}\,r_{,j}\right] \tag{9}$$

for two-dimensional plane strain problems. In Eqs. 8 and 9, r stands for the distance between $\underline{x}$ and $\underline{y}$, and the derivatives are at the field point $\underline{x}$, i.e.,

$$r_i = x_i - y_i$$

$$r = (r_i\, r_i)^{1/2} \tag{10}$$

$$r_{,i} = r_i / r$$

Stresses at a point $\underline{x}$ may also be written in the form

$$\sigma_{jk} = S_{ijk}(\underline{x},\underline{y})\, e_i \tag{11}$$

where the Kernel S_{ijk} may be easily obtained from U_{ij} with the aid of (2) and (3). Using Eq. 4, the tractions on a surface whose unit normal is $\underline{\eta}$ may be written as

$$t_j = T_{ij}(\underline{x},\underline{y})\, e_i \tag{12}$$

where

$$T_{ij}(\underline{x},\underline{y}) = -\frac{1}{8\pi(1-\nu)r^2}\left\{\frac{\partial r}{\partial \underline{\eta}}\left[(1-2\nu)\delta_{ij} + 3\, r_{,i}\, r_{,j}\right] + \right.$$

$$\left. + (1-2\nu)(\eta_i\, r_{,j} - \eta_j\, r_{,i})\right\} \tag{13}$$

for three-dimensional problems and

$$T_{ij}(\underline{x},\underline{y}) = -\frac{1}{4\pi(1-\nu)r}\left\{\frac{\partial r}{\partial \underline{\eta}}\left[(1-2\nu)\delta_{ij} + 2\, r_{,i}\, r_{,j}\right] + \right.$$

$$\left. + (1-2\nu)(\eta_i\, r_{,j} - \eta_j\, r_{,i})\right\} \tag{14}$$

for two-dimensional plane strain problems.

Basic Formulation of the Boundary Element Method

Betti's reciprocal work theorem, between two elastostatic states, σ_{ij}, ε_{ij}, u_j, b_j and σ^*_{ij}, ε^*_{ij}, u^*_j, b^*_j, defined over the domain Ω whose boundary is Γ, can be written as

$$\int_\Gamma t^*_j\, u_j\, d\Gamma + \int_\Omega b^*_j\, u_j\, d\Omega = \int_\Gamma t_j\, u^*_j\, d\Gamma + \int_\Omega b_j\, u^*_j\, d\Omega \qquad (15)$$

This theorem may be demonstrated from the equilibrium equations by means of a weighted residual statement (1). The procedure can be easily extend and integral reciprocal theorems obtained for phenomena governed by equations other than the elastic equilibrium equations.

Somigliana's Identity

In order to define the state "*", it may be assumed that the domain Ω is part of the infinite region and that "*" is due to a unit point load applied at a point $\underline{y}$ inside Ω (i.e. the fundamental solution). The reciprocal theorem then is established between the actual state and the fundamental solution.

Let us say that the load is applied along the i direction. The second term of Eq. 15 will become

$$\int_\Omega \delta(\underline{x},\underline{y})\, \delta_{ij}\, u_j(\underline{x})\, d\Omega(\underline{x}) = u_i(\underline{y}) \qquad (16)$$

and Eq. 15

$$u_i(\underline{y}) + \int_\Gamma T_{ij}(\underline{x},\underline{y})\, u_j(\underline{x})\, d\Gamma(\underline{x}) = \int_\Gamma U_{ij}(\underline{x},\underline{y})\, t_j(\underline{x})\, d\Gamma(\underline{x}) + $$
$$+ \int_\Omega U_{ij}(\underline{x},\underline{y})\, b_j(\underline{x})\, d\Omega(\underline{x}) \qquad (17)$$

where $(\underline{y})$, $(\underline{x})$ and $(\underline{x},\underline{y})$ have been used to indicate values that correspond to the point where the load is applied, values corresponding to general field points and values that depend on both. Eq. 17 is known as Somigliana's identity (28) and gives the value of the displacement at any internal point in terms of the boundary values of u_j and t_j, the body forces throughout the domain and the known fundamental solution.

When the body forces are zero, the displacements at any internal point may be obtained from the boundary displacements and tractions as

$$u_i(\underline{y}) = \int_\Gamma U_{ij}(\underline{x},\underline{y})\, t_j(\underline{x})\, d\Gamma(\underline{x}) - \int_\Gamma T_{ij}(\underline{x},\underline{y})\, u_j(\underline{x})\, d\Gamma(\underline{x}) \tag{18}$$

Zero body forces will be assumed from now on except in later sections where the treatment of those forces is studied.

Boundary Points

Somigliana's identity gives the displacements at any internal point once u_j and t_j are known at every boundary point. Only when the boundary value problem has been solved, Somigliana's identity for internal points will be useful. However, since Eq. 18 may be written for any point of the region including the boundary, an integral equation can obtained by taking point $\underline{y}$ to the boundary.

When $\underline{y}$ is a boundary point, the boundary integrals have a singular point. Consider that the boundary is smooth at $\underline{y}$. In order to analyze the behavior of the integrals at the point $\underline{y}$, the region will be supplemented by half of a sphere (half of a circle for two-dimensions) with centre at $\underline{y}$ and radius ε as shown in Fig. 1.

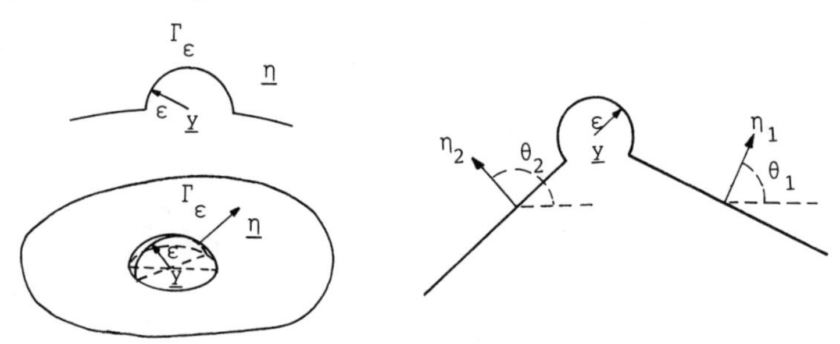

Figure 1. Boundary arround point $\underline{y}$.

Figure 2. Corner point.

The first integral of Eq. 18 can be written as

$$I_1 = \int_\Gamma U_{ij}(\underline{x},\underline{y})\ t_j(\underline{x})\ d\Gamma(\underline{x}) = \lim_{\varepsilon\to 0} \int_{\Gamma-\Gamma_\varepsilon} U_{ij}(\underline{x},\underline{y})\ t_j(\underline{x})\ d\Gamma(\underline{x}) +$$

$$+ \lim_{\varepsilon\to 0} \int_{\Gamma_\varepsilon} U_{ij}(\underline{x},\underline{y})\ t_j(\underline{x})\ d\Gamma(\underline{x}) = I_1' + I_1'' \tag{19}$$

The last integral of Eq. 19 becomes

$$I_1'' = t_j(\underline{y}) \lim_{\varepsilon\to 0} \int_{\Gamma_\varepsilon} U_{ij}(\underline{x},\underline{y})\ d\Gamma(\underline{x}) \tag{20}$$

It may be easily shown (8), by substitution of the values of U_{ij} given by Eqs. 8 or 9, that

$$\lim_{\varepsilon\to 0} \int_{\Gamma_\varepsilon} U_{ij}(\underline{x},\underline{y})\ d\Gamma(\underline{x}) = 0 \tag{21}$$

Integral I_1, may be written again as

$$I_1 = \int_\Gamma U_{ij}(\underline{x},\underline{y})\ t_j(\underline{x})\ d\Gamma(\underline{x}) \tag{22}$$

with the singular point being now excluded.

In a similar way, the second integral of Eq. 18 may be written as

$$I_2 = \int_\Gamma T_{ij}(\underline{x},\underline{y})\ u_j(\underline{x})\ d\Gamma(\underline{x}) = \lim_{\varepsilon\to 0} \int_{\Gamma-\Gamma_\varepsilon} T_{ij}(\underline{x},\underline{y})\ u_j(\underline{x})\ d\Gamma(\underline{x}) +$$

$$+ \lim_{\varepsilon\to 0} \int_{\Gamma_\varepsilon} T_{ij}(\underline{x},\underline{y})\ u_j(\underline{x})\ d\Gamma(\underline{x}) = I_2' + I_2'' \tag{23}$$

The last integral of Eq. 23 becomes

$$I_2'' = u_j(\underline{y}) \lim_{\varepsilon\to 0} \int_{\Gamma_\varepsilon} T_{ij}(\underline{x},\underline{y})\ d\Gamma(\underline{x}) \tag{24}$$

and by substitution of the values of T_{ij} given by Eq. 13 or Eq. 14

and integration over Γ_ε one may easily obtain

$$I_{ij} = \lim_{\varepsilon \to 0} \int_{\Gamma_\varepsilon} T_{ij}(\underline{x},\underline{y})\, d\Gamma(\underline{x}) = -\frac{1}{2}\, \delta_{ij} \tag{25}$$

Integral I_2 may now be written

$$I_2 = -\frac{1}{2}\, u_j(\underline{y})\, \delta_{ij} + \int_\Gamma T_{ij}(\underline{x},\underline{y})\, u_j(\underline{x})\, d\Gamma(\underline{x}) \tag{26}$$

where the integral on Γ is defined in the sense of the Cauchy Principal Value.

Therefore, for boundary points, Eq. 18 transforms into

$$C_{ij}(\underline{y})\, u_j(\underline{y}) + \int_\Gamma T_{ij}(\underline{x},\underline{y})\, u_j(\underline{x})\, d\Gamma(\underline{x}) = \int_\Gamma U_{ij}(\underline{x},\underline{y})\, t_j(\underline{x})\, d\Gamma(\underline{x}) \tag{27}$$

where the integrals are in the sense of Cauchy Principal Value and when Γ is smooth at $\underline{y}$, $C_{ij} = 1/2\ \delta_{ij}$. When $\underline{y}$ is a point where the boundary is not smooth, the integral of Eq. 25 will give a different result. For instance, in two-dimensional problems and with Fig.2 as a reference

$$I_{ij} = -\frac{1}{8\pi(1-\nu)} \begin{bmatrix} 4(1-\nu)(\pi+\theta_2-\theta_1)+\sin 2\theta_1-\sin 2\theta_2 & \cos 2\theta_2-\cos 2\theta_1 \\ \cos 2\theta_2-\cos 2\theta_1 & 4(1-\nu)(\pi+\theta_2-\theta_1)+\sin 2\theta_2-\sin 2\theta_1 \end{bmatrix} \tag{28}$$

so that $C_{ij} = \delta_{ij} + I_{ij}$. It is more complicated to obtain a general expresion for I_{ij} in three dimensions since the slope discontinuities may be of several types. However, one may always do the integration over the corresponding portion of a spherical surface. In Ref.(20) the I_{ij} integrals for the more frequent geometries of corners and edges may be found.

The boundary equation 27 that has been obtained permits to solve the general boundary value problem of elastostatics. If displacements are known over the whole boudary, Eq.27 will produce an integral equation of the first kind; if tractions are known over the whole boundary an integral equation of the second kind will be obtained and finally, if a combination of both types of boundary conditions are prescribed a mixed integral equation will be obtained.

Somigliana's identity Eq.18 for zero body forces and its boundary counterpart Eq.17 are also valid for the external region Ω^*. The integrals extend over the internal boundaries Γ^* provided that the regularity conditions are satisfied. This always happens, for three dimensions when the force resultant over the internal boundaries is bounded , and for two dimensions when this resultant is zero. If the force resultant of a plane problem is not zero and tractions are prescribed at every internal boundary point, a solution of the external region which is defined except for a rigid body motion is obtained. If displacement boundary conditions are prescribed in such a way that any rigid body motion is avoided, the right solution to the problem will be obtained considering only internal boundaries. A study of the uniqueness of the external region solution of static and steady state dynamic problems may be found in Ref.(16).

Boundary Discretization

In order to solve the integral equation numerically, the boundary will be discretized into a series of elements over which displacements and tractions are written in terms of their values at a series of nodal points. Writing the discretized form of Eq. 27 for every nodal point and coordinate, a system of linear algebraic equations is obtained. Once the boundary conditions are applied the system may be solved to obtain all the unknown nodal values and consequently an approximate solution of the boundary values.

For simplicity, matrix notation will be used. Vectors $\underline{u}$ and $\underline{t}$ have two or three components. $\underline{U}$ and $\underline{T}$ are 2x2 or 3x3 matrices. Displacements and tractions are approximated over the boundary as

$$\underline{u} = \underline{\psi}\, \underline{u}_n$$
$$\underline{t} = \underline{\psi}\, \underline{t}_n \tag{29}$$

where $\underline{u}_n$ and $\underline{t}_n$ are vectors containing the 3N (2N for two-dimensions) nodal displacements and tractions, and $\underline{\psi}$ is the 3x3N (2x2N for two-dimensions) interpolation functions matrix

$$\underline{\psi} = \begin{bmatrix} \psi^1 & 0 & 0 & \psi^2 & 0 & 0 & \dots & \psi^N & 0 & 0 \\ 0 & \psi^1 & 0 & 0 & \psi^2 & 0 & \dots & 0 & \psi^N & 0 \\ 0 & 0 & \psi^1 & 0 & 0 & \psi^2 & \dots & 0 & 0 & \psi^N \end{bmatrix} \tag{30}$$

The Cartesian coordinates of the boundary may also be written in terms of the nodal coordinates as

$$\underline{x} = \underline{\psi}\, \underline{x}_n \tag{31}$$

It is now assumed that the boundary is divided into elements as shown in Fig.3. Each function ψ^1 is associated with a different node 1 of the boundary and its support coincides with the surface or length of the elements to which the node belongs.

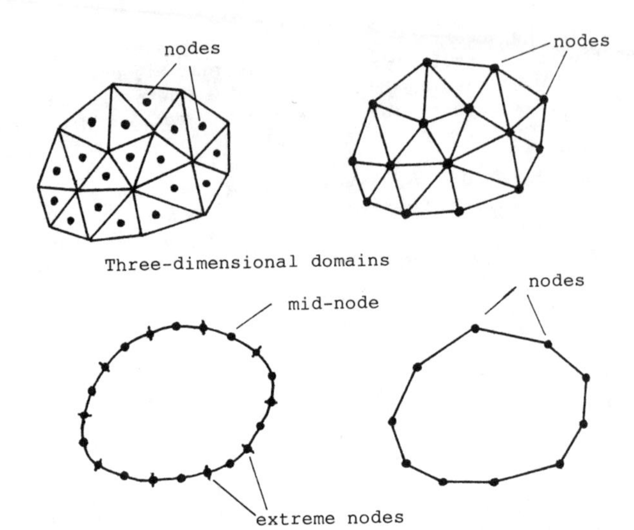

Two-dimensional domains

Figure 3. Boundary discretization.

Taking into account the boundary discretization and the functions representation, Somigliana's identity for node k may be written as

$$\underline{c}^k \underline{u}^k + \left[\sum_{e=1}^{M} \int_{\Gamma_e} \underline{T}\ \underline{\psi} d\Gamma_e \right] \underline{u}_n = \left[\sum_{e=1}^{M} \int_{\Gamma_e} \underline{U}\ \underline{\psi} d\Gamma_e \right] \underline{t}_n \tag{32}$$

where M is the total number of boundary elements and Γ_e is the surface or line domain corresponding to element e.

Many of the integrals of Eq.32 will be zero due to the fact that the functions ψ^l, that form $\underline{\psi}$,are different than zero only over the elements to which the node l belongs.

The integrals can be computed numerically or analytically over each boundary element domain. The kernels of the integrals depend not only on the element where the integration takes place but also on the node k for which equations are being written.

Eq.32 can be written for all the boundary nodes to produce

$$\underline{c}^k\ \underline{u}^k + \sum_{l=1}^{N} \underline{\hat{H}}^{kl}\ \underline{u}^l = \sum_{l=1}^{N} \underline{G}^{kl}\ \underline{t}^l \tag{33}$$

where $\underline{u}^l$ and $\underline{t}^l$ are the displacements and tractions vectors at the node l, $\underline{H}^{kl}$ is the summation of the integrals

$$\int_{\Gamma_e} \underline{T}\, \underline{\psi}^l \, d\Gamma_e \quad \text{being} \quad \underline{\psi}^l = \psi^l \, \underline{I} \tag{34}$$

over all the elements to which the node l belongs and $\underline{G}^{kl}$ is the summation over the same elements of the integrals

$$\int_{\Gamma_e} \underline{U}\, \underline{\psi}^l \, d\Gamma_e \tag{35}$$

Calling

$$\begin{aligned} \underline{H}^{kl} &= \underline{\hat{H}}^{kl} && \text{if} \quad k \neq l \\ \underline{H}^{kl} &= \underline{\hat{H}}^{kl} + \underline{C}^k && \text{if} \quad k = l \end{aligned} \tag{36}$$

the equation for node k becomes

$$\sum_{l=1}^{N} \underline{H}^{kl} \, \underline{u}^l = \sum_{l=1}^{N} \underline{G}^{kl} \, \underline{t}^l \tag{37}$$

and the complete system of equations may be written as

$$\underline{H}\, \underline{u}_n = \underline{G}\, \underline{t}_n \tag{38}$$

Boundary conditions are now applied to Eq.38. For each component of the nodal variables, either the displacement or the traction is known. Consequently, the system of equations can be reordered in such a way that all the unknowns are taken to the left hand side to obtain

$$\underline{A}\, \underline{X} = \underline{F} \tag{39}$$

where the matrix $\underline{A}$ and the vector $\underline{F}$ are known and $\underline{X}$ is a vector formed by all the unknowns.

It is worth to notice that the diagonal submatrices $\underline{H}^{kk}$ include terms of $\underline{\hat{H}}^{kk}$ and $\underline{C}^k$. Some difficulties frequently appear in conjunction with the explicit computation of those terms because of the singularity of the fundamental solution. However, there is an easy way to compute those terms (25). Assuming a rigid-body displacement along one of the Cartesian coordinates the traction vector must be zero and thus from Eq.38

$$\underline{H}\,\underline{I}^{j} = 0 \tag{40}$$

where $\underline{I}^{j}$ is a vector that for all the nodes has unit displacement along direction j and zero displacement in any other direction. Since Eq.40 must be satisfied for any rigid-body displacement

$$\underline{H}^{kk} = -\sum_{l=1}^{N} \underline{H}^{kl} \tag{41}$$

which gives the diagonal submatrices in terms of the rest of the terms of the H matrix.

When dealing with infinite or semi-infinite regions, Eq.41 must be modified. If a rigid body displacement is prescribed for a boundless domain the integral

$$\int_{\Gamma_\infty} \underline{T}\,\underline{I}^{j}\,d\Gamma = \underline{I}^{j} \int_{\Gamma_\infty} \underline{T}\,d\Gamma \tag{42}$$

over the external infinite boundary Γ_∞ will not be zero and since $\underline{T}$ are tractions due to a unit point load, this integral over a boundary enclosing the load must be

$$\int_{\Gamma_\infty} \underline{T}\,d\Gamma = -\underline{I} \tag{43}$$

where $\underline{I}$ is the 3x3 or 2x2 identity matrix. The diagonal submatrices are now

$$\underline{H}^{kk} = \underline{I} - \sum_{l=1}^{N} \underline{H}^{kl} \tag{44}$$

Displacements and Stresses at Internal Points

Somigliana's identity (Eq.18) gives the displacements at any internal point in terms of the known boundary displacements and tractions. Considering again the boundary discretization, the integral representation may be written in a way similar to Eq.32 as

$$\underline{u}(\underline{y}) = \{\sum_{e=1}^{M} \int_{\Gamma_e} \underline{U}(\underline{x},\underline{y})\ \underline{\psi}(\underline{x})\ d\Gamma_e(\underline{x})\}\ \underline{t}_n - \{\sum_{e=1}^{M} \int_{\Gamma_e} \underline{T}(\underline{x},\underline{y})\ \underline{\psi}(\underline{x})\ d\Gamma_e(\underline{x})\}\underline{u}_n \quad (45)$$

where Γ_e is once more the surface or line domain corresponding to element e and $\underline{y}$ is now an internal point. The internal point displacements in terms of the nodal displacements and tractions can be written in the same way as Eq.33, with $\underline{C}^k = \underline{I}$ and have the form

$$\underline{u}^y = \sum_{l=1}^{N} \underline{G}^{yl}\ \underline{t}^l - \sum_{l=1}^{N} \underline{\hat{H}}^{yl}\ \underline{u}^l \quad (46)$$

The terms $\underline{G}^{yl}$ and $\underline{\hat{H}}^{yl}$ consist of integrals over the elements to which the node l belongs. Those integrals do not contain any singularity and may be easily computed by numerical integration.

The stresses can be calculated from the stress-displacement relation. For a point $\underline{y}$ one has

$$\sigma_{ij}(\underline{y}) = \frac{2\mu\nu}{1-2\nu}\ \delta_{ij}\ \frac{\partial u_l(\underline{y})}{\partial y_l} + \mu\ (\frac{\partial u_i(\underline{y})}{\partial y_j} + \frac{\partial u_j(\underline{y})}{\partial y_i}) \quad (47)$$

and substituting Somigliana's identity for internal points

$$\sigma_{ij}(\underline{y}) = \int_\Gamma \{\frac{2\mu\nu}{1-2}\ \delta_{ij}\ \frac{\partial U_{lm}(\underline{x},\underline{y})}{\partial y_l} + \mu\ (\frac{\partial U_{im}(\underline{x},\underline{y})}{\partial y_j} + \frac{\partial U_{jm}(\underline{x},\underline{y})}{\partial y_i})\}t_m(\underline{x})\ d\Gamma(\underline{x}) -$$

$$- \int_\Gamma \{\frac{2\mu\nu}{1-2\nu}\ \delta_{ij}\ \frac{\partial T_{lm}(\underline{x},\underline{y})}{\partial y_l} + \mu\ (\frac{\partial T_{im}(\underline{x},\underline{y})}{\partial y_j} + \frac{\partial T_{jm}(\underline{x},\underline{y})}{\partial y_i})\}\ u_m(\underline{x})\ d\Gamma(\underline{x})$$

(48)

where all derivatives are taken at point $\underline{y}$.

The above equation may be written

$$\sigma_{ij}(\underline{y}) = \int_\Gamma D_{ijm}(\underline{x},\underline{y})\ t_m(\underline{x})\ d\Gamma(\underline{x}) - \int_\Gamma S_{ijm}(\underline{x},\underline{y})\ u_m(\underline{x})\ d\Gamma(\underline{x}) \quad (49)$$

where the third order tensors are

$$D_{ijm} = \frac{1}{r^{\alpha}} \{(1-2\nu)[\delta_{mi} r_{,j} + \delta_{mj} r_{,i} - \delta_{ij} r_{,m}] + $$

$$+ \beta r_{,i} r_{,j} r_{,m} \} \frac{1}{4\alpha\pi(1-\nu)} \quad (50)$$

$$S_{ijm} = \frac{2\mu}{r^{\beta}} \{\beta \frac{\partial r}{\partial \underline{n}} [(1-2\nu) \delta_{ij} r_{,m} + \nu (\delta_{im} r_{,j} + \delta_{jm} r_{,i}) -$$

$$- \gamma r_{,i} r_{,j} r_{,m}] + \beta \nu (\eta_i r_{,j} r_{,m} + \eta_j r_{,i} r_{,m}) +$$

$$+ (1-2\nu)(\beta \eta_m r_{,i} r_{,j} + \eta_j \delta_{im} + \eta_i \delta_{jm}) -$$

$$- (1-4\nu) \eta_m \delta_{ij}\} \frac{1}{4\alpha\pi(1-\nu)} \quad (51)$$

The above formulae are applicable for two and three dimensions. For the former case $\alpha=1$, $\beta=2$ and $\gamma=4$; for the latter one $\alpha=2$, $\beta=3$ and $\gamma=5$. All the derivatives indicated by comma are taken at the boundary point $\underline{x}$, i.e., $r_{,i} = \frac{\partial r}{\partial x_i} = - \frac{\partial r}{\partial y_i}$.

Eq. 49 can be discretized to obtain

$$\sigma_{ij} = \{ \sum_{e=1}^{M} \int_{\Gamma_e} \underline{D}_{ij}^{T}(\underline{x},\underline{y}) \underline{\psi}(\underline{x}) d\Gamma_e(\underline{x})\} \underline{t}_n -$$

$$-\{ \sum_{e=1}^{M} \int_{\Gamma_e} \underline{S}_{ij}^{T}(\underline{x},\underline{y}) \underline{\psi}(\underline{x}) d\Gamma_e(\underline{x})\} \underline{u}_n \quad (52)$$

where $\underline{D}_{ij}$ and $\underline{S}_{ij}$ are 3x1 or 2x1 vectors for two- and three-dimensions ,respectively. Eq.52 gives the stresses at internal points in terms of the known nodal displacements and tractions.

The values obtained for the internal stresses using the above formula are in general more accurate than those computed using other numerical methods and similar discretizations. The same may be said of the internal displacements computed through Eq.45 or Eq.46. However when the internal point is very close to the boundary (say less than 1/2 of the smallest length of the nearest element), because of the peak in the fundamental solution, special numerical integration schemes have to be used to obtain accurate stresses and displacements. The values of the displacements at the boundary are known from the integral equation solution. The same can not be said of the stresses since the stress tensor has more values to compute than the traction vector. This problem will be studied in a different section.

Indirect B.E.M. Formulation

Somigliana's identity, known as the direct integral representation, gives displacements at internal and boundary points in terms of boundary tractions and displacements. There are some other representations where the displacements are written in terms of certain quantities that are not boundary displacements or tractions in an explicit form. Those representations are called "Indirect Representations" and are the basic equations of the indirect BEM.

Let's assume that in addition to the state σ_{ij}, ε_{ij}, u_j, t_j defined over the domain Ω whose boundary is Γ, there is a "primed" state σ'_{ij}, ε'_{ij}, u'_j, t'_j defined over the complementary domain Ω^* with tractions over the external region referred to the normal $\underline{\eta}$ of the internal domain as shown in Fig.4.

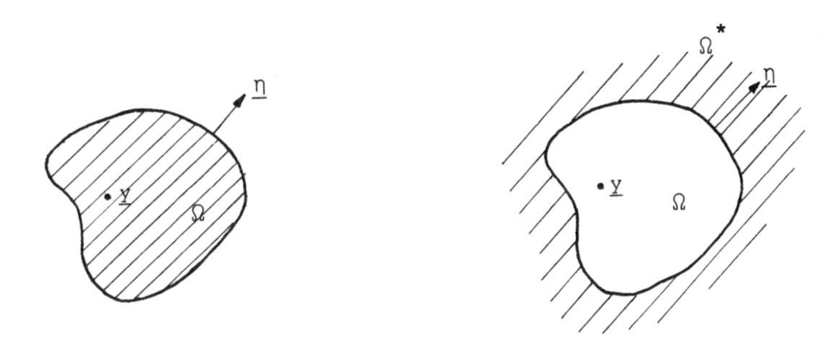

Figure 4. Internal and external domains.

A reciprocity relation between the "primed" state and the fundamental solution when the concentrated load is applied at a point inside Ω, may be written for the external region in the form

$$0 = \int_\Gamma U_{ij}(\underline{x},\underline{y})\ t'_j(\underline{x})\ d\Gamma(\underline{x}) - \int_\Gamma T_{ij}(\underline{x},\underline{y})\ u'_j(\underline{x})\ d\Gamma(\underline{x}) \qquad (53)$$

Subtracting Eq.53 from Eq.18, one obtains

$$u_i(\underline{y}) = \int_\Gamma U_{ij}(\underline{x},\underline{y}) \{t_j(\underline{x}) - t'_j(\underline{x})\} d\Gamma(\underline{x}) + \int_\Gamma T_{ij}(\underline{x},\underline{y}) \{u'_j(\underline{x}) - u_j(\underline{x})\} d\Gamma(\underline{x}) \tag{54}$$

Since the "primed" state may be defined arbitrarily, it will be assumed that it is such that its displacements on the boundaries are the same with those on the internal boundaries, i.e.,

$$u'_j(\underline{x}) = u_j(\underline{x}) \tag{55}$$

Calling $s_j = t_j(\underline{x}) - t'_j(\underline{x})$, Eq.54 can be written as

$$u_i(\underline{y}) = \int_\Gamma U_{ij}(\underline{x},\underline{y}) \, s_j(\underline{x}) \, d\Gamma(\underline{x}) \tag{56}$$

It may be easily shown that

$$U_{ij}(\underline{x},\underline{y}) = U_{ji}(\underline{x},\underline{y}) = U_{ji}(\underline{y},\underline{x}) \tag{57}$$

and Eq.56 may be written as

$$u_i(\underline{y}) = \int_\Gamma U_{ji}(\underline{y},\underline{x}) \, s_j(\underline{x}) \, d\Gamma(\underline{x}) \tag{58}$$

that may be interpreted as follows: the displacements at a point $\underline{y}$ inside Ω may be obtained by the summation of those due to loads $s_j(\underline{x})$ $d\Gamma(\underline{x})$ applied at every $d\Gamma(\underline{x})$ when Ω is considered to be part of the complete region. The integral representation given by Eq.56 is known as the single layer potential with density $\underline{s}$. As may be seen from Eq.55, the displacements u_j are continuous at the boundary Γ while the tractions are discontinuous.

Tractions on a point $\underline{y}$, referred to a certain normal, $\underline{\eta}$, may be obtained from Eq.58 by appropriate differentiations and use of Eqs.2 and 3 in the form

$$t_i(\underline{y}) = \int_\Gamma T_{ji}(\underline{y},\underline{x}) \, s_j(\underline{x}) \, d\Gamma(\underline{x}) \tag{59}$$

If the "primed" state is assumed to be such that the tractions on the boundaries are

$$t'_j(\underline{x}) = t_j(\underline{x}) \tag{60}$$

calling $d_j(\underline{x}) = u'_j(\underline{x}) - u_j(\underline{x})$, Eq.54 can be written as

$$u_i(\underline{y}) = \int_\Gamma T_{ij}(\underline{x},\underline{y})\ d_j(\underline{x})\ d\Gamma(\underline{x}) \tag{61}$$

The above integral representation is known as double layer potential with density $\underline{d}$. As may be seen from the definition of $\underline{d}$, the double layer potential produces displacements that are discontinuous at Γ while the tractions as given by Eq.60 are continuous. A more meaningful interpretation of the double layer potential representation may be given as follows. It may be easily shown that (11)

$$T_{ij}(\underline{x},\underline{y}) = Z_{ji}(\underline{y},\underline{x}) \tag{62}$$

where $Z_{ji}(\underline{y},\underline{x})$ represents the i displacement at $\underline{y}$ when a unit j displacement dislocation is applied at $\underline{x}$, both being points within the infinite domain. Thus, Eq.61 can be written as

$$u_i(\underline{y}) = \int_\Gamma Z_{ji}(\underline{y},\underline{x})\ d_j(\underline{x})\ d\Gamma(\underline{x}) \tag{63}$$

which means that the displacement at an internal point $\underline{y}$ may be obtained as the summation of those produced by dislocations $d_j(\underline{x})\ d\Gamma(\underline{x})$ applied at every $d\Gamma(\underline{x})$ when Ω is considered to be part of the complete region.

Tractions **at** a point $\underline{y}$ referred to normal $\underline{\eta}$ may be obtained from

$$t_i(\underline{y}) = \int_\Gamma Q_{ji}(\underline{y},\underline{x})\ d_j(\underline{x})\ d\Gamma(\underline{x}) \tag{64}$$

where $Q_{ji}(\underline{y},\underline{x})$ is derived from $Z_{ji}(\underline{y},\underline{x})$ by appropriate differentiations.

It has been shown how the internal displacements may be represented by a single layer potential (Eq.56), a double layer potential (Eq.61), or a combination of both (Eq.18). Internal stresses are obtained by differentiation. In general, as it was said above for Somigliana's representation, the internal stresses and displacements computed by means of the discretized form of the integral representations are more accurate than those obtained using other numerical methods and similar discretizations. This is a consequence of the fact that the internal values are obtained by integration of fundamental solutions that are exact and only the boundary densities of the potentials are approximated. According to St. Venant principle the local errors of this approximation may be expected to damp quickly.

When the point of the single layer potential is taken to the boundary, Eq.56 does not change, being now the integral in the sense of Cauchy Principal Value

$$u_i(\underline{y}) = \int_\Gamma U_{ij}(\underline{x},\underline{y})\, s_j(\underline{x})\, d\Gamma(\underline{x}) \;;\; \underline{y} \in \Gamma \tag{65}$$

However, the traction representation given by Eq.59 for a point $\underline{y}$ where the boundary is smooth becomes

$$t_i(\underline{y}) = \frac{1}{2}\,\delta_{ij}\, s_j(\underline{y}) + \int_\Gamma T_{ji}(\underline{y},\underline{x})\, s_j(\underline{x})\, d\Gamma(\underline{x}) \;;\; \underline{y} \in \Gamma \tag{66}$$

On the other hand, when the point $\underline{y}$ of the double layer potential is taken to the boundary, the following relation is obtained for the displacement at a point where the boundary is smooth

$$u_i(\underline{y}) = -\frac{1}{2}\,\delta_{ij}\, d_j(\underline{y}) + \int_\Gamma T_{ij}(\underline{x},\underline{y})\, d_j(\underline{x})\, d\Gamma(\underline{x}) \;;\; \underline{y} \in \Gamma \tag{67}$$

The tractions of the double layer potential do not present any discontinuity at the boundary, i.e.,

$$t_i(\underline{y}) = \int_\Gamma Q_{ji}(\underline{y},\underline{x})\, d_j(\underline{x})\, d\Gamma(\underline{x}) \;;\; \underline{y} \in \Gamma \tag{68}$$

When boundary conditions are applied to Eqs.65-66 or 67-68 an integral equation is obtained with unknowns $s_j(\underline{x})$ or $d_j(\underline{x})$, respectively. The integral equation may be solved numerically and once $s_j(\underline{x})$ or $d_j(\underline{x})$ are known, displacements and tractions at any point may be easily computed.

The methods based on the solution of the above integral equations by boundary discretization are known as Indirect Boundary Element Methods. Some times this name is used only for the single layer potential representation (5), while the method based on the double layer potential is called the Displacement Discontinuity Method (11). The indirect formulation that has been summarized in this section may be found in more detail in the book by Jaswon and Symm (22). Discussion of the indirect method may also be found in the work by Hartmann (20).

Applications

A few examples describing applications of the direct BEM will be shown next.

Example 1: Hollow cylinder under internal pressure.

The problem is represented by the plane-strain model shown in Fig. 5. The internal pressure is assumed to be p = 100 Nw/mm^2, while the internal and external radii are r_1 = 10 mm and r_2 = 25 mm, re-

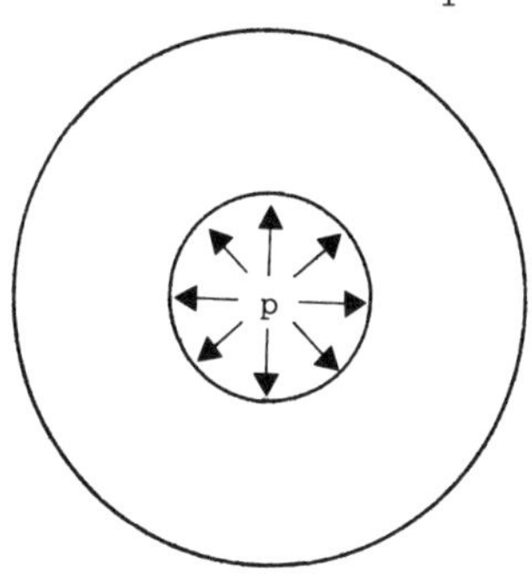

Figure 5. Hollow cylinder under internal pressure.

spectively. The elastic constants of the material are: Elasticity modulus, E = 200,000 Nw/mm^2 and Poisson's ratio, ν = 0.25.

Due to the symmetry of the problem only one quarter of the section has been discretized. Three node elements are used to obtain a quadratic representation of the tractions, displacements and geometry. The boundary conditions are shown in Fig. 6. Three different discretizations consisting of 4, 10 and 15 elements are used as shown in Fig. 6.

Table 1 shows the radial displacements computed for the nodes along the radius. The displacements are given in $\mu \equiv 10^{-3}$ mm and are compared with the exact values given by the liner elasticity theory (see, for instance, Ref. 18).

It should be noticed that using a coarse boundary discretization like (a), results within 2% of the exact solution are obtained. When 15 elements are used, the results are whitin a range of 0.1% of the exact solution.

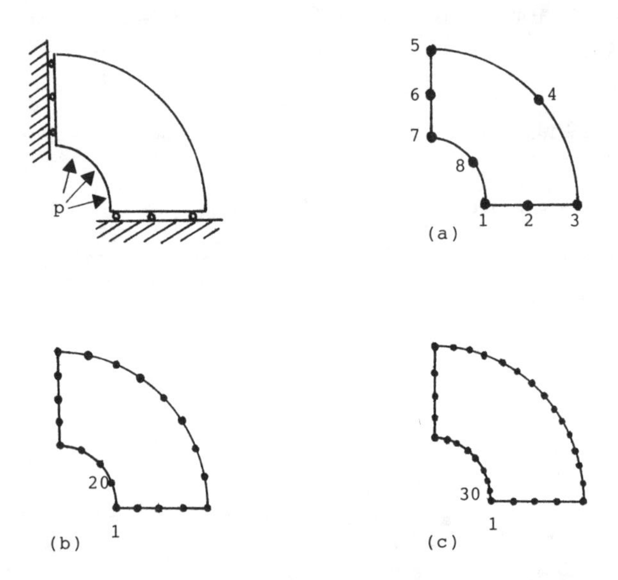

Figure 6. Boundary element discretization.

NODE	EXACT	DISCRETIZATION		
		a	b	c
1	8.0325	7.8781	8.0246	8.0350
2	6.2272	-	6.2193	6.2303
3	5.2912	5.1668	5.2845	5.2928
4	4.7644	-	4.7593	4.7652
5	4.4625	4.3896	4.4570	4.4631

Table 1. Radial displacements for hollow cylinder under internal pressure.

Example 2: Square plate with a circular hole.

A square plate with a circular hole under uniform tractions along two opposite sides is considered. Fig. 7 shows the plate, the boundary conditions and the boundary discretization using 11 quadratic elements. The geometry of the plate is defined by a radius of the hole $r_1 = 8.25$ mm, a lenght of the sides $l = 165$ mm and a value of the normal tractions $\sigma = 100$ Nw/mm^2. The elastic constants are the same as in the previous example.

In order to check the accuracy of the BEM solution the stress concentration factor around the hole is computed. This factor is defined by

$$K = \frac{\sigma_{yy\ max}}{\sigma} \tag{69}$$

and in the case of the problem of Fig. 8, a normal traction of $\sigma_{yy} = 303.66$ Nw/mm^2 was computed at node A. Thus, the stress concentration factor is K = 3.0366. Peterson (33) gives an experimental value for this problem, K = 3.03, which shows a very good agreement with the BEM solution.

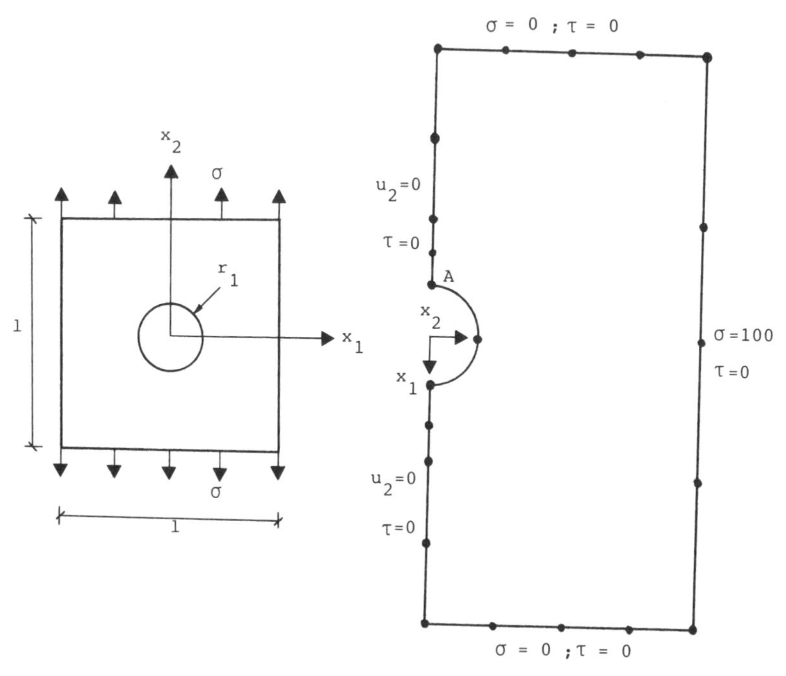

Figure 7. Square plate with circular hole.

The BEM using quadratic elements is very well suited not only to problems with non singular stress concentrations but also to problems that include cracks or any other kind of singularity. Singular quarter point elements that reproduce the stress singular behavior have been developed (6). Using those elements, several procedures to compute the stress intensity factors were studied by Blandford et al. (6) and Martínez and Domínguez (29). No more attention is paid here to the use of BEM for elastic stress intensity factor computations since there is a whole chapter dedicated to that topic in this book.

Example 3: Anchor plate.

The last example corresponds to an anchor plate of a sheet-pile buried in a soil which is assumed to be linear elastic. The anchor plate is assumed to be infinitely rigid compared to the surrounding soil. The plate is loaded by a wire that pulls it along the x_2 direction.

Fig. 8 shows the boundary discretization using constant elements with a single node. The boundary conditions are: a constant displacement along x_2 and zero displacement along x_1 for the elements at the side of the plate where there are compressions; zero tractions in both directions for the rest of the elements in the plate-soil interface; zero tractions in both directions along the soil free surface and zero displacements for the elements that close the boundary through the soil. The geometry and position of the plate are defined by l=1350 mm, b = 300 mm and d = 450 mm.

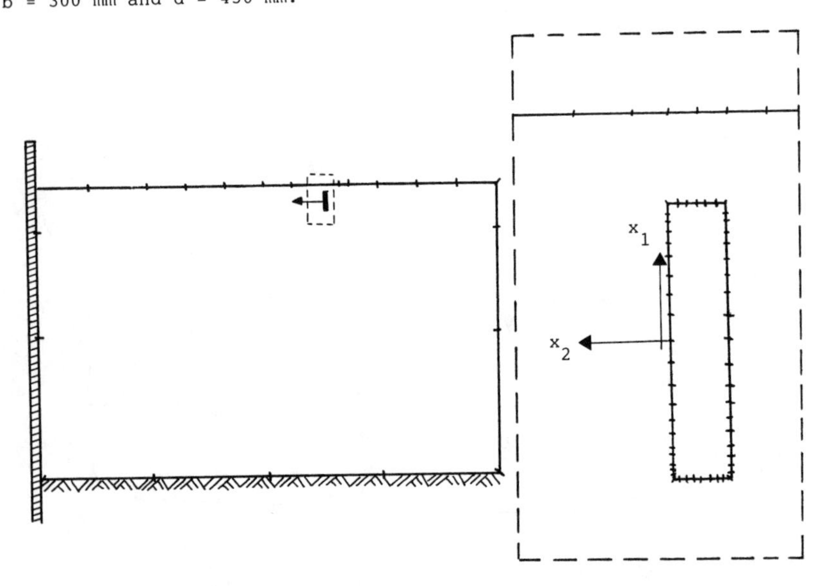

Figure 8. Discretization of the soil arround the anchor plate.

Fig. 9 shows the distribution of the stress component σ_{22} along two lines that are 100 mm and 500 mm far from the anchor plate loaded surface. The figure shows the peaks in the vicinity of the plate -ends and how these peaks damp quickly with the distance. The effects of the close soil free surface makes the stress distribution to be non-symmetric with respect to x_2. No exact solution is known for this

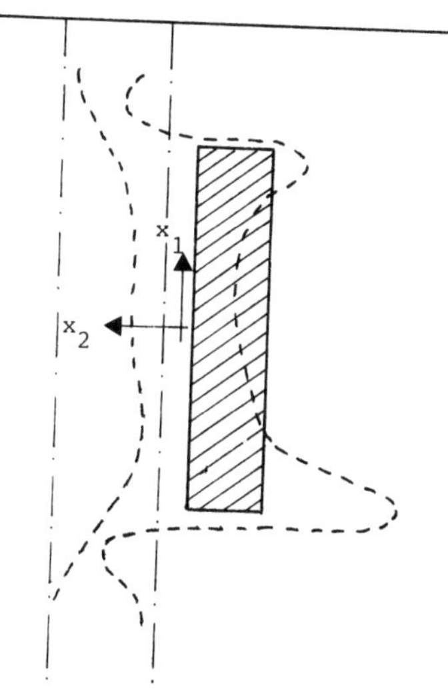

Figure 9. σ_{22} stress component along two lines close to the anchor plate.

problem. It may be said that the results are in quantitative agreement with the stress distribution around a loaded rigid plate included in a complete space that was studied by Muskhelishvili (32).

Stresses on the Boundary.

As a result of the numerical solution of the boundary integral equation, values of the tractions all over the boundary are known. However, in many cases, the complete stress tensor for boundary points is desired. The immediate idea, would be to establish the stress representation given by Eq. 52 for boundary points. This procedure was studied by Cruse (12) but it is complicated and time consuming because of the order of the singularity of the kernels $\underline{D}_{ij}$ and $\underline{S}_{ij}$. For three-dimensional problems $\underline{D}_{ij}$ and $\underline{S}_{ij}$ contain singularities of order $1/r^2$ and $1/r^3$, respectively. For two-dimensional problems the order of the singularities is reduced by one. In spite of the singularities, Eq. 52 may be taken to the boundary (12); however, the computation of the principal value of the integrals is not efficient in terms of speed and accuracy.

The easiest way of determining the stress tensor at boundary points is to compute its components from the known boundary tractions and displacements (9), (20). Let's assume a three-dimensional domain and a local Cartesian system of coordinates at the boundary point whose stress tensor is to be computed as shown in Fig. 10. It is immediate that

$$\begin{aligned} \hat{\sigma}_{13} &= \hat{\sigma}_{31} = \hat{t}_1 \\ \hat{\sigma}_{23} &= \hat{\sigma}_{32} = \hat{t}_2 \\ \hat{\sigma}_{33} &= \hat{t}_3 \end{aligned} \tag{70}$$

where the symbol ^indicates local coordinates. In addition to the tractions, a discrete expresion of the boundary displacements is also known from Eq. 29. That expresion may be written in local coordinates as

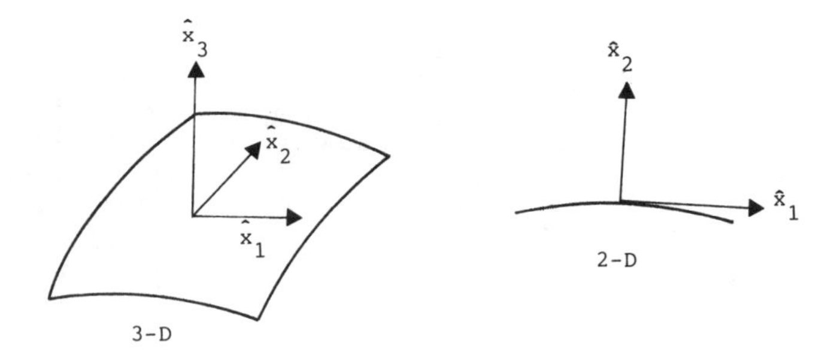

Figure 10. Local systems of coordinates.

$$\underline{\hat{u}} = \underline{L}^T \underline{\psi}\, \underline{u}_n \tag{71}$$

where $\underline{L}$ is the coordinates transformation matrix obtained from Eq. 31. The four components of the strain tensor are computed by differentiation of $\underline{\hat{u}}$ in accordance with

$$\hat{\varepsilon}_{ij} = \frac{1}{2} \left(\frac{\partial \hat{u}_j}{\partial x_i} + \frac{\partial \hat{u}_i}{\partial x_j} \right) \forall\ i,j = 1,2 \tag{72}$$

Thus, the terms $\hat{\varepsilon}_{ij}$ will depend on the derivatives of the shape functions and on the nodal displacements. If constant elements are used, the displacement derivatives should be computed using a finite difference appoximation between adjacent nodes.

The rest of the terms of the stress tensor can now be computed from Hooke's law as follows:

$$\hat{\sigma}_{12} = \hat{\sigma}_{21} = 2\mu\, \hat{\varepsilon}_{12}$$

$$\hat{\sigma}_{11} = \frac{1}{1-\nu} [\nu\, \hat{\sigma}_{33} + 2\mu\, (\hat{\varepsilon}_{11} + \nu\, \hat{\varepsilon}_{22})] \tag{73}$$

$$\hat{\sigma}_{22} = \frac{1}{1-\nu} [\nu\, \hat{\sigma}_{33} + 2\mu\, (\hat{\varepsilon}_{22} + \nu\, \hat{\varepsilon}_{11})]$$

For two-dimensional problems, the procedure is analogous. Three components of the stress tensor are obtained from the tractions, i.e.,

$$\hat{\sigma}_{12} = \hat{\sigma}_{21} = \hat{t}_1$$

$$\hat{\sigma}_{22} = \hat{t}_2 \tag{74}$$

one component of the strain tensor is computed from the surface displacements from

$$\hat{\varepsilon}_{11} = \frac{\partial \hat{u}_1}{\partial \hat{x}_1} \tag{75}$$

and the last stress component is computed from

$$\sigma_{11} = \frac{1}{1-\nu} (\nu\, \sigma_{22} + 2\mu\, \varepsilon_{11}) \tag{76}$$

by using Hooke's law for plane strain.

Traction Discontinuities at Corner Points.

When a node is located at a point where the boundary is not smooth, i.e., it is characterized by corners for two-dimensions and corners and edges for three-dimensions, a discontinuity in the traction occurs at that node. This fact implies that when the nodal tractions are unknown, the number of equations at that node is smaller than the number of unknowns. The problem has been treated by different authors (10), (2), (3), (37).

Two-dimensional domains will be considered first for simplicity with reference Fig.11. When the tractions are known at both sides of the corner node, only the two components of the nodal displacements are unknown and no special treatment of the corner node is needed. The same can be said of any situation where only two of the six variables are unknown; however, when two tractions and one displacement, three tractions or four tractions are unknown at the corner, additional

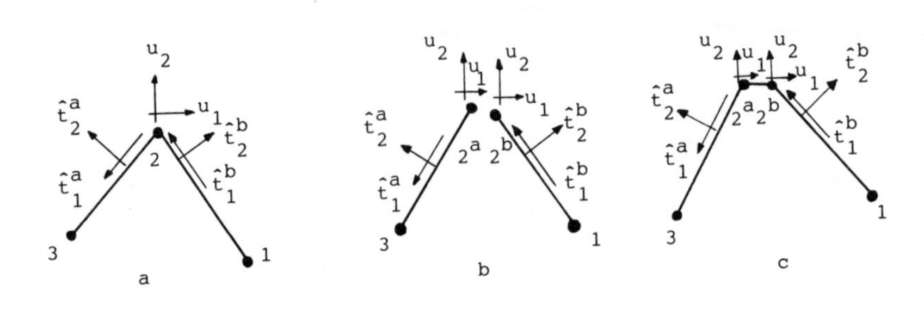

Figure 11. Corner tractions and displacements.

equations are needed. An easy way of solving the problem (7), (8) is by duplicating the corner node. The geometry of the problem is slightly modified and only two traction component are assigned to each node (Fig. 11). The problem may now be solved by the standard procedure. The distance between the two corner nodes must be very small and it is limited by the numerical problems that may be originated by the existence of two equations whose coefficients are very close to each other. Good results are obtained with distances that do not cause numerical problems. When the corner node is duplicated, a "leak" may be left between the two nodes (Fig. 11.b) or a small element may be assumed between the two (Fig. 11c). In the latter case, tractions over the small elements are assumed to be $\hat{t}_1^b$, $\hat{t}_2^b$ and $\hat{t}_1^a$, $\hat{t}_2^a$ when the equations for nodes 2^b and 2^a, respectively, are written.

The corner problem may also be treated by obtaining additional equations from the theory of elasticity (10), (2), (37). An extra equation is obtained by the condition of uniqueness of the stress tensor at the corner. Only three components of the stress tensor are independent and consequently an equation relating the four tractions can be written in the form

$$\eta_1^a \, t_1^b - \eta_1^b \, t_1^a = \eta_2^b \, t_2^a - \eta_2^a \, t_2^b \tag{77}$$

known as Cauchy's equation and where $\underline{\eta}^b$ and $\underline{\eta}^a$ are the unit normals before and after the corner respectively. Using local coordinates Eq. 77 becomes

$$(\hat{t}_1^a - \hat{t}_1^b) \cos \phi = (\hat{t}_2^a - \hat{t}_2^b) \sin \phi \tag{78}$$

where ϕ is the angle between the two normals.

Cauchy's equation solves the problem when two tractions and one displacement or three tractions are unknown. When the four tractions are unknown one more equation is needed. Chaudonneret (10) proposed to use the condition of invariance of the trace of the strain tensor. The strains are written in terms of nodal displacements by finite differences and related to the stresses and tractions through Hooke's law. This gives an equation that relates tractions and displacements.

The above two procedures for the treatment of the corner problem imply an increase of the number of equations in order to be able to compute more than two unknowns at corner nodes. When dealing with three -dimensional problems the duplication of nodes must be done along edges and a node must be located at each face for corners. If the second procedure is used, all tractions may be written in terms of six using Cauchy's equations. Three more equations relating tractions and displacements may be necesary.

The aforementioned difficulties or enlargement of the system of equations may be avoided using a simple procedure which is consistent with the treatment of the boundary stresses studied in the previous section and follows an idea presented by Alarcón (3) for one of the possible corner boundary conditions. Again a two-dimensional domain is assumed and the boundary displacements are written in global coordinates as

$$\underline{u} = \underline{\psi}\, \underline{u}_n \tag{79}$$

The derivatives of $\underline{u}$ can be easily computed from Eq. 79 at the elements before and after the corner node, as shown in Fig. 12, and take the form

$$\begin{Bmatrix} \dfrac{\partial \underline{u}}{\partial \hat{x}_1^b} \\[2ex] \dfrac{\partial \underline{u}}{\partial \hat{x}_1^a} \end{Bmatrix} = \begin{Bmatrix} \dfrac{\partial \underline{\psi}}{\partial \hat{x}_1^b} \\[2ex] \dfrac{\partial \underline{\psi}}{\partial \hat{x}_1^a} \end{Bmatrix} \underline{u}_n \tag{80}$$

while the derivatives of $\underline{u}$ with respect to the global coordinates at the corner point may be written as

$$\begin{Bmatrix} \dfrac{\partial \underline{u}}{\partial x_1} \\[2ex] \dfrac{\partial \underline{u}}{\partial x_2} \end{Bmatrix} = (J_C^{-1}) \begin{Bmatrix} \dfrac{\partial \underline{u}}{\partial \hat{x}_1^b} \\[2ex] \dfrac{\partial \underline{u}}{\partial \hat{x}_1^a} \end{Bmatrix} \tag{81}$$

Where (J_C^{-1}) is the inverse of the Jacobian

$$(J) = \begin{bmatrix} \dfrac{\partial x_1}{\partial \hat{x}_1^b} & \dfrac{\partial x_2}{\partial \hat{x}_1^b} \\ \\ \dfrac{\partial x_1}{\partial \hat{x}_1^a} & \dfrac{\partial x_2}{\partial \hat{x}_1^a} \end{bmatrix} \tag{82}$$

particularized at the corner point, that may be easily computed from the definition of the Cartesian coordinates of the boundary elements, i.e.,

$$\underline{x} = \underline{\psi}\, \underline{x}_n \tag{31}$$

where $\underline{\psi}$ is a function of $\hat{x}_1^b$ and $\hat{x}_1^a$.

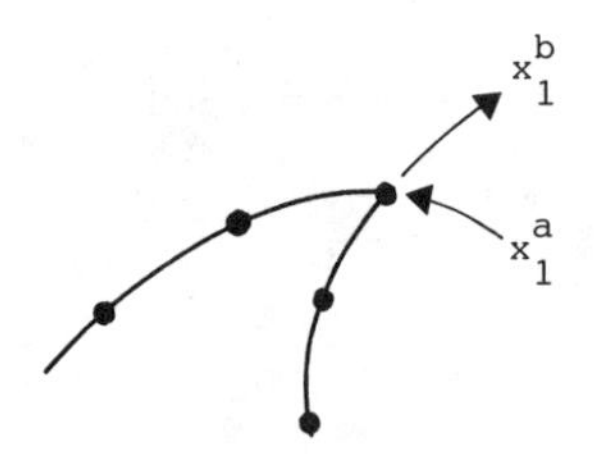

Figure 12. Local coordinates at corner node.

Equation 31 substituted into Eq. 82 permits one to compute the Jacobian that may be particularized at the corner point and inverted. Substituting the value of (J_C^{-1}) and Eq. 80, also particularized at the corner point, into Eq. 81, one obtains an expresion of the displacement global derivatives at the corner point in terms of the nodal displacements. The stress tensor at the corner and consequently any desired tractions may be computed using the strains definition and Hooke's law. Thus, $\hat{t}_1^b$, $\hat{t}_2^b$, $\hat{t}_1^a$ and $\hat{t}_2^a$ are written in term of the nodal displacements and if the known boundary conditions at the corner are the two displacements, no BEM equation will be needed for that point. One only has to use as many BEM equations at the corner as unknown displacements exist at that point according to the prescribed boundary conditions. Obviously more than two BEM equations will never be needed.

Zoned Media

In the previous sections the BEM for homogeneous linear elastic isotropic media has been formulated. In many cases media that are not homogeneous but consist of several zones each one being homogeneous must be analysed. Those problems may be studied using the above formulation of the BEM and continuity conditions. All the boundaries of the body have to be discretized, including internal boundaries that separate homogeneous zones within the medium. The BEM equations formulated for every homogeneous zone plus the displacement and traction continuity conditions over the internal boundaries produce a system of equations that may be solved once the external boundary conditions are taken into account.

Consider the problem shown in Fig. 13 consisting of three zones of different elastic materials. A two-dimensional domain has been represented for simplicity; however, the equations and reasoning below apply to both two and three-dimensional problems.

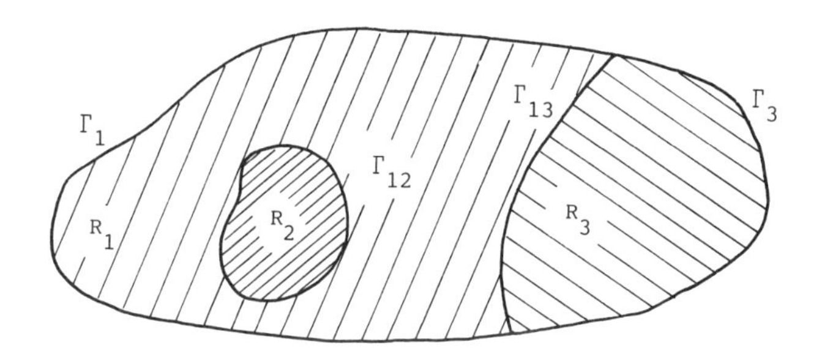

Figure 13. Zoned medium.

The following symbols are used

Γ_i : external part of the boundary of zone R_i

Γ_{ij} : boundary between zones R_i and R_j

$\underline{u}^i$, $\underline{t}^i$: nodal displacements and tractions at nodes on the boundary Γ_i of zone R_i

$\underline{u}^{ij}$, $\underline{t}^{ij}$: nodal displacements and tractions at nodes on Γ_{ij} as part of zone R_i

$\underline{H}^i$, $\underline{G}^i$: parts of the H and G matrices obtained for zone R_i that multiply $\underline{u}^i$ and $\underline{t}^i$, respectively

$\underline{H}^{ij}$, $\underline{G}^{ij}$: parts of the H and G matrices obtained for zone R_i that multiply $\underline{u}^{ij}$ and $\underline{t}^{ij}$, respectively.

The BEM equations for the three homogeneous zones are

$$[\underline{H}^1 \ \underline{H}^{12} \ \underline{H}^{13}] \begin{Bmatrix} \underline{u}^1 \\ \underline{u}^{12} \\ \underline{u}^{13} \end{Bmatrix} = [\underline{G}^1 \ \underline{G}^{12} \ \underline{G}^{13}] \begin{Bmatrix} \underline{t}^1 \\ \underline{t}^{12} \\ \underline{t}^{13} \end{Bmatrix} \tag{83}$$

for zone R_1,

$$\underline{H}^{21} \ \underline{u}^{21} = \underline{G}^{21} \ \underline{t}^{21} \tag{84}$$

for zone R_2 and

$$[\underline{H}^3 \ \underline{H}^{31}] \begin{Bmatrix} \underline{u}^3 \\ \underline{u}^{31} \end{Bmatrix} = [\underline{G}^3 \ \underline{G}^{31}] \begin{Bmatrix} \underline{t}^3 \\ \underline{t}^{31} \end{Bmatrix} \tag{85}$$

for zoned R_3. The traction equilibrium conditions and displacement compatibility conditions over the internal boundaries Γ_{ij} are

$$\begin{aligned} \underline{u}^{12} &= \underline{u}^{21} \\ \underline{u}^{13} &= \underline{u}^{31} \\ \underline{t}^{12} &= -\underline{t}^{21} \\ \underline{t}^{13} &= -\underline{t}^{31} \end{aligned} \tag{86}$$

that transform Eq. 84 and 85 into

$$[\underline{H}^{21} \ \underline{G}^{21}] \begin{Bmatrix} \underline{u}^{12} \\ \underline{t}^{12} \end{Bmatrix} = 0 \tag{87}$$

$$[\underline{H}^3 \; \underline{H}^{31} \; \underline{G}^{31}] \begin{Bmatrix} \underline{u}^3 \\ \underline{u}^{13} \\ \underline{t}^{13} \end{Bmatrix} = \underline{G}^3 \, \underline{t}^3 \tag{88}$$

The last two Eqs. plus Eq. 83 can be rearranged into

$$\begin{bmatrix} \underline{H}^1 & \underline{0} & \underline{H}^{12} & -\underline{G}^{12} & \underline{H}^{13} & -\underline{G}^{13} \\ \underline{0} & \underline{0} & \underline{H}^{21} & \underline{G}^{21} & \underline{0} & \underline{0} \\ \underline{0} & \underline{H}^3 & \underline{0} & \underline{0} & \underline{H}^{31} & \underline{G}^{31} \end{bmatrix} \begin{Bmatrix} \underline{u}^1 \\ \underline{u}^3 \\ \underline{u}^{12} \\ \underline{t}^{12} \\ \underline{u}^{13} \\ \underline{t}^{13} \end{Bmatrix} = \begin{bmatrix} \underline{G}^1 & \underline{0} \\ \underline{0} & \underline{0} \\ \underline{0} & \underline{G}^3 \end{bmatrix} \begin{Bmatrix} \underline{t}^1 \\ \underline{t}^3 \end{Bmatrix} \tag{89}$$

The above system of equations may be solved once the boundary conditions on Γ_1 and Γ_3 are prescribed. The total number of unknowns is equal to the number of nodal d.o.f. over the external boundaries plus twice the number of nodal d.o.f. over the internal boundaries.

The subdivision of the region into several zones may be used also in homogeneous media as a way of avoiding numerical problems or improving computational efficiency. For instance, problems that include cracks or notches, as the one shown in Fig. 14, present numerical difficulties when the boundary is discretized due to the proximity of

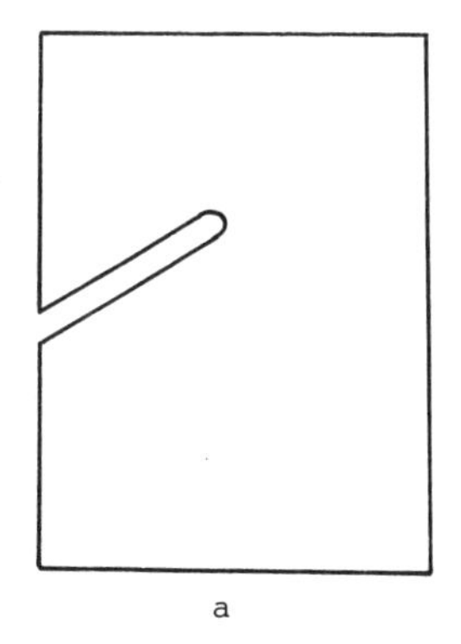

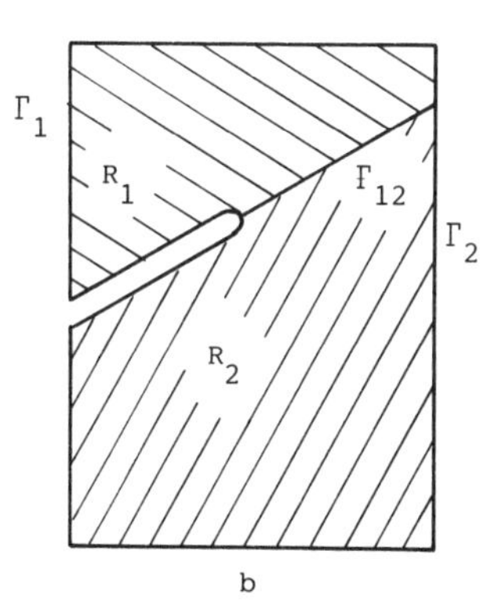

Figure 14. Subdivision to avoid numerical difficulties.

some of the nodes. The difficulties disappear if the region is divided into two zones in a way that the nodes that are very close belong to different regions. Another situation where the subdivision of a homogeneous region may be useful corresponds to problems with a large number of unknowns, as the one shown in Fig. 15. In those cases the subdivision transforms the fully propulated matrices into banded matrices, which are more convenient from a computational point of view. In those cases, the increase in the number of unknowns because of the internal boundaries must be small, otherwise the subdivision will be worthless.

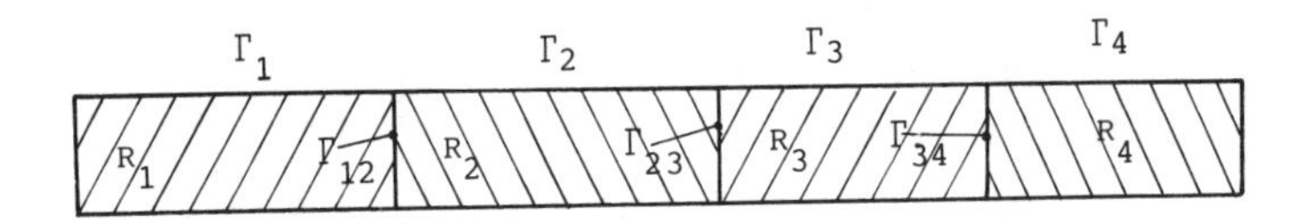

Figure 15. Subdivision to increase computational efficiency.

Body Forces

In the previous sections, the BEM has been studied for problems with zero body forces. When body forces exist, in addition to the boundary integrals, Somigliana's identity includes the domain integral (Eq. 17)

$$\int_{\Omega} U_{ij} (\underline{x},\underline{y})\, b_j(\underline{x})\, d\Omega(\underline{x}) \tag{90}$$

This integral doesn't include any unknown value but in order to compute it directly by numerical procedures, the whole domain has to be discretized into cells, which represents an important difficulty and reduces the greatest advantage of the BEM over domain techniques. Fortunately, the domain integration may be avoided in many practical situations.

The domain integral can be transformed into a boundary integral always that the body forces b_j may be obtained from a function Φ such that

$$b_j = \Phi_{,j} \tag{91}$$

where

$$\Phi_{,jj} = K_o = \text{Constant} \tag{92}$$

Then

$$\int_\Omega U_{ij}\, b_j\, d\Omega = \int_\Omega U_{ij}\, \Phi_{,j}\, d\Omega = \int_\Omega (U_{ij}\, \Phi)_{,j}\, d\Omega - \int_\Omega U_{ij,j}\, \Phi\, d\Omega \qquad (93)$$

and using the divergence theorem

$$\int_\Omega U_{ij}\, b_j\, d\Omega = \int_\Omega U_{ij}\, \Phi\, \eta_j\, d\Gamma - \int_\Omega U_{ij,j}\, \Phi\, d\Omega \qquad (94)$$

where $\underline{\eta}$ is the normal to Γ at the integration point. In order to take the last integral to the boundary, the fundamental solution is written in terms of its Galerkin vector G_{ij} in the form (18)

$$U_{ij} = G_{ij,kk} - \frac{1}{2(1-\nu)}\, G_{ik,jk} \qquad (95)$$

It is worth noting that the subindex i doesn't change neither in U_{ij} nor in G_{ij} and because of that G_{ij} is being called "vector". Eq. 95 may be written as

$$\underline{U}_i = \nabla^2\, \underline{G}_i - \frac{1}{2(1-\nu)}\, \underline{\nabla}\, \underline{\nabla}\, \underline{G}_i \qquad (96)$$

and

$$\underline{U}_i = \nabla^2\, \underline{G}_i - \frac{1}{2(1-\nu)}\, (\underline{\nabla} \wedge \underline{\nabla} \wedge \underline{G}_i + \nabla^2\, \underline{G}_i) \qquad (97)$$

Applying the divergence operator to both sides of Eq. 97

$$\underline{\nabla}\, \underline{U}_i = \frac{1-2\nu}{2(1-\nu)}\, \nabla^2\, \underline{\nabla}\, \underline{G}_i \qquad (98)$$

Taking into account Eq. 98, the last integral of Eq. 94 becomes

$$\int_\Omega U_{ij,j}\, \Phi\, d\Omega = \frac{1-2\nu}{2(1-\nu)} \int_\Omega G_{ij,jkk}\, \Phi\, d\Omega \qquad (99)$$

Now Green's theorem may be written between $G_{ij,j}$ and Φ

$$\int_{\Omega} (G_{ij,j}\ \Phi_{,kk} - G_{ij,jkk}\ \Phi) d\Omega = \int_{\Gamma} G_{ij,j}\ \Phi_{,k}\ \eta_k\ d\Gamma - \int_{\Gamma} G_{ij,jk}\ \Phi\ \eta_k\ d\Gamma \tag{100}$$

The integral of the first term may be written as

$$\int_{\Omega} G_{ij,j}\ \Phi_{,kk}\ d\Omega = K_o \int_{\Omega} G_{ij,j}\ d\Omega = K_o \int_{\Gamma} G_{ij}\ \eta_j\ d\Gamma \tag{101}$$

and

$$-\int_{\Omega} G_{ij,jkk}\ \Phi\ d\Omega = \int_{\Gamma} G_{ij,j}\ \Phi_{,k}\ \eta_k\ d\Gamma - \int_{\Gamma} G_{ij,jk}\ \Phi\ \eta_k\ d\Gamma - K_o \int_{\Gamma} G_{ij}\ \eta_j\ d\Gamma \tag{102}$$

The transformation of the domain integral into boundary integrals is completed by substitution of Eqs. 99 and 102 into Eq. 94. Thus,

$$\int_{\Omega} U_{ij}\ b_j\ d\Omega = \int_{\Gamma} U_{ij}\ \Phi\ \eta_j\ d\Gamma + \frac{1-2\nu}{2(1-\nu)} \Big[\int_{\Gamma} G_{ij,j}\ \Phi_{,k}\ \eta_k\ d\Gamma - \int_{\Gamma} G_{ij,jk}\ \Phi\ \eta_k\ d\Gamma - K_o \int_{\Gamma} G_{ij}\ \eta_j\ d\Gamma\Big] \tag{103}$$

Galerkin Vector

The Galerkin vector for the solution of three-dimensional elastostatic problems is well known. In fact, one of the easiest ways of computing the fundamental solution is using the Galerkin vector (see, for instance, Ref. 12).

$$G_{ij} = \frac{1}{8\pi\mu}\ r\ \delta_{ij} \tag{104}$$

For two-dimensional problems

$$G_{ij} = \frac{1}{8\pi\mu}\ r^2\ \ln\left(\frac{1}{r}\right)\ \delta_{ij} \tag{105}$$

It must be noticed that when the two-dimensional Galerkin vector is derived, the following fundamental solution is obtained

$$U_{ij} = \frac{1}{8\pi\mu(1-\nu)} \left[(3-4\nu)\ \ln\frac{1}{r}\ \delta_{ij} - \frac{7-8\nu}{2}\ \delta_{ij} + r_{,i}\ r_{,j}\right] \tag{106}$$

This solution differs from the fundamental solution given in Eq. 9 by a constant term that was dropped in that case because a rigid body motion doesn't change the solution of the system of equations. However, when body forces exist and the Galerkin vector is used to take the domain integral to the boundary, the fundamental solution given by Eq. 106 must be used.

Gravitational Loads

As an example of body forces, gravitational loads may be considered. For three-dimensional problems

$$\Phi = -\rho\ g\ x_3 \ ; \ \underline{b} = \begin{Bmatrix} 0 \\ 0 \\ -\rho g \end{Bmatrix} \text{ and } \Phi_{,jj} = 0 \tag{107}$$

where ρ is the density and g the gravitational acceleration.

The domain integral becomes

$$\int_\Omega U_{ij}\ b_j\ d\Omega = -\int_\Gamma U_{ij}\ \rho\ g\ x_3\ d\Gamma - \frac{(1-2\nu)\ \rho\ g}{16\pi\mu(1-\nu)} \int_\Gamma (r_{,i}\ \eta_3 - x_3\ r_{,ik}\ \eta_k) d\Gamma \tag{108}$$

that can be integrated over the boundary elements.

Internal Stresses

The integral representation for the internal stresses as given by Eq. 49 will have an additional term when body forces exist and it will be of the form

$$\sigma_{ij}(\underline{y}) = \int_\Gamma D_{ijm}(\underline{x},\underline{y})\ t_m(\underline{x})\ d\Gamma(\underline{x}) - \int_\Gamma S_{ijm}(\underline{x},\underline{y})\ u_m(\underline{x})\ d\Gamma(\underline{x}) +$$

$$+ \int_\Omega D_{ijm}(\underline{x},\underline{y})\ b_m(\underline{x})\ d\Omega(\underline{x}) \tag{109}$$

The last integral can be transformed into a boundary integral following the same procedure shown above. According to Eq. 103, it can be written as

$$\int_\Omega D_{ijm}\ b_m\ d\Omega = \int_\Gamma D_{ijm}\ \Phi\ \eta_m\ d\Gamma + \frac{1-2\nu}{2(1-\nu)} [\int F_{ijm,m}\ \Phi_{,k}\ \eta_k\ d\Gamma -$$

$$- \int_\Gamma F_{ijm,mk}\ \Phi\ \eta_k\ d\Gamma - K_o \int_\Gamma F_{ijm}\ \eta_m\ d\Gamma] \tag{110}$$

where F_{ijm} is obtained from G_{im} by means of the same transformation used to obtain D_{ijm} from U_{im}.

Thermoelastic Problems

It is well known that the linear steady-state thermoelastic problem may be formulated using Duhamel-Neumann law (see, for instance, Ref. 38) that says that the effect of a temperature change θ in an elastic body is equivalent to adding a body force $-\gamma\ \theta_{,i}$ at each point and increasing the boundary tractions by $\gamma\ \theta\ \eta_j$, where $\gamma = 2\mu\ \alpha\ (1+\nu)/(1-2\nu)$ and α is the coefficient of thermal expansion. Thus, the thermoelastic problem is a particular case of the elastostatic problem with body forces.

The boundary integral equation for the thermoelastic problem in the absence of actual body forces is

$$C_{ij}\ u_j + \int_\Gamma T_{ij}\ u_j\ d\Gamma = \int_\Gamma U_{ij}\ t_j\ d\Gamma + \int_\Gamma U_{ij}\ \gamma\ \theta\ \eta_j\ d\Gamma - \\ - \int_\Omega U_{ij}\ \gamma\ \theta_{,j}\ d\Omega \qquad (111)$$

where the points on which each variable depends are implicit.

It is obvious that the potential of the equivalent body forces is

$$\Phi = -\gamma\ \theta$$

and the domain integral in Eq. 111 may be written as

$$-\int_\Omega U_{ij}\ \gamma\ \theta_{,j}\ d\Omega = \int_\Gamma -\gamma\ \theta\ U_{ij}\ \eta_j + \frac{1-2\nu}{2(1-\nu)}\ [\int_\Gamma G_{ij,j}\ (-\gamma\ \theta_{,k})\ \eta_k\ d\Gamma - \\ - \int_\Gamma G_{ij,jk}\ (-\gamma\ \theta)\eta_k\ d\Gamma - K_o \int_\Gamma G_{ij}\ \eta_j\ d\Gamma] \qquad (112)$$

Taking into account that for steady-state heat conduction $\Phi_{,jj} \neq 0$ and substituting Eq. 112 into Eq. 111 the following integral equation is obtained

$$C_{ij}\ u_j + \int_\Gamma T_{ij}\ u_j\ d\Gamma = \int_\Gamma U_{ij}\ t_j\ d\Gamma + \frac{(1-2\nu)\gamma}{2(1-\nu)} \int_\Gamma (G_{ij,jk}\ \theta\ \eta_k - G_{ij,j}\theta_{,k}\ \eta_k)d\Gamma \qquad (113)$$

The above integral equation may be discretized in the usual way. All the functions in the last integral of Eq. 113 are known and may be easily integrated by numerical procedures over the boundary elements.

Foundations and Underground Structures

The analysis of foundations and underground structures taking into account the interaction with the soil makes necessary the modeling of the structure and the soil. When the structure is deeply buried (for instance, tunnels) the soil is sometimes considered as a complete boundless space, while for many other cases of buried structures as well as for the analysis of foundations the soil is represented by the more realistic half-space model. Problems including boundless domains constitute one of the types for which the BEM is more advantageous.

Fig. 16 shows a very simple BE model used to study the stresses in the soil arround a circular cavity under internal pressure. The soil is assumed to be a linear elastic homogeneous space and thus, only the boundary of the cavity must be discretized. To do so, 24 constant elements are used. Tractions are prescribed for all the boundary nodes and stresses computed at 7 internal points. The values of the diagonal term of the stress tensor corresponding to the radial direction are shown in Table 2 where they compare well with the exact values known from the elasticity theory.

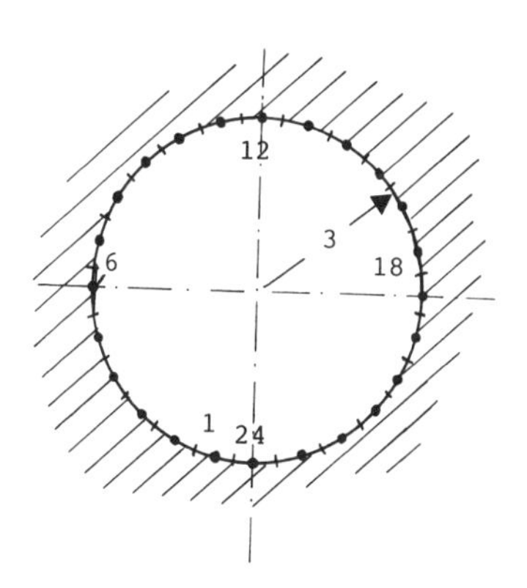

Figure 16. Circular cavity discretization (1)

One of the important aspects for the design of foundations is the computation of their stiffness (impedance) matrix, i.e., the matrix $\underline{K}$ that relates the vector of loads applied to the foundation and the

Distance to the centre of the cavity	24 Constant Elements	Exact values
4	-57.23	-56.25
6	-25.29	-25
10	-9.1	-9
20	-2.27	-2.25
50	-0.36	-0.36
200	-0.227×10^{-1}	-0.225×10^{-1}
1000	-0.911×10^{-3}	-0.9×10^{-3}

Table 2. Radial stress at internal points.

vector of displacements due to the loads in the form

$$\underline{F} = \underline{K}\ \underline{u} \tag{114}$$

The analysis of the response of foundations under the action of forces is not only an important problem in statics but also in dynamics where the time or frequency domain stiffness matrix is used for the design of foundations of vibrating machinery or other structures that are subjected to dynamic excitations and where the soil-structure interaction effect may be important. Since the dynamic formulation of the BEM and its use in dynamic soil-structure interaction are treated in other chapters, only the problem of static stiffnesses of foundations will be analyzed in this section.

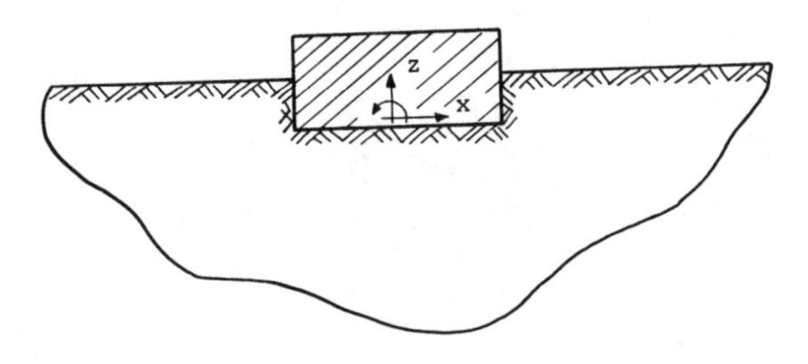

Figure 17. Rigid foundation embedded in the soil.

Fig. 17 shows a transversal section of a square foundation embedded in the soil. The most simple and probably most frequent assumption about the behavior of the soil is considering it to be a linear isotropic elastic half-space. The foundation is assumed to behave as a rigid body when excited by a force or a moment.

If the standard BEM with Kelvin's fundamental solution is used, the modelling of the soil requires the discretization of the soil-foundation interface and also of the soil boundless free surface. This fact introduces an approximation either because the discretization of the soil free surface extends only to a finite zone arround the foundation or because some kind of approximate infinite element is used. Fig. 18 shows one quarter of the boundary element discretization used for the analysis of an embeded square foundation. The level of embedment E/B = 4/3 and the amount of free-field taken into account is defined by A/B = 2.5. Constant displacement and traction elements are used. The boundary conditions are the displacements corresponding to a unit rigid body motion of the foundation over the interface elements and zero tractions over the free soil surface.

The BE mesh for the soil-foundation interface does not have to be very dense (15) since the stress resultants over the foundation and not the stress distributions are needed. The variation of the computed stiffnesses with the number of elements along half of the side of the bottom of an embedded foundation is shown in Fig. 19. The discretization of the lateral walls is varied consistently. It may be seen that values of $N \geq 6$ give accurate stiffnesses when compared with those extrapolated for $N = \infty$. The terms of the stiffness matrix shown in

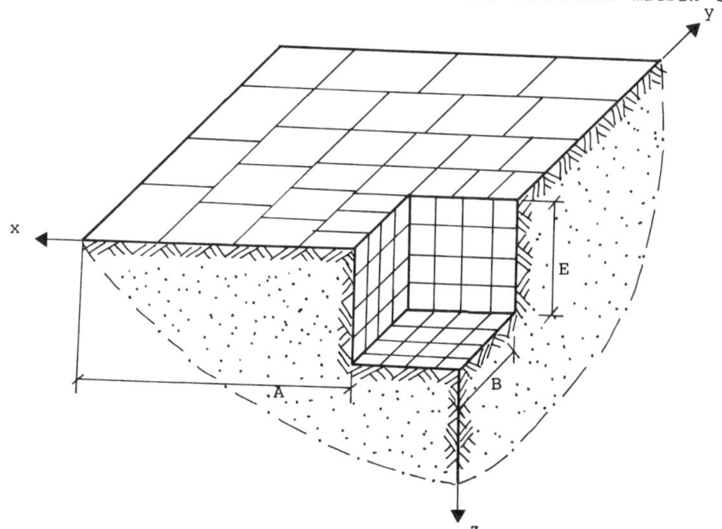

Figure 18. Discretization for one quarter of a square embedded foundation.

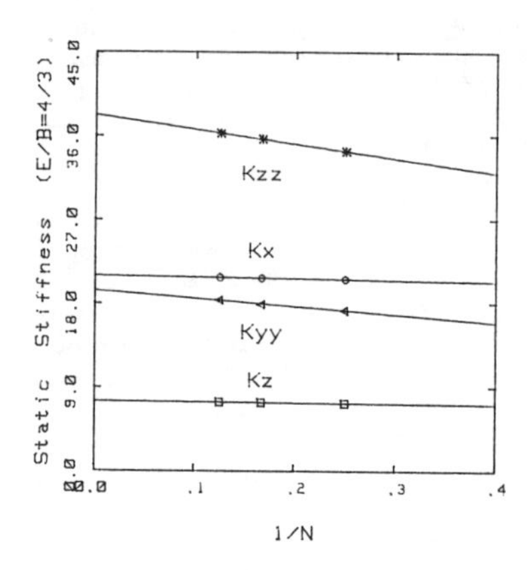

Figure 19. Effect of the interface discretization on the foundation stiffnesses.

Fig. 19 are normalized as follows: Horizontal stiffness: $K_{x\ norm}$ = $= K_x\ (2-\nu)/GB$; Vertical stiffness: $K_{z\ norm} = K_z\ (1-\nu)/GB$; Rocking stiffness: $K_{yy\ norm} = K_{yy}\ (1-\nu)/GB^3$; and Torsional stiffness: $K_{zz\ norm}$ $= K_{zz}/GB^3$.

The effect of the amount of soil free surface that is discretized may be seen in Fig. 20 for the same embedded foundation (E/B = 4/3). The study is done adding successively lines of constant elements arround the previous discretization of the soil free surface. The stiffnesses computed using Kelvin's fundamental solution for the complete space converge rapidly.

The discretization of the soil free surface may be avoided by using Mindlin's solution (31), corresponding to the point load in an elastic half-space, as fundamental solution. The BEM formulation remains the same with the only difference being that the discretization extends only over those parts of the boundary that are not soil free surface. The use of Mindlin's solution reduces the number of elements but the computer time per integration over a boundary element grows due to the greater amount of terms involved in the fundamental solution. A more efficient procedure is based on the use of the complete space fundamental solution plus another point load following the same direction located at the point image of the first with respect to the soil surface. The use of this solution makes the fundamental solution traction tensor $\underline{T}$, for points of the soil free surface, to have all the terms equal to zero except T_{13}, T_{23}, T_{31} and T_{32} and consequently, the effect on any soil free surface elements may be expected to be very small. Fig. 20 shows that when the double fundamental solution is used, almost identical values of the foundation stiffnesses are obtained for any amount of soil free surface discretized. The same behavior is observed for different levels of embedment and thus, it may be concluded that when the double force fundamental solution is used, no elements are needed over the soil free surface.

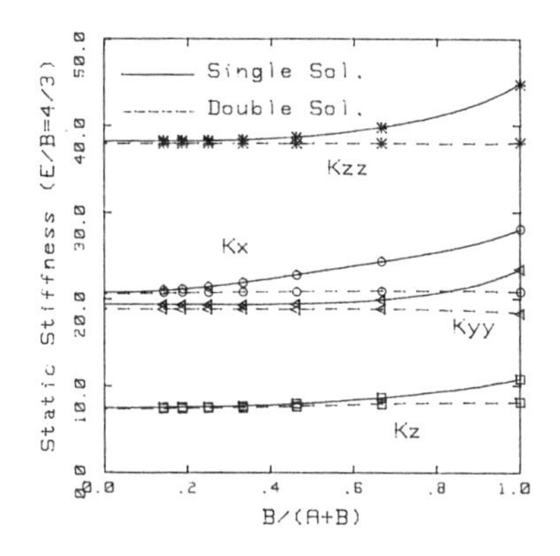

Figure 20. Effect of the free surface discretization on the foundation stiffnesses.

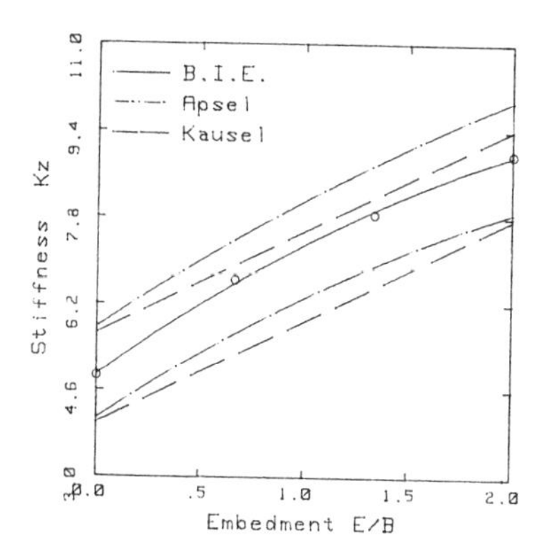

Figure 21. Normalized vertical stiffness for embedded square foundations.

As an example of the use of the BEM for computation of stiffnesses of three-dimensional foundations, Fig. 21 shows the value of the normalized vertical stiffness of square foundations versus the level of

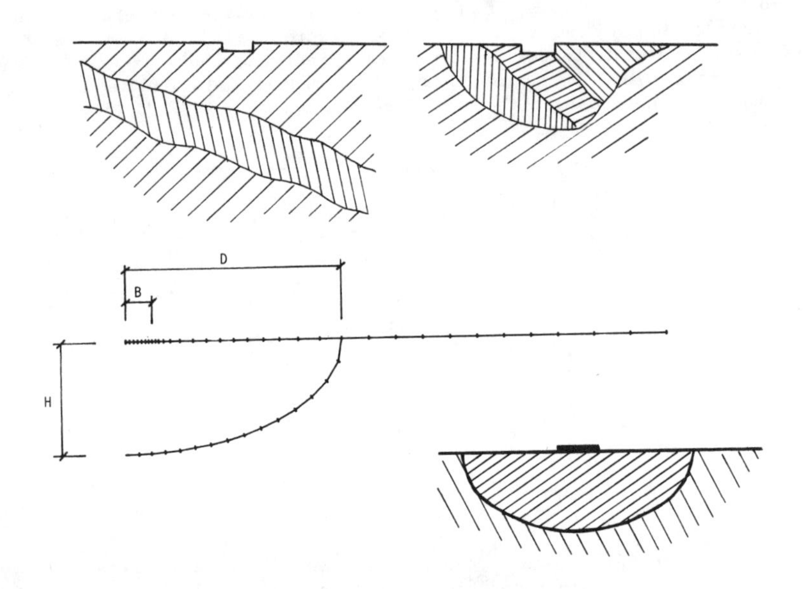

Figure 22. Zoned soil models and boundary discretization.

embedment. The values have been computed using the double fundamental solution and discretizing one quarter of the soil-foundation interface into constant rectangular elements, 16 for the bottom and elements of about the same size for the lateral walls. As may be seen in Fig. 21 the values computed for the static stiffness are between those corresponding to the inscribed and the circunscribed cylindrical foundations given by Apsel (4) and Kausel and Ushijima (23).

An example of the kind of soil models that may be studied using the BEM is shown in Fig. 22. A boundary element discretization for a non-homogeneous soil is also shown there.

Axisymmetric Problems

There are many elastostatics problems that present an axisymmetric geometry and very frequently also axisymmetric loading conditions. In the following, it is first explained how the BEM may be applied to axisymmetric problems with respect to both the geometry and the loading by taking advantage of the symmetry that reduces a three-dimensional analysis to two uncoupled plane domain problems; one with two degrees of freedom per point (radial and axial) and another with one degree of freedom per point (tangential). Nonaxisymmetric loads are studied later. The first formulation of the BEM for axisymmetric problems was done in

1975 (24), (30). It is based on the Somigliana's identity obtained from the application of the reciprocity theorem between the actual axisymmetric problem and the fundamental solutions corresponding to a radial ring load and an axial ring load, for one part of the problem, and a tangential ring load for the other (torsion) part of the problem, as indicated in Fig. 23.

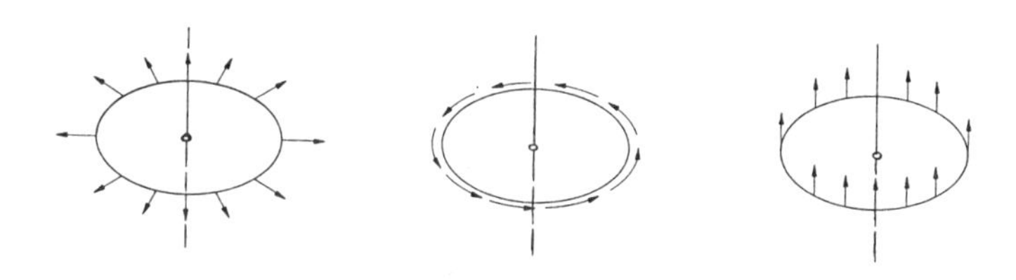

Figure 23. Ring loads.

Somigliana's identity has now the same expresion as in previous cases (Eqs. 17, 18 and 27) with U_{ij} and T_{ij} being the displacements and tractions due to the ring loads. Those fundamental solutions were obtained by different procedures by Kermanidis (24), Mayr (30), Cruse et al. (13), and Domínguez and Abascal (16). They are written in terms of Legendre functions or complete eliptic integrals, which makes their integration along the boundary elements rather involved. Explicit expressions of the ring loads fundamental solutions may be found, for instance, in Refs. (24), (13).

The BEM using Kelvin 's fundamental solution may also be applied to the analysis of axisymmetric problems. Integration of tensors $\underline{U}$ and $\underline{T}$ through axisymmetric elements must then be performed.

The basic BEM equation is written in cylindrical coordinates using matrix notation as,

$$\underline{C}^c(\underline{y})\ \underline{u}^c(\underline{y}) = \int_\Gamma \underline{Q}^T(\underline{y})\ \underline{U}(\underline{x},\underline{y})\ \underline{Q}(\underline{x})\ \underline{t}^c(\underline{x})\ d\Gamma(\underline{x}) - \\ - \int_\Gamma \underline{Q}^T(\underline{y})\ \underline{T}(\underline{x},\underline{y})\ \underline{Q}(\underline{x})\ \underline{u}^c(\underline{x})\ d\Gamma(\underline{x}) \tag{115}$$

where the superscriptc c stands for cylindrical,

$$\underline{u}(\underline{x}) = \underline{Q}(\underline{x})\ \underline{u}^c(\underline{x}) \\ \underline{u}(\underline{y}) = \underline{Q}(\underline{y})\ \underline{u}^c(\underline{y}) \tag{116}$$

the same being applicable for $\underline{t}(\underline{x})$, $\underline{t}(\underline{y})$,

$$\underline{C}^c(\underline{y}) = \underline{Q}^T(\underline{y})\ \underline{C}(\underline{y})\ \underline{Q}(\underline{y}) \tag{117}$$

and

$$\underline{Q}(\underline{x}) = \begin{bmatrix} \cos\theta(\underline{x}) & -\sin\theta(\underline{x}) & 0 \\ \sin\theta(\underline{x}) & \cos\theta(\underline{x}) & 0 \\ 0 & 0 & 1 \end{bmatrix} \tag{118}$$

The transformation of tensors $\underline{U}$ and $\underline{T}$ is in agreement with a more general study of the transformations of these tensors done by Rizzo and Shippy (36).

Since $\underline{y}$ is the collocation point selected to apply the unit point load, it may be assumed that $\theta(\underline{y}) = 0$ (Fig. 24), which makes the transformation matrix $\underline{Q}(\underline{y}) = \underline{I}$. The kernels of the integrals will have the form

$$\underline{U}^c = \underline{U}(\underline{x},\underline{y})\ \underline{Q}(\underline{x}) = \begin{bmatrix} U_{11}\cos\theta + U_{12}\sin\theta & -U_{11}\sin\theta + U_{12}\cos\theta & U_{13} \\ U_{21}\cos\theta + U_{22}\sin\theta & -U_{21}\sin\theta + U_{22}\cos\theta & U_{23} \\ U_{31}\cos\theta + U_{32}\sin\theta & -U_{31}\sin\theta + U_{32}\cos\theta & U_{33} \end{bmatrix} \tag{119}$$

where $\theta \equiv \theta(\underline{x})$.

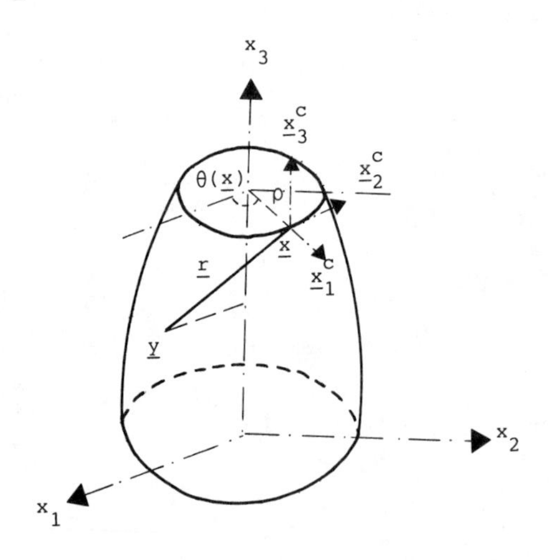

Figure 24. Cylindrical coordinates.

Equation 115 may be written in a short form as

$$\underline{c}^c \, \underline{u}^c(\underline{y}) = \int_\Gamma \underline{U}^c \, \underline{t}^c(\underline{x}) \, d\Gamma - \int_\Gamma \underline{T}^c \, \underline{u}^c(\underline{x}) \, d\Gamma \tag{120}$$

One half of a meridional section of the body is discretized into elements (Fig. 25) and Eq. 120 may be written for the boundary node k as

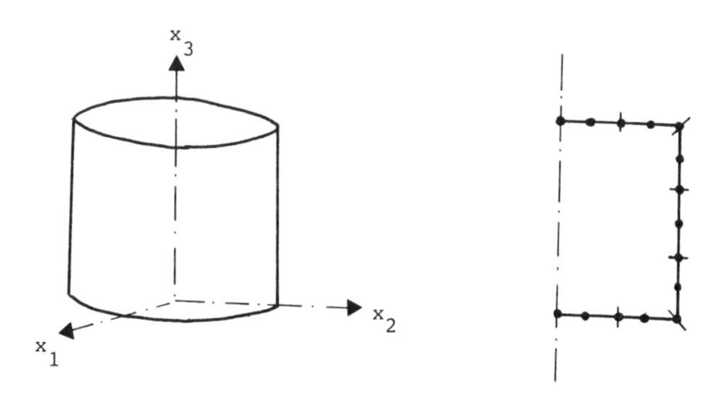

Figure 25. Quadratic BE discretization for axisymmetric problem.

$$\underline{c}^k \, \underline{u}^k + [\sum_{e=1}^{M} \int_{\Gamma_e} \rho \, \underline{\psi} \, d\Gamma_e \int_0^{2\pi} \underline{T}^c \, d\theta] \, \underline{u}_n^c = [\sum_{e=1}^{M} \int_{\Gamma_e} \rho \, \underline{\psi} \, d\Gamma_e \int_0^{2\pi} \underline{U}^c \, d\theta] \, \underline{t}_n^c \tag{121}$$

where Γ_e are boundary elements of the section and $\underline{\psi}$ are shape functions along line elements. The integration along θ may be easily done by numerical procedures. The submatrices $\underline{H}^{kl}$ and $\underline{G}^{kl}$ that relate two nodes k and l have the pattern

$$\begin{vmatrix} * & 0 & * \\ 0 & * & 0 \\ * & 0 & * \end{vmatrix} \begin{matrix} \leftarrow \rho \\ \leftarrow \theta \\ \leftarrow z \end{matrix} \tag{122}$$

where the zeros denote elements that are null due to the skewsymmetry of the corresponding terms in Eq. 119. It is clear that the torsion and the radial-axial problems are uncoupled and both may be studied on a plane domain.

It should be noticed that since a rigid body motion along the radial coordinate of an axisymmetric body is meaningless, the terms of $\underline{c}^c$ cannot be computed from the rest of the terms of the $\underline{H}$ matrix. However, they may be computed using the analytical expression of Eq. 28 given for plane problems.

When the boundary conditions are not axisymmetric, the problem may still be analysed by means of a plane model. The problem is divided into a number of uncoupled plane problems by representing the prescribed loading or displacement by a Fourier series along the tangential coordinate (43). Each term of the series produces displacements and stresses in the same Fourier mode and if the prescribed values do not vary very rapidly around the axis, a few modes will be enough for an accurate solution. The Fourier expansion is of the form

$$
\begin{aligned}
u_\rho &= \sum_{n=0}^{\infty} (u^s_{n\rho} \cos n\,\theta + u^a_{n\rho} \sin n\,\theta) \\
u_\rho &= \sum_{n=0}^{\infty} (-u^s_{n\rho} \sin n\,\theta + u^a_{n\rho} \cos n\,\theta) \\
u_z &= \sum_{n=0}^{\infty} (u^s_{nz} \cos n\,\theta + u^a_{nz} \sin n\,\theta)
\end{aligned}
\tag{123}
$$

where "s" indicates the symmetric terms and "a" the antisymmetric ones.

For each Fourier mode amplitude a discretized boundary equation like Eq. 121 may be written with the only difference being that $\underline{T}^c$ and $\underline{U}^c$ are now weighted by a sine or a cosine function and are the integrals around the axis of the form

$$
\int_0^{2\pi} \underline{U}^c \sin n\theta \; d\theta \quad ; \quad \int_0^{2\pi} \underline{U}^c \cos n\,\theta \; d\theta \tag{124}
$$

It is worth noting that since $\sin n\,\theta$ has zero value at $\theta = 0$, the $\underline{y}$ point cannot be located at $\theta(\underline{y}) = 0$ to compute the amplitude of those terms of the Fourier series. One only has to move $\underline{y}$ to a point, for instance $\theta(\underline{y}) = -\pi/2n$, where the amplitude is not zero. This change is easily taken into account by a shift of the origin of θ in Eq. 119 and consequently in Eq. 124.

Computational Aspects

There are many computational aspects that must be taken into account when writing a BEM program. Since a chapter of this book is dedicated to computer programs, only a brief description of the procedures used for computation of the elements of the $\underline{G}$ and $\underline{H}$ matrices will be presented in this section. Those terms are obtained by integration of the products of the $\underline{U}$ and $\underline{T}$ tensors and the shape functions over the boundary elements (Eq. 45).

The immediate procedure for computing the integrals is the use of a standard numerical quadrature. However, it is clear that due to the existence of a singularity in the functions of the integrand, even though the integrals are used in the Cauchy Principal Value sense, the standard numerical quadrature will be inadequate when the distance between the collocation point $\underline{y}$ and the integration point $\underline{x}$ tends to zero. One way of solving this problem is using a special integration approach for those elements to which the collocation point belongs.

The integration over the elements including the collocation point may be done in several ways. First of all, an analytical integration may be carried out. This scheme implies a somewhat tedious analytical work for developing the program and is mostly used for two-dimensional programs (8), (1). A second procedure, which reduces the analytical work for three-dimensional programs, is the use of a semianalytical integration scheme. An analytical integration is performed for one of the coordinates of the surface element and a standard numerical quadrature is used for the other coordinate along boundaries of the element where no singularity exist (15). The third procedure for the integration over the elements that contain the collocation point is the use of a special numerical integration scheme. For two-dimensional problems (one-dimensional integration), the integrals including $\underline{U}$ are singular only in a weak sense and may be accurately computed using a numerical quadrature weighted by a logarithmic function (42), (39). For three-dimensional domains, there are some schemes that are based on a subdivision of the boundary element into sublements which become smaller as they get closer to the collocation point (Fig. 26). A standard numerical quadrature is used for each subelement (26), (27). It is reminded that the integrals including $\underline{T}$ and containing a singularity can be determined in an implicit form by means of a rigid body motion.

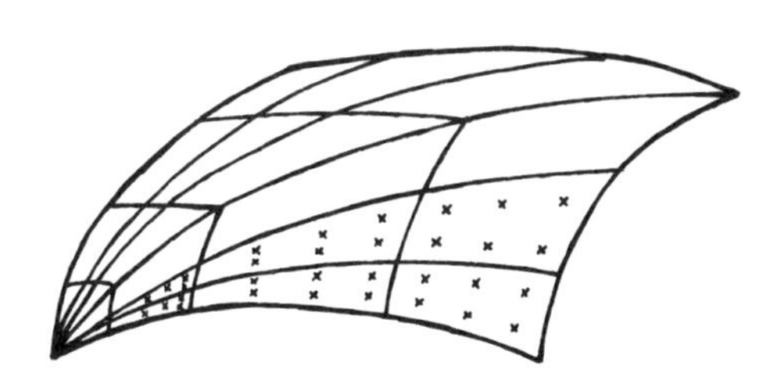

Figure 26. Element subdivision.

The efficiency of the integrations over the elements that do not include the collocation point may be increased by using an adaptive numerical integration scheme (27) that either uses a variable number of integration points depending on the distance between the collocation point and the integration element or does a subdivision of the integration element where the number of subelements depends on that same distance.

A special numerical integration scheme is succesfully used for axisymmetric problems (19). When Kelvin's solution is integrated around the axis, a Gauss quadrature may be applied to every semiring; however, its accuracy is easily improved by increasing the density of integration points arround the collocation point by means of a parabolic transformation of the circumferencial coordinate

$$\theta = \frac{\pi}{4}\,(\xi + 1)^2 \quad ; \quad -1 \leq \xi \leq 1 \tag{125}$$

A complete study of computational aspects of the BEM for different applications may be found in the work by Doblaré (14).

Combination with the Finite Element Method

Since the BEM began to be known by the engineering community some papers have been written presenting so many advantages of the method that it would seem like the FEM was coming to its end. On the other hand some of the people working with the FEM would say that there was no future for the BEM. After more than fifteen years of living together it is widely recognized that both methods have advantages and drawbacks, that there are types of problems for which one of them is more appropriate than the other and that codes including the possibility of FE models, BE models and a combination of both promise to be extremely efficient.

Among the advantages of the BEM, when compared with the FEM, the following may be emphasized:

- The dimension of the problem is reduced by one which implies a smaller number of degrees of freedom. This fact reduces not only the size of the systems of equations but what is more important, the amount of time needed for preparation of the data. This time, as opposed to computer time, is every day more expensive.

- Infinite domains are modelled by discretization only of the internal boundaries.

- Stress singularities are accurately represented.

Among the drawbacks of the BEM the following may be enumerated:

- The matrices of the system of equations are fully populated and non-symmetric.

- As a general rule, the computer time required to determine each coefficient of the system is larger than it is in the FEM.

- The material properties must be homogeneous or at least piece-wise homogeneous; however, if the medium consists of small homogeneous zones the number of degrees of freedom will grow very rapidly.

Advantages and drawbacks, within other fields, like elastodynamics or plasticity, are not mentioned.

The idea of representing certain parts of a problem using BE and others using FE is immediate, trying to use each model where it is more appropriate. To do so, a stiffness matrix of the BE zones, as shown in Fig. 27, relating nodal displacements and forces is obtained from the BEM system of equations. The matrix is assembled to the FEM stiffness matrix of the rest of the domain.

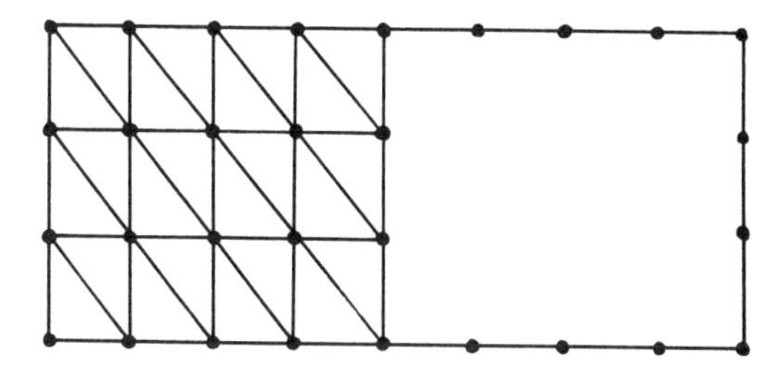

Figure 27. BE-FE coupling.

Assuming the domain of the problem is that of Fig. 27, one may write for the BE region the following system of equations:

$$\underline{H}\ \underline{u}_n = \underline{G}\ \underline{t}_n \tag{38}$$

In order to obtain a stiffness matrix, one first has to transform tractions into forces. This may be done by using the shape functions. Thus

$$\underline{F}' = \underline{N}^T\ \underline{t}_n \tag{126}$$

where

$$\underline{N}^T = \int_\Gamma \begin{vmatrix} \underline{\psi}^1 & & \underline{0} \\ & \underline{\psi}^2 & \\ \underline{0} & & \underline{\psi}^N \end{vmatrix} d\Gamma \tag{127}$$

and

$$\underline{\psi}^i = \begin{bmatrix} \psi^i & 0 & 0 \\ 0 & \psi^i & 0 \\ 0 & 0 & \psi^i \end{bmatrix} \tag{128}$$

By inversion of $\underline{G}$, it is obtained

$$(\underline{N}^T \underline{G}^{-1} \underline{H}) \underline{u}_n = \underline{N}^T \underline{t}_n \tag{129}$$

which is of the form

$$\underline{K}' \underline{u}_n = \underline{F}' \tag{130}$$

where $\underline{K}'$ is the stiffness matrix of the BE zone.

This stiffness matrix may be directly assembled to the FE matrix and the resulting system solved. The main difficulty that is found in the matrix given by Eq. 130 is its lack of symmetry. This fact reduces to an important extent the efficiency of the solution process and it is very convenient that the stiffness matrix $\underline{K}'$ may be transformed into a symmetric form.

The idea of a nonsymmetric stiffness matrix is difficult to accept because it seems to be in disagreement with the reciprocal theorem. However, the lack of symmetry may be explained. When a unit force (Eq. 130) is prescribed at a certain boundary point, what is really prescribed is a traction distribution that depends on the boundary shape function and local geometry. When the unit force is prescribed at a different point the traction distribution will be different and there is no reason to obtain a displacement reciprocal of the first. It might be said that FE point loads at the boundary also imply a traction distribution; however, the main difference is that the FE stiffness matrix is obtained from displacement patterns consistent with the load distribution, while in the BEM the displacements due to certain tractions, and vice-versa, are determined from the fundamental solution that doesn't follow any particular shape. As may be expected the accuracy of the model does not change very much if the $\underline{K}'$ matrix is symmetrized.

Several justifications and methods of symmetrization have been proposed. Error minimization (8) or energy consideration (44) justify in a more elegant and rigurous way what is the immediate procedure to make $\underline{K}'$ become symmetric; i.e., its substitution by

$$\underline{K}'_s = \frac{1}{2}\left(\underline{K}' + \underline{K}'^T\right) \tag{131}$$

which is ready to be coupled to the FEM stiffness matrix. Several other procedures for symmetrization have been presented (41) with their efficiency being similar to the one above.

Examples of problems solved using a combination of the FEM and the BEM may be found, for instance, in Refs. (44), (9). The results obtained are very promising.

References

1. Alarcón, E., Brebbia, C., and Domínguez, J., "The Boundary Element Method in Elasticity", International Journal of Mechanical Sciences, Vol. 20, 1978, pp. 625-639.

2. Alarcón, E., Martín, A., and París, F., "BE in Potential and Elasticity Theory", Computeres & Structures, Vol. 10, 1979, pp. 351-362.

3. Alarcón, E., Letters to the editor, International Journal for Numerical Methods in Engineering, Vol. 19, 1983, p. 1105.

4. Apsel, R.J., "Dynamic Green's Function for Layered Media and Applications to Boundary Value Problems", Ph. D. Thesis, Univ. Calif. San Diego, 1979.

5. Banerjee, P.K., and Butterfield, R., "Boundary Element Methods in Engineering Science", McGraw-Hill, London, 1981.

6. Blandford, G.E., Integraffea, A.R., and Liggett, J.A., "Two-Dimensional Stress Intensity Factor Computations Using the Boundary Element Method", International Journal for Numerical Methods in Engineering, Vol. 17, 1981, pp. 387-404.

7. Brebbia, C.A., and Domínguez, J., "Boundary Element Method for Potential Problems", Applied Mathematical Modelling, Vol. 1, 1977, pp. 372-378.

8. Brebbia, C.A., "The Boundary Element Method for Engineers", Pentech Press, London, 1978.

9. Brebbia, C.A., Telles, J.C.F., and Wrobel, L.C., "Boundary Element Techniques", Springer-Verlag, Berlin, 1984.

10. Chaudonneret, M., "On the Discontinuity of the Stress Vector in the Boundary Integral Equation Method for Elastic Analysis", Recent Advances in Boundary Element Methods, C.A. Brebbia, Ed. , Pentech Press, London, 1978.

11. Crouch, S.L., "Solution of Plane Elasticity Problems by the Displacement Discontinuity Method", International Journal for Numerical Methods in Engineering, Vol. 10, 1976, pp. 301-343.

12. Cruse, T.A., "Mathematical Foundations of the Boundary Integral Equation Method in Solid Mechanics", Report No. AFOSR-TR-77-1002, Pratt and Whitney Aircraft Group, 1977.

13. Cruse, T.A., Snow, D.A., and Wilson, R.B., "Numerical Solutions in Axisymmetric Elasticity", Computeres & Structures, Vol. 8, 1977, pp. 445-451.

14. Doblaré, M., "Computational Aspects of the Boundary Element Method" Topics in Boundary Element Research, C.A. Brebbia, Ed. , Vol. 2, 1985.

15. Domínguez, J., "Dynamic Stiffness of Rectangular Foundations", M.I.T. Research Report R-78-20, Civil Eng. Dept., 1978.

16. Domínguez, J., and Abascal, R., "On Fundamental Solutions for the Boundary Integral Equations Method in Static and Dynamic Elasticity", Engineering Analysis, Vol. 1, 1984, pp. 128-134.

17. Fredholm, L., "Sur une Classe D'equations Fonctionelles", Acta Mathematica, Vol. 27, 1903, pp. 365-390.

18. Fung, Y.C., "Foundations of Solid Mechanics", Prentice Hall, Englewood Cliffs, 1969.

19. Gómez-Lera, M.S., Domínguez, J., and Alarcón, E., "On the Use of a 3-D Fundamental Solution for Axisymmetric Steady-State Dynamic Problems", Boundary Elements VII, Proceedings of the 7th International Conference, Italy, C.A. Brebbia and G. Maier, Eds., Springer-Verlag, Berlin, 1985.

20. Hartmann, F., "Elastostatics", Progress in Boundary Element Method, C.A. Brebbia Ed., Pentech Press, London, 1981.

21. Jaswon, M.A., "Integral Equation Methods in Potential Theory I", Proceedings of the Royal Society, Vol. 275, 1963, pp. 23-32.

22. Jaswon, M.A., and Symm, G.T., "Integral Equation Methods in Potential Theory and Elastostatics", Academic Press, London, 1977.

23. Kausel, E., and Ushijima, R., "Vertical and Torsinal Stiffness of Cylindrical Footing", M.I.T. Research-Report, R-79-63, Civil Eng. Dept., 1979.

24. Kermanidis, T.A., "Numerical Solution for Axially Symmetrical Elasticity Problems", Intern. J. of Solids and Structures, Vol. 11, 1975, pp. 493-500.

25. Lachat, J.C., "A Futher Development of the Boundary Integral Technique for Elastostatics", Ph. D. Thesis, University of Southampton 1975.

26. Lachat, J.C., and Watson, J.O., In: "Boundary Integral Equation Method: Computational Applications in Applied Mechanics", Cruse and Rizzo Eds., AMD 11, ASME, New York, 1975.

27. Lachat, J.C., and Watson, J.O., "Effective Numerical Treatment of Boundary Integral Equations", International Journal for Numerical Methods in Engineering, Vol. 10, 1976, pp. 991-1005.

28. Love, A.E.H., "A Treatise on the Mathematical Theory of Elasticity" Dover, New York, 1944.

29. Martínez, J., and Domínguez, J., "On the Use of Quarter-Point Boundary Elements for Stress Intensity Factor Computations", International Journal for Numerical Methods in Engineering, Vol. 20, 1984, pp. 1941-1950.

30. Mayr, M., "Ein Integralgleichungsverfahren Losung Rotationssymetrischer Elastizitatsprobleme", Dissertation, Technical University München, 1975.

31. Mindlin, R.D., "Force at a Point in The Interior of a Semi-Infinite Solid", Physics, Vol. 7, 1936, pp. 195-202.

32. Muskhelishvili, N.I., "Some Basic Problems of the Mathematical Theory of Elasticity", Noordhoff, Groningen, 1953.

33. Peterson, R.E., "Stress Concentration Factors", John Wiley, New York, 1974.

34. Rizzo, F.J. "An Integral Equation Approach to Boundary Value Problems of Clasical Elastostatics", Quarterly of Applied Mathematics, Vol. 25, 1967, pp. 83-95.

35. Rizzo, F.J., and Shippy, D.J., "An Advanced Boundary Integral Equation Method for 3-D Thermoelasticity", International Journal for Numerical Methods in Engineering, Vol. 11, 1977, pp. 1753-1768.

36. Rizzo, F.J., and Shippy, D.J., "Some Observations on Kelvin's Solution in Clasical Elastostatics as a Double Tensor Field with Implications for Somigliana's Integral", Journal of Elasticity, Vol. 13, 1983, pp. 91-97.

37. Rudolphi, T.J., "An Implementation of the BEM for Zoned Media with Stress Discontinuities", International Journal for Numerical Methods in Engineering, Vol. 19, 1983, pp. 1-15.

38. Sokolnikoff, I.S., "Mathematical Theory of Elasticity", McGraw-Hill, New York, 2nd edt., 1956.

39. Stroud, A.H., and Secrest, D., "Gaussian Quadrature Formulas", Prentice-Hall, Englewood Cliffs, NJ, 1966.

40. Symm, G.T., "Integral equation methods in potential theory. II", Proceedings of the Royal Society, Vol. 275, 1963, pp. 33-46.

41. Tullberg, O., and Bolteus, L., "A Critical Study of Different Boundary Element Stiffness Matrices", Boundary Element Method in Engineering, C.A. Brebbia ed., Springer-Verlag, Berlin, 1982.

42. Watson, J.O., "Advanced Implementation of the Boundary Element Method for Two and Three-Dimensional Elastostatics", Developments in Boundary Element Method, Banerjee and Butterfield,Eds., Applied Science, London, 1979.

43. Wilson, E., "Structural Analysis of Axisymmetric Solids", AIAA Journal, Vol. 3, 1965, pp. 2269-2274.

44. Zienkiewicz, O.C., Kelly, D.W., and Bettess, P., "The Coupling of FEM and Boundary Solution Procedures", International Journal for Numerical Methods in Engineering, Vol. 11, 1977, pp. 355-375.

Stability Analysis of Beams and Plates

George D. Manolis,*

The linear elastic stability analysis of Bernoulli-Euler beams and Kirchhoff (thin) plates by using the direct boundary element method is presented in this chapter. The formulation is based on the reciprocal work theorem of Betti although other techniques, such as the method of weighted residuals, can also be used. Two classes of fundamental solutions are used; namely, fundamental solutions which correspond to pure flexure and ones that incorporate the effects of axial (in-plane) forces on bending. In the former case, discretization of the boundary as well as of the interior of the problem is necessary, while in the latter case discretization of the boundary only is sufficient. The price that is paid for dispensing with the interior discretization is that the fundamental solutions in the latter case are more complicated than the ones in the former case. Numerical examples are subsequently presented to illustrate the methodology and to demonstrate its applicability to stability problems. The basic conclusion drawn is that both classes of fundamental solutions yield perfectly adequate results.

Introduction

Linear elastic stability analysis of beams and plates has been studied analytically (22) and numerically (11,12) for simple as well as complicated geometries, loading and boundary conditions, respectively. Among the numerical methods that have been used for the determination of the elastic critical load of beams and plates one can mention the Finite Difference Method (FDM) and the Finite Element Method (FEM). A comprehensive exposition with applications can be found in Ghali and Neville (12) for the former method and in Gallagher (11) for the latter method.

During the last fifteen years, the Boundary Element Method (BEM) has been successfully used to formulate in integral equation form and numerically solve a wide variety of problems in engineering science as the recent treatises of Banerjee and Butterfield (3) and Brebbia et al (6) clearly demonstrate. The primary advantage of this technique is that normally only a boundary discretization of the domain of interest is needed, instead of a boundary as well as an interior discretization required by other 'domain type' methods, such as the FDM and the FEM, provided that the correct fundamental solutions are employed in the formulation of the problem. Furthermore, the BEM can easily accommodate complex geometries and boundary conditions and handle

*Associate Professor, Department of Civil Engineering, State University of New York, Buffalo, NY 14260

infinite or semi-infinite domains without any difficulty.

The bending of beams using the BEM is elaborated in Ref. 3. Use of boundary elements instead of finite elements does not carry any advantage for this class of problems. Perhaps because of this, there is a paucity of information on the solution of frames by boundary element (or boundary integral) techniques. Plate flexure using integral equation methods was first studied by Jaswon and Maiti (14), who employed an indirect formulation. Indirect formulations (2,15) usually do not involve the natural variables of the problem, such as nodal displacements, but instead rely on source distribution densities which define harmonic potentials. The displacements are then related to the derivatives of these harmonic potentials. Direct formulations, which are based on the reciprocal work identity and involve the natural boundary variables (displacement, slope, moment, and shear), were first used by Forbes and Robinson (9) and later by others (4,19,23). A review of the general theory and applications of the boundary integral method to thin plate bending can be found in Stern (20). Brunet (7), by employing an integral equation - finite difference scheme, was able to numerically study the stability of thin wall tapered beams. Sekiya and Katayama (18) and Tai, Katayama and Sekiya (21) employed an integral equation formulation to numerically determine the elastic critical load of beams and plates for which the fundamental solution was obtained either by the FDM or experimentally from surface strain measurements. Niwa, Kobayashi and Kitahara (17) in their comprehensive study of determining the natural frequencies of plates by both the direct and indirect BEM, also very briefly indicated how a stability analysis can be done, and determined the critical load of a circular clamped plate as a byproduct of their eigenfrequency analysis. Very recently, Gospodinov and Ljutskanov (13) using an indirect BEM, were able to determine the critical load of a square plate. In both Refs. 18 and 21, the authors do not take full advantage of the potential of their integral equation formulations and, by introducing fundamental solutions in discrete form, restrict the range of applicability of their method and introduce inaccuracies. In one case (18) for instance, the fundamental solution is obtained by solving the corresponding flexural problem a number of times with the point load at different locations using FDM. In the other case (21), the surface strains of typical specimens under flexure must be experimentally measured before the fundamental solution can be synthesized. References 13 and 17 describe a more general methodology, but the presentation is very brief. Very recently, two references (5,8) describe the buckling of simply supported rectangular and triangular plates. Both use fundamental solutions for flexure and this necessitates area integrations. Furthermore, Ref. 5 includes, in an iterative manner, the effect of an elastic foundation on the plates' buckling loads.

This chapter presents a general way of treating stability problems, namely, the direct BEM in conjunction with fundamental solutions for point loads obeying 'radiation type' of boundary conditions. Numerical discretization is necessary in order to reproduce the boundary conditions and loading pattern. The formulation is based on the reciprocal work theorem of Betti (16), although other ways are possible. Either exact fundamental solutions of the general stability problem are utilized, in which case only boundary discretization is required, or approximate fundamental solutions of the corresponding static problem are utilized, in which case both boundary and interior discretization is necessary. This last characteristic allows recovery of modal shapes from values of the displacements at boundary and interior nodes. Numerical examples drawn from various references illustrate the technique and demonstrate its advantages.

Beam Stability

This section presents the formulation and solution of the classical Bernoulli-Euler beam stability problem by the BEM. The sign convention used follows that of Timoshenko and Gere (22). Consider the beam element in axial compression shown in Fig. 1(a) and under a uniformly distributed load q. The governing equation is

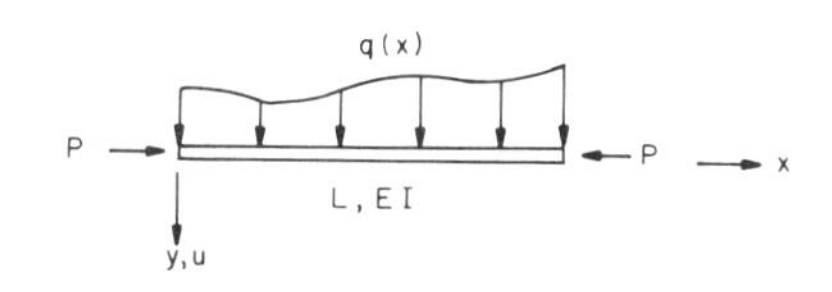

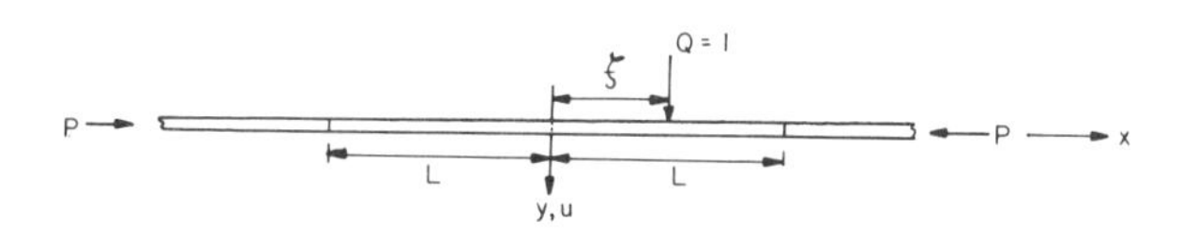

Figure 1. (a) Beam-column Element and (b) Beam-column of infinite extent under a point load

$$EIu''''(x) + Pu''(x) = q(x) \tag{1}$$

where $'$ denotes the operation d/dx and

$$\theta(x) = u' \tag{2}$$

$$m(x) = -EIu'' \tag{3}$$

$$s(x) = -EIu''' - Pu' \tag{4}$$

In the above, θ is the rotation, m the moment, s the shear and EI the flexural rigidity. The fundamental solution U is the solution of Eq. 1 for a concentrated unit load Q replacing the distributed load q as shown in Fig. 1(b), i.e.,

$$EIU''''(x,\xi) + PU''(x,\xi) = \delta(x,\xi) \tag{5}$$

where ξ is the location of Q and $\delta(x,\xi)$ is the delta function. The solution for $U(r)$, where $r = x - \xi$, is (16)

$$U(r) = (1/(2Pk))(-\sin k|r| + \tan kL \cos k|r| + k(|r| - L)) \tag{6}$$

where $k^2 = P/(EI)$. This solution is applicable to a beam of 'infinite' extent and the boundary conditions are that the transverse displacement decays to zero past a length L from the origin, i.e.,

$$U(+L) = U(-L) = 0 \tag{7}$$

Let $\Theta(x,\xi) = U'(x,\xi) = (dU(r)/dr)(dr/dx)$, where $dr/dx = \text{sgn}(r)$ which is equal to +1 if $r>0$ and -1 if $r<0$. Similarly, $M(x,\xi)$ and $S(x,\xi)$ are computed in view of Eqs. 3 and 4, respectively. We define u, θ, m and s to be state 1 and U, Θ, M and S to be state 2. The expressions for state 2 are collected in Appendix II.

To derive the BEM integral identity we multiply Eq. 1 by U and integrate from 0 to L, i.e.,

$$\int_0^L EIu'''' \, Udx + \int_0^L Pu''Udx = \int_0^L Uqdx \tag{8}$$

After integrating by parts the first expression on the left hand side four times and the second expression two times, using Eqs. 2-4 for both elastic states, and employing Eq. 5, we get

$$\int_0^L u(x)\, \delta(x,\xi)dx = u(\xi) = [Us - \Theta m + \theta M - uS]|_0^L + \int_0^L qUdx \tag{9}$$

Equation 9 is the Betti reciprocal theorem relating the unknown elastic state 1 with state 2. An additional equation can be derived by differentiating Eq. 9 with respect to ξ, where $d/d\xi = -d/dx$, as

$$\theta(\xi) = [U's - \Theta'm + \theta M' - uS']\Big|_0^L + \int_0^L qU'dx \tag{10}$$

where U', Θ', M' and S' are also given in Appendix II. All that remains is to evaluate Eqs. 9 and 10 at $\xi = L - \varepsilon$ and at $\xi = 0 + \varepsilon$ as $\varepsilon \to 0$. This yields a 4x4 system of equations

$$\begin{Bmatrix} \hat{u} \\ \hat{\theta} \end{Bmatrix} = \begin{bmatrix} \hat{U} & \hat{\Theta} \\ \hat{U}' & \hat{\Theta}' \end{bmatrix} \begin{Bmatrix} \hat{s} \\ \hat{m} \end{Bmatrix} + \begin{bmatrix} \hat{S} & \hat{M} \\ \hat{S}' & \hat{M}' \end{bmatrix} \begin{Bmatrix} \hat{u} \\ \hat{\theta} \end{Bmatrix} + \int_0^L \begin{Bmatrix} U \\ U' \end{Bmatrix} qdx \tag{11}$$

where column vector $\hat{u} = \lfloor u(L), u(0) \rfloor^T$, the submatrix

$$[\hat{U}] = \begin{bmatrix} U(L,L) & -U(0,L) \\ U(L,0) & -U(0,0) \end{bmatrix} \tag{12}$$

and similarly for the remaining expressions.

The method of solution is similar to the one employed in the FEM (3), where a generalized displacement vector is related to a generalized load vector through the usual flexural stiffness matrix plus a geometric stiffness matrix, in which the axial load P appears as a parameter. Equation 11 is therefore written for all beams composing the structure and a global matrix equation is thus obtained by superposition. For the computation of buckling loads, the term containing the distributed load q is neglected. Upon application of the boundary conditions a composite matrix is obtained whose determinant must vanish. This leads to a transcendental equation whose first root is the critical load and subsequent roots yield the remaining buckling loads. For small size problems, an iterative approach using the false position method to minimize the number of iterations is adequate (10).

In the above formulation, one may use the fundamental solution $\bar{U}$ of the flexural problem as an approximation, i.e., the solution of

$$EI\bar{U}''''(x,\xi) = \delta(x,\xi) \tag{13}$$

in lieu of Eq. 5. For this case (3)

$$\bar{U}(r) = (L^3/(12EI))(2 + (|r|/L)^3 - 3(|r|/L)^2) \tag{14}$$

which has the same boundary conditions as before, i.e., $\bar{U}(L) =$

$\bar{U}(-L) = 0$. The corresponding rotation, moment, and shear associated with $\bar{U}(r)$, as well as the first derivatives of these quantities that define elastic state 2 are also collected in Appendix II. As far as the BEM formulation using $\bar{U}(r)$ is concerned, the difference comes in deriving Eq. 9 , where we have the term

$$\int_0^L u(EI\bar{U}'''' + P\bar{U}'')dx = u(\xi) + \int_0^L Pu\bar{U}''dx \tag{15}$$

instead of just $u(\xi)$ as before. This implies that the right hand sides of Eqs. 9 and 10 need now to be augmented by the terms

$$\left.\begin{array}{l} P\int_0^L u(x)\ \bar{U}''(x,\xi)dx \\ \\ \text{and} \\ \\ P\int_0^L u(x)\ \bar{U}'''(x,\xi)dx \end{array}\right\} \tag{16}$$

respectively. It is precisely these terms that give rise to the interior discretization. We now need to consider beam elements that are 'small' enough to allow for a linear variation of the transverse displacement between the end nodes. Thus, the above integrals of $\bar{U}''(r) = (L/(2EI))(|r/L|-1)$ and $\bar{U}'''(r) = (1/(2EI))\text{sgn}(r)$ over the length can easily be evaluated analytically. Other than that, the numerical procedure used in conjunction with the exact fundamental solutions of the buckling problem remains unchanged and can be used here as well.

Plate Stability

This section develops the BEM as applied to thin (Kirchhoff) plate theory. Once more, the sign convention of Ref. 1 is adopted. The governing equation for plate buckling in cartesian coordinates is

$$D\nabla^4 w - (N_{xx}\partial^2 w/\partial x^2 + N_{yy}\partial^2 w/\partial y^2 + 2N_{xy}\partial^2 w/\partial x\partial y) = q \tag{17}$$

where $w(x,y)$ is the transverse deflection, D the plate's flexural rigidity, $N_{ij}(x,y)$ the in-plane forces, and $q(x,y)$ the distributed load. Furthermore, ∇ is the del operator. The moments per unit length and the shears are defined as

$$M_{ij} = -D[(1 - \nu)w,_{ij} + \nu\, w,_{kk}\, \delta_{ij}] \tag{18}$$

and

$$Q_i = M_{ij,j} \tag{19}$$

respectively, where ν is Poisson's ratio and δ_{ij} the Kronecker delta. Also, repeated symbols in the indicial notation used above imply summation and the index following a comma denotes spatial differentiation. The in-plane forces are assumed to remain unchanged during bending and obey the following equations of equilibrium

$$\partial N_{xx}/\partial x + \partial N_{xy}/\partial y = \partial N_{xy}/\partial x + \partial N_{yy}/\partial y = 0 \tag{20}$$

For simplicity, we assume that $N_{xx} = N_{yy} = N$, $N_{xy} = 0$. For generality, we introduce an orthogonal system (n,t), n being the normal and t the tangent to the plate's perimeter S. A typical plate plus the coordinate systems mentioned are shown in Fig. 2.

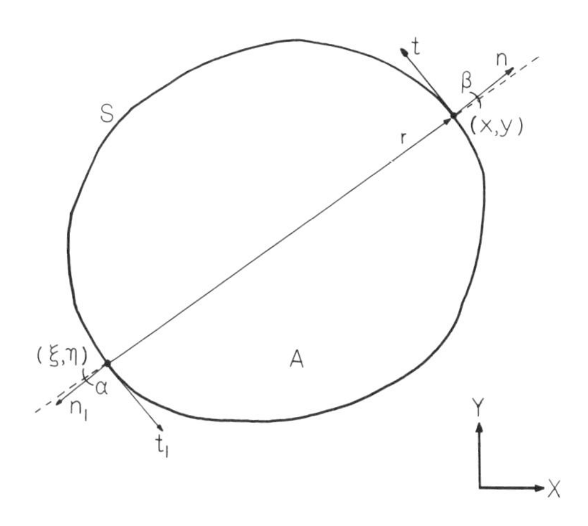

Figure 2. Plate Geometry Details

Equation 17 can now be condensed to read

$$D \nabla^4 w - N \nabla^2 w = q \tag{21}$$

Betti's reciprocal theorem for two elastic states has the form

$$\iint_A qw^* dA + \int_S (Vw^* - M\theta^*) ds + \int_S N\theta w^* ds + \sum_{l=1}^{L} T_l w_l^*$$

$$= \iint_A q^* w dA + \int_S (V^* w - M^* \theta) ds + \int_S N\theta^* w ds + \sum_{l=1}^{L} T_l^* w_l \tag{22}$$

where

$$\theta = \partial w / \partial n$$

$$M = (\partial^2 w / \partial n^2 + \nu\, \partial^2 w / \partial t^2) \tag{23}$$

$$V = Q_n + \partial M_{nt} / \partial_t = (\partial^3 w / \partial n^3 + (2 - \nu) \partial^3 w / \partial n \partial t^2)$$

$$T = D(1 - \nu)(\partial^2 w / \partial n_2 \partial t_2 - \partial^2 w / \partial n_3 \partial t_3)$$

In Eq. 23, θ is the rotation, V is known as the Kirchhoff shear, and T denotes the jump in the twisting couple at corner l. Furthermore, the normals and tangents on one side of the corner are denoted by subscript 2 and on the other side by subscript 3. For plates with a smooth perimeter, there is no summation over L, the total number of corners, in Eq. 22. If the starred state (*) is identified as the solution of Eq. 21 for a unit point force $\delta(\xi,\eta)$ replacing q, then Eq. 22 becomes

$$cw(\xi,\eta) = \int_S (-wV^* + \theta M^* - M\theta^* + Vw^*) ds$$

$$+ \int_S N(-w\theta^* + \theta w^*) ds + \iint_A qw^* \, dA \tag{24}$$

If point (ξ,η) is on the perimeter S, $c = 0.5$ and if (ξ,η) is inside the area A, $c = 1.0$. A second integral equation is generated by taking $w'(\xi,\eta)$, $' = \partial / \partial n_1$, where n_1 is the normal at the load point (ξ,η). Thus,

$$c\theta(\xi,\eta) = \int_S (-wV^{*\prime} + \theta M^{*\prime} - M\theta^{*\prime} + Vw^{*\prime})ds$$

$$+ \int_S N(-w\theta^{*\prime} + \theta w^{*\prime})ds + \iint_A qw^{*\prime}\, dA \tag{25}$$

Equations 24 and 25 are the counterparts of Eqs. 9 and 10 in plate buckling. The difference is that the boundary S must now be discretised into a number of line elements, while in beam buckling the beam segment between two nodes plays the role of a line element. Note that the area integrals in both Eqs. 24 and 25 are not considered if only critical loads are sought. Finally, if $\bar{w}^*$ is the solution of the flexure problem, i.e., $D\nabla^4 \bar{w}^* = \delta$, then the right hand sides of Eqs. 24 and 25 must be augmented by the terms

$$\left.\begin{array}{ll} \iint_A N \nabla^2 \bar{w}^* dA & \text{and} \\ \iint_A N \nabla^2 \bar{w}^{*\prime} dA, & \text{respectively.} \end{array}\right\} \tag{26}$$

Again, it is these area integrals that give rise to an interior discretization using area elements.

The fundamental solution for pure flexure is obtained from solving for the transverse displacement of a circular plate with a concentrated load at the center. This is an axisymmetric problem and from Ref. 22 we have

$$\bar{w}^*(r) = (1/(16\pi D))(r^2 \ln r^2 - r^2) \tag{27}$$

where $r^2 = (x - \xi)^2 + (y - \eta)^2$. Note that the displacement is finite at $r = 0$ and decays to zero past $r = \sqrt{e}$, where e is the base of the natural logarithm. Similarly, the exact solution w^* is obtained from the same problem with the addition of an in-plane compressive radial force N at the perimeter of the plate. Consulting Ref. 17 we have

$$w^*(r) = (D/(2\pi N))(\ln r + K_o(\sqrt{Nr^2/D})) \tag{28}$$

where K_o is the modified Bessel function of order zero. Expressions for $\bar{w}^*$, $\bar{\theta}^*$, $\bar{M}^*$, $\bar{V}^*$ and $\bar{w}^{*\prime}$, $\bar{\theta}^{*\prime}$, $\bar{M}^{*\prime}$, $\bar{V}^{*\prime}$ are collected in Appendix III. Similar expressions can be derived for w^*, θ^*, and so on, by noting the recurrence relations for Bessel functions and their derivatives, i.e.,

$$\left.\begin{array}{l} dK_o(r)/dr = -K_1(r),\ d^2K_o(r)/dr^2 = K_2(r), \\ d^3K_o(r)/dr^3 = -K_3(r),\ d^4K_o(r)/dr^4 = K_4(r) \\ \text{and} \\ K_2(r) = (2/r)K_1(r) + K_o(r),\ K_3(r) = (4/r)K_2(r) + K_1(r), \\ K_4(r) = (6/r)K_3(r) + K_2(r) \end{array}\right\} \quad (29)$$

Note that the subscript denotes the order of the Bessel function and that it is possible to express all the derivatives of $K_o(r)$ in terms of $K_o(r)$ and $K_1(r)$ only. The expressions for the elastic state w^* are also collected in Appendix III.

Numerical Implementation

A few general comments about the numerical implementation of the BEM are collected here. As far as beam buckling is concerned, the only remark is that no numerical integrations are required, since all quantities of interest can be obtained analytically. For the plate buckling case, two types of line elements are commonly used for the discretization of the plate's perimeter: (a) Constant elements, where all boundary quantities (displacement, rotation, moment, and shear) are assumed to remain constant over the element and are collocated at a node in the center, and (b) Linear elements, which defined by two nodes placed at the endpoints of the element and over which the boundary quantities vary in a linear fashion. For the case of a closed boundary, both types of elements result in the same number of nodes and the only difference is in the coefficient matrices [A], [B], and [C] that result from a standard nodal collocation, i.e.,

$$c\left\{\frac{w}{\theta}\right\} = [A]\left\{\frac{w}{\theta}\right\} + [B]\left\{\frac{M}{V}\right\} + [C(N)]\left\{\frac{w}{\theta}\right\} + \{D(N)\} + \{E(q)\} \quad (30)$$

In the above equation, the column matrices $\lfloor w,\theta \rfloor^T$ and $\lfloor M,V \rfloor^T$ contain the nodal values of the obvious boundary quantities and coefficient matrices [A], [B], and [C] result from integrating the appropriate fundamental solutions (see Eqs. 24 and 25) over the perimeter S. Also, if the approximate fundamental solutions are to be employed, column matrix {D} must be included (see Eq. 26), and if the plate supports any concentrated or distributed loads, column matrix {E} must be included.

For the case of linear elements and in conjunction with the approximate (flexural) fundamental solutions, Tottenham (23) was able to exactly integrate the kernels over an element. In the case of constant elements, it is standard practice to use (four point) gaussian quadrature for the numerical integration of the kernels over an element. When singular cases arise, i.e., when a receiving node coincides with the node defining the element over which integration is being performed, the integrals remain finite. Thus, no special precautions are needed other than to subdivide the constant element into two parts, one to the left and one to the right of the singular node. A standard four point quadrature can then be applied to each subelement and the results are added. It was shown in Ref. 16 that the answers obtained from some plate flexure trial problems using the same number of either constant or linear elements were nearly identical.

In order to compute {D} and/or {E}, area elements must be introduced. The simplest approach is to use three noded triangles and collocated at the center of each triangle. Both quadratic and quintic quadrature schemes can then be employed. A singular area element arises when one side of a triangle coincides with the singular line element. No problems are encountered in this case as long as the quadrature scheme used for the triangle does not include gauss points that coincide with the singular node.

Numerical Examples - Beams

As far as beam-column stability is concerned, the following examples are drawn from Ref. 16:

(a) Simply supported beam: The critical load is $P_{cr} = \pi^2(EI/L^2)$. For a beam with $EI/L^2 = 1$, $P_{cr} = \pi^2 = 9.8696$. Using the exact fundamental solution we get a $P_{cr} = 9.8707$ after 8 iterations, an answer that is 0.01% in error. The same error level is obtained using the approximate fundamental solution if the beam is subdivided into 6 segments.

(b) Cantilever beam: For the same material and geometric properties as above, $P_{cr} = \pi^2/4 = 2.4674$. After 17 iterations the formulation with the exact fundamental solution gave $P_{cr} = 2.4829$, which is 0.6% in error. Using 4 elements in conjunction with the approximate fundamental solution gave an error of 0.01%.

(c) Propped cantilever: Here $P_{cr} = 20.19$ and the exact fundamental solution formulation gave $P_{cr} = 20.188$ after 7 iterations, which is 0.05% in error. The other formulation required 5 elements in order to obtain the same error level. For such small problems, computer execution time requirements are minor.

Reference 18 reports three significant figure agreement with the exact solution for the Euler buckling load of a simply supported beam, if the beam is divided into 50 equal segments. Similar results, i.e., 50 elements are necessary to achieve better than 0.1% accuracy, are obtained in Ref. 21 for a simply supported beam, a pinned cantilever, and a simply supported beam with a linearly varying moment of inertia. Similar studies concerning the FEM can be found in Ref. 11. It is concluded that the levels of accuracy attained by the BEM and the FEM are comparable.

Numerical Examples - Plates

All the examples presented are for clamped plates. Clamped edges considerably simplify the solution scheme because the displacements and rotations are zero around the perimeter.

(a) Circular plate: Reference 16 considers a plate of radius 10 in. (25.4 mm), of flexural rigidity 1000 lbf-in. (113 N-mm), and a Poisson's ratio of 0.30 . Preliminary work on the flexure of this plate under (i) a concentrated point load $P = 1000$ lbf (4450 N) applied at the center and (ii) a uniformly distributed load $q = 100$ psi (689 kPa) served to investigate mesh discretization questions. Note that for case (ii) an interior discretization is required to compute integrals of the form $\int_A\int q\bar{w}^* dA$ and $\int_A\int q\bar{w}^{*\prime} dA$, while for case (i) these area integrals are easily evaluated analytically. In case (i), a disctretization of the perimeter into 24 constant elements gave a moment at the boundary $M = -81.28$ lbf (-361.7 N) and a discretization into 24 linear elements gave $M = -79.24$ lbf (-352.6 N). The analytic solution (22) is $M = -(P/4\pi) = -79.58$ lbf (-354.1 N). Thus, the first answer is 2.1% in error and the second one is 0.4% in error. In the second case, two interior discretization patterns employing 3-noded triangles were used: the coarse one (6 elements per quadrant) is shown in Fig. 3(a), and the fine one (12 elements per quadrant) is shown in Fig. 3(b). Compared to the analytical solution for a normal edge moment of $M = -qa^2/8 = -1250$ lbf (-5562 N), the 24 constant element discretization with the coarse and fine interior mesh gave answers 4.2% and 2.9% in error, respectively, while the 24 linear element discretization with the coarse and fine interior mesh gave answers 3.4% and 1.8% in error, respectively. Obviously, the plate discretization patterns which gave the best results in the flexural problems were used for the buckling problem. The exact critical load (radially applied) is $N_{cr} = 14.68\ D/a^2 = 146.8$ lbf/in. (25.72N/mm) (22). Compared to that, the approximate fundamental solution formulation was 3.2% in error and the exact one was 2.1% in error. In both cases over 15 iterations were required for convergence. It should be added that the quadratic scheme was used for integrating over triangles and constant line elements were employed in the buckling problem. Also, computer time and memory requirements for these problems are very modest. Additionally, Ref. 17 presents a first buckling load for a circular clamped plate that is no more than 0.84% in error, and subsequent buckling loads are found to be no more than 1.9% in error. These results were obtained for a perimeter discretization of 28 arc segments and an 8

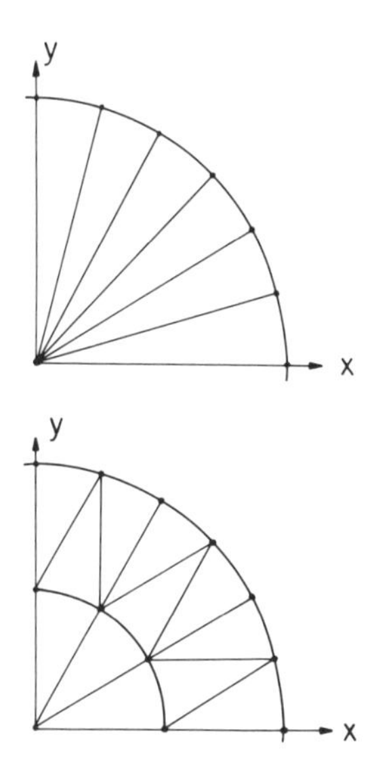

Figure 3. Typical Boundary and Interior Discretization Patterns for a Circular Plate Quadrant.

point Gaussian quadrature. The same exact fundamental solution (Eq. 28) was used.

(b) Square plate: By adjusting the strength of the fictitious loads applied along the fictitious contours surrounding the perimeter of a square plate in their indirect formulation, Gospodinov and Ljutskanov (13) obtained the following results: for a total of 8, 12, and 20 constant elements, the critical load was 6.1%, 1.0%, and 0.4 %, respectively, in error. The applied in-plane forces were $N_{xx} = N_{yy} = N$ and the exact fundamental solution was similar to the one in Eq. 28. There is a loss of accuracy when numerically constructed fundamental solutions for flexure are used (18). By modelling one quarter of a square plate under uniaxial compression ($N_{xx} = N$, $N_{yy} = 0$) and using 16 rectangular area elements and 8 boundary elements, the critical load was 4.4 % in error. When the same problem as above was solved (21) using experimental surface strain influence data measured at 25 points, the error was either +6 % or -3 % , depending on the integration rule used to numerically integrate the system equations.

Concluding Remarks

This chapter presented the buckling of slender beams and thin plates by the direct BEM. Two approaches were discussed, one using fundamental solutions of the differential equation governing buckling and one using fundamental solutions of the differential equation governing pure flexure. As far as beam problems are concerned, no quadrature is required since all relevant quantities are analytically evaluated and both approaches yield very accurate results. As far as plate problems are concerned, quadrature is required and the integrands exhibit mild (integrable) singularities. If the plate interior is adequately discretized, then the results obtained using the approximated fundamental solutions have the same level of accuracy as the ones obtained from the exact fundamental solutions, in which case only a boundary discretization is necessary.

The BEM formulations presented in this chapter are applicable to steady-state dynamics as well. Furthermore, they can be extended to problems involving material as well as geometric type of non-linearities.

Acknowledgements

The author would like to thank Prof. D.E. Beskos and Mr. M.F. Pineros for their cooperation. Thanks are also due to Mrs. I. Isihara for typing the manuscript.

Appendix I - References

1. Abramowitz, M., and Stegun, I., Handbook of Mathematical Functions, Dover, New York, 1972.

2. Altiero, N.J., and Sikarskie, D.L., 'A Boundary Integral Method Applied to Plates of Arbitrary Plan Form,' Computers and Structures, Vol. 9, 1978, pp. 163-168.

3. Banerjee, P.K., and Butterfield, R., Boundary Element Methods in Engineering Science, McGraw-Hill, London, 1981.

4. Bézine, G., 'Boundary Integral Formulation for Plate Flexure with Arbitrary Boundary Conditions,' Mechanics Research Communications, Vol. 5, 1978, pp. 197-206.

5. Bézine, G., Cimetiere, A., and Gelbert, J.P., 'Unilateral Buckling of Thin Elastic Plates by the Boundary Integral Equation Method,' International Journal for Numerical Methods in Engineering, Vol. 21, 1985, pp. 2189-2199.

6. Brebbia, C.A., Telles, J.C.F., and Wrobel, L.C., Boundary Element Techniques, Theory and Application in Engineering, Springer-Verlag, Berlin, 1984.

7. Brunet, M., 'An Integral Equation Method for Buckling Analysis of Thin Walled Non-Uniform Tapered Members,' in C.A. Brebbia, Editor, Recent Advances in Boundary Element Methods, Pentech Press, London, 1979, pp. 317-325.

8. Costa, J.A., and Brebbia, C.A., 'Elastic Buckling of Plates Using the Boundary Element Method,' in C.A. Brebbia and G. Maier, Editors, Boundary Elements VII, Vol. I, Springer-Verlag, Berlin, 1985, pp. 4-29 to 4-42.

9. Forbes, D.J., and Robinson, A.R., Numerical Analysis of Elastic Plates and Shallow Shells by an Integral Equation Method, Structural Research Series Report 345, University of Illinois, Urbana, 1969.

10. Froberg, C.E., Introduction to Numerical Analysis, 2nd Edition, Addison-Wesley, Reading, Massachusetts, 1973.

11. Gallagher, R.H., Finite Element Analysis: Fundamentals, Prentice Hall, Englewood Cliffs, 1975.

12. Ghali, A., and Neville, A.M., Structural Analysis: A Unified Classical and Matrix Approach, Chapman and Hall, London, 1978.

13. Gospodinov, G., and Ljutskanov, D., 'The Boundary Element Method Applied to Plates,' Applied Mathematical Modelling, Vol. 6, 1982, pp. 237-244.

14. Jaswon, M.A., and Maiti, M., 'An Integral Equation Formulation of Plate Bending Problems,' Journal of Engineering Mathematics, Vol. 2, No. 1, 1968, pp. 83-93.

15. Maiti, M., and Chakrabarty, S.K., 'Integral Solutions for Simply-supported Polygonal Plates,' International Journal of Engineering Science, Vol. 12, 1978, pp. 793-806.

16. Manolis, G.D., Beskos, D.E., and Pineros, M.F., 'Beam and Plane Stability by Boundary Elements,' Computers and Structures, Vol. 22, No. 6, 1986, pp. 917-923.

17. Niwa, Y., Kobayashi, S., and Kitahara, M., 'Determination of Eigenvalues by Boundary Element Methods,' in P.K. Banerjee and R.P. Shaw, Editors , Developments in Boundary Element Methods - 2, Applied Science Publishers, London, 1982, pp. 143-176.

18. Sekiya, T., and Katayama, T., 'Analysis of Buckling Using the Influence Function,' in Proceedings, 29th Japan National Congress of Applied Mechanics, University of Tokyo Press, Tokyo, 1981, pp. 25-31.

19. Stern, M., 'A General Boundary Integral Formulation for the Numerical Solution of Plate Bending Problems,' International Journal of Solids and Structures, Vol. 15, 1979, pp. 769-782.

20. Stern, M., 'Boundary Integral Equations for Bending of Thin Plates,' in C.A. Brebbia, Editor, Progress in Boundary Element Methods, Vol. 2, Pentech Press, London, 1983.

21. Tai, H., Katayama, T., and Sekiya, T., 'Buckling Analysis by Influence Function,' Mechanics Research Communications, Vol. 9, No. 3, 1982, pp. 139-144.

22. Timoshenko, S.P., and Gere, J.M., Theory of Elastic Stability, McGraw-Hill, New York, 1961.

23. Tottenham, H., 'The Boundary Element Method for Plates and Shells,' in Banerjee, P.K., and Butterfield, R., Editors, Developments in Boundary Element Methods - 1, Applied Science Publishers, London, 1979, pp. 173-205.

Appendix II - Beam Fundamental Solutions

The approximate fundamental solutions that constitute state 2 for the case of beam buckling are listed below (3,16):

$$\begin{aligned}
\bar{U}(r) &= (L^3/(12EI))(2 + |r/L|^3 - 3|r/L|^2) \\
\bar{\theta}(r) &= (L^2/(4EI))|r/L|(|r/L| - 2)\ \mathrm{sgn}(r) \\
\bar{M}(r) &= (L/2)(1 - |r/L|) \\
\bar{S}(r) &= -(1/2)\ \mathrm{sgn}(r) \\
\bar{U}'(r) &= -\bar{\theta}(r) \\
\bar{\theta}'(r) &= -(6L/(12EI))(|r/L| - 1) \\
\bar{M}'(r) &= (1/2)\ \mathrm{sgn}(r) \\
\bar{S}'(r) &= 0
\end{aligned} \tag{A.1}$$

where $'$ here is $d/d\xi = -d/dx$

The exact fundamental solutions for the same problem are (16):

$$\begin{aligned}
U(r) &= (1/(2Pk))(-\sin k|r| + \tan kL \cos k|r| + k(|r| - L)) \\
\theta(r) &= (1/(2P))(-\cos k|r| - \tan kL \sin k|r| + 1)\ \mathrm{sgn}(r) \\
M(r) &= (-1/(2k))(-\sin k|r| + \tan kL \cos k|r|) \\
S(r) &= -(1/2)\ \mathrm{sgn}(r) \\
U'(r) &= -\theta(r) \\
\theta'(r) &= (k/(2P))(-\sin k|r| + \tan kL \cos k|r|) \\
M'(r) &= (1/2)(\cos k|r| + \tan kL \sin k|r|)\ \mathrm{sgn}(r) \\
S'(r) &= 0
\end{aligned} \tag{A.2}$$

Appendix III - Plate Fundamental Solutions

For plate buckling, the approximate fundamental solutions are (22, 16):

$$w^* = (1/(16\pi D))\,|r^2 \log r^2 - r^2|$$
$$\theta^* = (1/(16\pi D))\; 2r \log r^2 \cos\beta$$
$$M^* = (1/(8\pi))\,|(1 + v)(1 + \log r^2) + (1 - v)\cos 2\beta|$$
$$V^* = (1/(8\pi r))\,|(5 - v)\cos\beta + (1 - v)\cos 3\beta|$$
$$w^{*\prime} = -(1/(8\pi D))\; r \log r^2 \cos\alpha \tag{A.3}$$
$$\theta^{*\prime} = -(1/(8\pi D))\,|(1 + \log r^2)\cos(\alpha - \beta) + \cos(\alpha + \beta)|$$
$$M^{*\prime} = -(1/(8\pi r))\,|2(1 + v)\cos\alpha + (1 - v)(\cos(2\beta - \alpha) - \cos(2\beta + \alpha))|$$
$$V^{*\prime} = +(1/(8\pi r^2))\,|(5 - v)\cos(\alpha + \beta) + 2(1 - v)\cos(3\beta + \alpha) - (1 - v)\cos(3\beta - \alpha)|$$

The above equations are taken from Ref. 21. As shown in Fig. 2, β is the angle between the normal at receiver (x,y) and the line connecting (x,y) with the source (ξ,η), while α is the angle between the normal at (ξ,η) and that line.

The exact fundamental solutions for plate buckling can be synthesized from the following expressions (17, 16):

$$w = a\,[\ln r + K_o]$$
$$\partial w/\partial r = a\,[1/r - z K_1]$$
$$\partial^2 w/\partial r^2 = a\,[-1/r^2 - z^2(K_o + (2/(z r))K_1)] \tag{A.4}$$
$$\partial^3 w/\partial r^3 = a\,[2/r^3 + z^3((4/(z r))K_o + (1 + 8/(z r)^2)K_1)]$$
$$\partial^4 w/\partial r^4 = a\,[-6/r^4 + z^4((1 + 24/(z r)^2)K_o + (8/(z r) + 48/(z r)^3)K_1)]$$

where $z^2 = N/D$, $a = 1/(2\pi z^2)$, and $K_o = K_o(z r)$, $K_1 = K_1(z r)$ are evaluated from expressions given in Ref. 1. Furthermore,

$$\partial w/\partial n = (\partial w/\partial r)(\partial r/\partial n)$$

$$\partial^2 w/\partial^2 n = (\partial^2 w/\partial r^2)(\partial r/\partial n)^2 + (\partial w/\partial r)(\partial^2 r/\partial n^2) \quad (A.5)$$

$$\partial^3 w/\partial^3 n = (\partial^3 w/\partial r^3)(\partial r/\partial n)^3 + 3(\partial^2 w/\partial r^2)(\partial^2 r/\partial n^2)(\partial r/\partial n) + + (\partial w/\partial r)(\partial^3 r/\partial n^3)$$

where, for instance,

$$\partial r/\partial n = (\partial r/\partial x)(\partial x/\partial n) + (\partial r/\partial y)(\partial y/\partial n) = \cos\beta$$

$$\partial^2 r/\partial^2 n = (1/r)\sin^2\beta \quad (A.6)$$

and β was identified above. The expressions involving the tangential derivatives $\partial w/\partial t$, etc, are derived from the ones involving the normal derivative ($\partial w/\partial n$, etc) by simply replacing n by t. The only difference is that

$$\partial r/\partial t = \sin\beta$$

$$\partial^2 r/\partial t^2 = (1/r)\cos^2\beta \quad (A.7)$$

Using all the above expressions in Eq. 23 yields Θ, M, and V. The remaining expressions are obtained as $w' = \partial w/\partial n_1$, $\Theta' = -\partial\Theta/\partial n_1$, $M' = -\partial M/\partial n_1$, and $V' = -\partial V/\partial n_1$, where n_1 is the normal at the source (ξ,η).

Dynamic Analysis of Beams, Plates and Shells

Dimitri E. Beskos*, M. ASCE

This chapter deals with the linear elastic dynamic analysis of beams, plates and shells by using the boundary element method. Both free and forced vibration problems of Bernoulli-Euler beams, Kirchhoff thin plates and thin shallow shells are considered. In addition, the free vibration analysis of membranes is also considered in this chapter. The discussion is not restricted to the conventional (direct or indirect) boundary element method in the frequency or the time domain, but also includes other boundary methods as well as a hybrid technique that combines boundary and finite elements. In general, the subject of this chapter is not characterized by an extensive literature. A lot of work is presently under way, however, and more interesting results are expected in the near future.

Introduction

Linear elastic dynamic analysis of beams, plates and shells has been studied both analytically and numerically. And whereas an analytic treatment is only possible for simple geometries, loading and boundary conditions, a numerical treatment permits one to consider beam, plate and shell dynamic problems with complicated geometries, loading and boundary conditions. Among the plethora of books dealing with the dynamic analysis of beams, plates and shells one can mention those of Love (25), Graff (16) and Ghali and Neville (15), Kraus (22) and Bathe (3) for their analytical and numerical treatment, respectively. A comprehensive review work on vibrations of plates and shells can be found in the two NASA reports of Leissa (23, 24).

The most popular numerical methods for dynamic analysis of beams, plates and shells are the Finite Difference Method (FDM) and especially the Finite Element Method (FEM). During the last fifteen years the Boundary Element Method (BEM) has emerged as an accurate and efficient numerical method for the solution of a wide variety of structural analysis problems as it is evident in the books of Banerjee and Butterfield (2), Brebbia, Telles and Wrobel (9) and Beskos (5). In general, the main advantage of the BEM is the dimensionality reduction of the problem, which results in a discretization that is restricted on the surface of the domain. Thus, no interior discretization is required, as it is the case with "domain" type of methods, such as the FDM and the FEM.

The present chapter deals with the dynamic analysis of beams, plates and shells by the BEM. A comprehensive review article on the application of the BEM in dynamic analysis has recently been prepared by Beskos (6). The treatment by the BEM of one-dimensional structures, such

* Professor, Department of Civil Engineering, University of Patras, Patras 261 10, Greece.

as Bernoulli-Euler beams, does not correspond to a numerical scheme more efficient than the FEM or the FDM, because for this type of structures the BEM loses its basic advantage of dimensionality reduction over the other two methods. A presentation of the BEM as applied to beam dynamics is given here, however, because it clearly describes in a nice tutorial manner the fundamentals of the method.

The treatment of plate dynamics by the BEM does correspond to an efficient numerical scheme that retains the dimensionality reduction advantage over the FEM and the FDM. Even when the load is applied in the interior and not at the edges of the plate, one simply has to compute a surface integral with a known integrand in addition to the usual BEM computations. There is so little available in the literature concerning shell dynamics by the BEM, that no definite statement about the efficiency of the method can be made at the present.

The BEM as applied to the free and forced vibration problems of Bernoulli-Euler beams, Kirchhoff thin plates and thin shallow shells is discussed in the following sections of this chapter. For reasons of completeness the interesting and related problem of free vibrations of membranes is also discussed in this chapter. Both frequency and time domain BEM formulations are considered in a direct or indirect manner. The discussion is not restricted to the conventional BEM that employs the Green's function in its formulation, but includes additionally other boundary methods that simply employ solutions of the homogeneous part of the governing equation instead of Green's functions, as well as a hybrid scheme that essentially combines boundary and finite elements.

Dynamic Analysis of Beams

This section is based exclusively on the work of Providakis and Beskos (33). The governing differential equation of flexural vibrations of a uniform Bernoulli-Euler elastic beam of length l and flexural rigidity EI is

$$EI \frac{d^4 y}{dx^4} + m \frac{d^2 y}{dt^2} = q \tag{1}$$

where $y = y(x,t)$ is the lateral deflection, m is the mass per unit length, $q = q(x,t)$ is the distributed lateral dynamic load, x is the longitudinal axis of the beam and t represents time.

For free vibrations for which $q = 0$ one can assume $y(x,t) = Y(x) \sin\omega t$ and reduce Eq. 1 to the eigenvalue problem

$$\frac{d^4 Y}{dx^4} - \lambda^4 Y = 0, \qquad \lambda^4 = m\omega^2/EI \tag{2}$$

that has to be solved for the natural frequencies ω and the corresponding modal shapes $Y = Y(x)$ for given boundary conditions. Multiplication of both sides of Eq. 2_1 by a suitably continuous function G, integration over the length of the beam and subsequent integration by parts four times in conjunction with the understanding that primes indicate differentiation with respect to x, produce the integral reciprocity relation

$$\int_0^1 Y(G^{IV} - \lambda^4 G)\,dx + \left[Y'''G - Y''G' + Y'G'' - YG'''\right]_0^1 = 0 \tag{3}$$

between the functions Y and G which forms the basis of the direct BEM. The function G is now taken to be the infinite beam Green's function (33)

$$G(x,\xi) = (1/4\lambda^3 EI)\left[\sec\lambda l \,\sin\lambda(l - |r|) - \operatorname{sech}\lambda l \,\sinh\lambda(l - |r|)\right] \tag{4}$$

which is the solution of the equation

$$G^{IV} - \lambda^4 G = (1/EI)\,\delta(x,\xi) \tag{5}$$

where $|r| = |x-\xi|$ and $\delta(x,\xi)$ is the Dirac's delta function. In view of Eq. 5, Eq. 3 reduces to

$$Y(\xi) = \left[sG - mF + \Theta E - YD\right]_0^1 \tag{6}$$

where

$$\Theta = \Theta(x) = Y', \; m = m(x) = -EIY'', \; s = s(x) = -EIY'''$$

$$F = F(x,\xi) = G', \; E = E(x,\xi) = -EIG'', \; D = D(x,\xi) = -EIG''' \tag{7}$$

and where the dependence on λ of G and its derivatives with respect to x has been dropped for notational simplicity. Differentiation of Eq. 6 with respect to ξ results in

$$\Theta(\xi) = \left[sG' - mF' + \Theta E' - YD'\right]_0^1 \tag{8}$$

where

$$G' = \frac{dG}{d\xi}, \quad D' = \frac{dD}{d\xi}, \quad E' = \frac{dE}{d\xi}, \quad F' = \frac{dF}{d\xi} \tag{9}$$

Eqs 6 and 8 represent essentially the integral representation of the solution of Eq. 2 for any interior point ξ, in terms of G and its derivatives which are given explicitly in (33). If the field point ξ is taken to the boundaries 0 and 1 through a limiting process one can obtain in matrix form the boundary integral equation

$$[A]\{v\} + [B]\{u\} = \{0\} \tag{10}$$

where the 4x4 influence matrices $[A(\lambda)]$ and $[B(\lambda)]$ consist of entries which are the values of G and its derivatives at 0 and 1 and

$$\{v\} = \{s(1),\ s(o),\ m(1),\ m(o)\}^T$$
$$\{u\} = \{Y(1),\ Y(o),\ \theta(1),\ \theta(o)\}^T \tag{11}$$

After application of the boundary conditions and recordering, Eq. 10 takes the form

$$[H\ (\lambda)]\{w\} = \{0\} \tag{12}$$

which has nontrivial solutions for $\{w\}$ iff

$$\det\ [H(\lambda)] = 0 \tag{13}$$

The real roots λ of the frequency equation 13 in conjunction with Eq. 2_2 provide the ω's, while solution of Eq. 12 in conjunction with Eq. 6 the Y's. It should be noted that the above method provides the exact values of all (infinitely many) natural frequencies and corresponding modal shapes, because the beam is treated as a continuous body and no discretization is involved. From this viewpoint the above BEM is equivalent to the FEM that is based on the concept of the dynamic stiffness method (10).

The forced vibration problem of beams is treated with the aid of the Laplace transform with respect to time which for a function f(x,t) is defined by

$$\bar{f}(x,k) = \int_o^\infty f(x,t)\ e^{-kt}\ dt \tag{14}$$

where the Laplace transform parameter k is, in general, a complex number. Application of Laplace transform on Eq. 1 under zero initial conditions yields

$$\frac{d^4\bar{y}}{dx^4} + \mu^4\bar{y} = \frac{\bar{q}}{EI}\ , \quad \mu^4 = mk^2/EI \tag{15}$$

Following the same procedure as before one can derive the following two equations for an interior point ξ:

$$\bar{y}(\xi) = \left[\bar{s}\bar{G} - \bar{m}\bar{F} + \bar{\theta}\bar{E} - \bar{y}\bar{D}\right]_o^1 + \int_o^1 \bar{q}\bar{G}\,dx \tag{16}$$

$$\bar{\theta}(\xi) = \left[\bar{s}\bar{G}' - \bar{m}\bar{F}' + \bar{\theta}\bar{E}' - \bar{y}\bar{D}'\right]_o^1 + \int_o^1 \bar{q}\bar{G}'\,dx \tag{17}$$

where the Green's function $\bar{G}$ satisfies the equation

$$\bar{G}^{IV} + \mu^4\bar{G} = (1/EI)\ \delta(x,\xi) \tag{18}$$

and its explicit expression is given by Eq. 4 with ω being replaced by ik (33). If the field point ξ is taken to the boundaries, then Eqs 16 and 17 can be used to form the matrix equation

$$[\bar{A}]\ \{\bar{v}\} + [\bar{B}]\ \{\bar{u}\} = \{\bar{Q}\} \tag{19}$$

which after application of the boundary conditions and reordering becomes

$$[\bar{H}]\ \{\bar{w}\} = \{\bar{P}\} \tag{20}$$

This equation is solved for a sequence of values of the parameter k to obtain the boundary data $\{\bar{w}\}$ which is subsequently used in Eq. 16 to obtain the transformed solution. The response y(x,t) is finally obtained by a numerical inversion of the transformed solution. If q(x,t) is a complicated function of time its direct Laplace transform is also computed numerically. Both direct and inverse numerical Laplace transforms can be very accurately computed by the method of Durbin (14) based on the sine-cosine transform, which has been proven to be the most accurate and general method for structural dynamics in a recent detailed comparison study by Narayanan and Beskos (26). In the special case the forcing function is harmonic in time the response will be also harmonic and its amplitude can be obtained directly from Eq. 20 if k is replaced by $-i\omega$ without any need for numerical inversion (33). The above formulation is capable of taking into account the effects of a constant axial force, external viscous or internal viscoelastic damping and an elastic foundation on the response of a beam, with the aid of some minor modifications (33). This direct BEM formulation that employs the Laplace transform provides theoretically the exact response of a beam structure, which is treated as a continuum, and from this viewpoint is equivalent with analogous FEM schemes that also utilize the Laplace transform (4). Fig. 1, which is taken from (33), shows the history of the deflection at the midspan of a simply supported beam due to a suddenly applied load there for various values of the internal viscoelastic damping coefficient f.

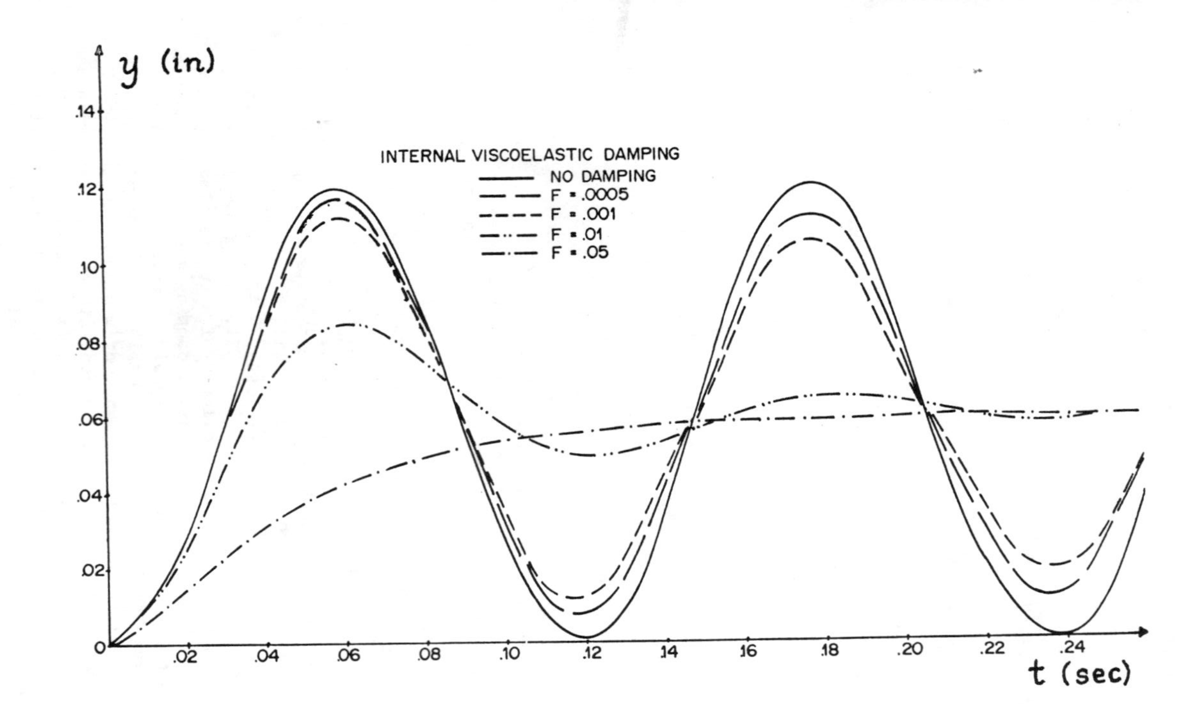

Figure 1. History of deflection at the midspan of a simply supported beam due to a suddenly applied load there for various values of internal viscoelastic damping; Ref. (33).

Dynamic Analysis of Plates

Apparently Vivoli and Filippi (42, 43) were the first to consider free flexural vibrations of plates by the indirect BEM and provide numerical results. Niwa, Kobayashi and Kitahara (28, 30) and Kitahara (21) presented the most comprehensive treatment of free vibration analysis of plates by both the direct and indirect BEM including detailed numerical results. Hutchinson and Wong (19, 44) also presented a theoretical treatment of the problem by employing the direct BEM. The following brief outline of the direct BEM formulation of the problem is based on Refs (28, 44).

The governing differential equation of flexural vibrations of thin elastic Kirchhoff plates is of the form

$$D\nabla^4 w + m\,(d^2 w/dt^2) = p \tag{21}$$

where $w = w(x,y,t)$ is the lateral deflection, m is the mass per unit surface, D is the flexural rigidity of the plate and $p = p(x,y,t)$ is the distributed lateral dynamic load. For the free vibration case for which $p = 0$, one can assume $w = W(x,y)\sin\omega t$ and reduce Eq. 21 to the eigenvalue problem

$$\nabla^4 W - \beta^4 W = 0, \qquad \beta^4 = m\omega^2/D \tag{22}$$

that has to be solved for the natural frequencies ω and the corresponding modal shapes $W = W(x,y)$ for given boundary conditions. The Green's function for an infinite plate $G = G(x,y,\xi,\eta)$ satisfies the equation

$$\nabla^4 G - \beta^4 G = \delta(x,y,\xi,\eta) \tag{23}$$

and has the explicit form (28)

$$G(x,y,\xi,\eta) = (i/8\beta^2)\left[H_o^{(1)}(\beta r) - H_o^{(1)}(i\beta r)\right] \tag{24}$$

where $r^2 = (x-\xi)^2 + (y-\eta)^2$ and $H_o^{(1)}(.)$ is the zero order Hankel function of the first kind. Eq. 22 can be reduced to an integral form by using the Rayleigh-Green identity with K corners of the static case (40), i.e.,

$$\int_S (G\nabla^4 W - W\nabla^4 G)\,dS = \frac{1}{D}\int_C \Big[V_n(G)W - M_n(G)\frac{dW}{dn} + \frac{dG}{dn}M_n(W) - G\,V_n(W)\Big]dC - \frac{1}{D}\sum_{k=1}^{K} \langle M_t(W)\,G - M_t(G)\,W \rangle \tag{25}$$

where $V_n(W)$, $M_n(W)$, $M_t(W)$, dW/dn and n are the effective shear, normal bending moment, twisting moment, normal slope and outward normal vector, respectively, $<\cdot>$ represents the discontinuity jump in the direction of increasing arc length and S and C stand for the surface and the perimeter of the plate, respectively. It is easy to see that due to Eqs 22 and 23, the volume integral in Eq. 25 simply reduces to $-W(\xi,\eta)$ and Eq. 25 becomes

$$cW(\xi,\eta) = \frac{1}{D}\int_C \left[-V_n(G)W + M_n(G)\frac{\partial W}{\partial n} - \frac{\partial G}{\partial n} M_n(W) + G\, V_n(W)\right] dC + \frac{1}{D}\sum_{k=1}^{K} < M_t(W)\, G - M_t(G)\, W > \qquad (26)$$

where

$$c = 1 \text{ for } (\xi,\eta) \in S, \quad c = 1/2 \text{ for } (\xi,\eta) \in C, \quad c = 0 \text{ for } (\xi,\eta) \not\in S \cup C \qquad (27)$$

A second independent integral equation is obtained by taking the directional derivative of Eq. 25, which reads

$$c\frac{\partial W}{\partial \nu} = \frac{1}{D}\int_C \left[-\frac{\partial V_n}{\partial \nu}(G)W + \frac{\partial M_n}{\partial \nu}(G)\frac{\partial W}{\partial n} - \frac{\partial^2 G}{\partial \nu \partial n} M_n(W) + \frac{\partial G}{\partial \nu} V_n(W)\right] dC + \frac{1}{D}\sum_{k=1}^{K} < M_t(W)\frac{\partial G}{\partial \nu} - \frac{\partial M_t}{\partial \nu}(G)W > \qquad (28)$$

Explicit expressions for the various derivatives of G can be found in (21).

For the numerical solution of the system of the two boundary integral equations 26 and 28 for $c = 1/2$, the boundary C is divided into a finite number N of boundary elements, which may be straight or curved with a constant or higher order variation of displacements and forces, respectively. Thus, Eqs 26 and 28 in their discretized form can be written as

$$[A]\{W\} + [B]\{\Theta\} + [R]\{M\} + [T]\{V\} = \{0\} \qquad (29)$$

where the 2Nx2N influence matrices $[A]$, $[B]$, $[R]$ and $[T]$ contain integrals of G and its derivatives and the vectors $\{W\}$, $\{\Theta\}$, $\{M\}$ and $\{V\}$ stand for boundary values of the deflection W, the normal slope $\Theta = dW/d\nu$, the bending moment M_n and the shear force V_n, respectively. Application of the N boundary conditions of the problem of interest in Eq. 29 and rearrangement yield the equation

$$[Q(\beta)]\ \{\psi\} = \{0\} \qquad (30)$$

which has nontrivial solutions for $\{\psi\}$ iff

$$\det\left[Q(\beta)\right] = 0 \tag{31}$$

The real roots β of the frequency equation 31 in conjunction with Eq.22$_2$ provide the natural frequencies ω, while solution of Eq. 30 the modal shapes $\{\psi\}$ of the free plate vibration problem.

The above procedure constitutes the direct BEM. An indirect BEM has also been proposed and used by Niwa et al (28) for free vibrations of plates. This approach starts with the definition of layer potentials W_o and W_1 of order zero and one, respectively, which are given in terms of the corresponding densities μ_o and μ_1 by the equations

$$W_o = \pi_o\mu_o = \int_C G\,\mu_o\,dC \tag{32}$$

$$W_1 = \pi_1\mu_1 = \int_C (\partial G/\partial n)\,\mu_1 dC \tag{33}$$

These potentials experience discontinuities at the boundary as discussed in (28). Noting that boundary conditions for plates enforce two conditions on a boundary point, one can assume that W can be represented as the sum of the two potentials W_o and W_1 of Eqs 32 and 33, respectively, i.e.,

$$W = \pi_o\mu_o - \pi_1\mu_1 \tag{34}$$

Thus, for example, for clamped all around boundaries one has

$$W = \pi_o\mu_o - \pi_1\mu_1 = 0 \tag{35}$$

$$\frac{\partial W}{\partial n} = \left(\frac{\partial \pi_o}{\partial n}\right)\mu_o - \left(\frac{\partial \pi_1}{\partial n}\right)\mu_1 = 0 \tag{36}$$

which in view of Eqs 32 and 33 become the boundary integral equations

$$\int_C \left[G\mu_o - (\partial G/\partial n)\mu_1\right] dC = 0 \tag{37}$$

$$\int_C \left[(\partial G/\partial n)\mu_o - (\partial^2 G/\partial n\partial \nu)\mu_1\right] dC = 0 \tag{38}$$

These equations are solved numerically by a boundary element discretization procedure analogous to the one employed in the direct BEM to obtain the values of μ_o and μ_1 at the boundary. This process results again in an equation that has the same form as Eq. 30. This is solved for the eigenvalues β and the eigenvectors $\{\psi\}$ with entries values of μ_o and μ_1 at the boundary. Use of Eqs. 22_2 and 35 finally yields the natural frequencies ω and modal shapes W. This indirect BEM was used by Niwa et al (28) for plate vibration problems characterized by uniform boundary conditions in preference of the direct BEM. For mixed boundary conditions, however, use was made of the direct BEM because of its generality. In Refs (30, 21) the effect of constant in-plane forces on the natural frequencies was also studied by the BEM, thereby providing indirectly plate buckling loads when the frequencies tend to zero.

Table 1 presents numerical values for the first 6 natural frequencies of a clamped circular plate and a simply supported square plate as obtained by the BEM in Ref. (28). These values show an excellent agreement with the exact solution taken from Leissa (23).

Bézine (7) in an effort to reduce the computational difficulties associated with the use of the complicated Green's function of the problem and its derivatives, introduced a hybrid scheme that combines the direct BEM in static form to take care of the elasticity of the plate with the FEM to take care of the inertia of the plate, and was able to study free flexural vibrations of plates. This hybrid scheme requires, of course, an interior discretization of the domain in addition to the boundary one but this disadvantage is compensated by the much simpler entries of the influence matrices in the BEM formulation. Thus, for example, for the case of no corners, one has the equations

$$cW(\xi,\eta) = \int_S G\beta^4 W dS - \frac{1}{D}\int_C \left[V_n(G)W - M_n(G)\frac{\partial W}{\partial n} + \frac{\partial G}{\partial n}M_n(W) - GV_n(W)\right] dC \qquad (39)$$

$$c\frac{\partial W}{\partial \nu} = \int_S \frac{\partial G}{\partial \nu}\beta^4 W dS - \frac{1}{D}\int_C \left[\frac{\partial V_n(G)}{\partial \nu}W - \frac{\partial M_n(G)}{\partial \nu}\frac{\partial W}{\partial n} + \frac{\partial^2 G}{\partial n \partial \nu}M_n(W) - \frac{\partial G}{\partial \nu}V_n(W)\right] dC \qquad (40)$$

with G and its derivatives being those of the static case given explicitly, e.g., in Stern (40). In Eqs. 39 and 40 the integrals over the surface S of the plate are discretized with the aid of the FEM, while the integrals over the perimeter C of the plate are discretized with the aid of the static direct BEM. Finally, the free vibration problem by taking first Eqs. 39 and 40 for c = 1/2 and then Eq 39 for c = 1 takes the matrix form (7)

$$\begin{aligned} [G_c]\{I\} - \beta^4 [J_s]\{W_s\} &= \{0\} \\ [G_s]\{I\} - \beta^4 [J_s^*]\{W_s\} &= \{0\} \end{aligned} \qquad (41)$$

where the subscripts **c** and **s** correspond to quantities on the perimeter and the surface of the plate, respectively, $\{I\}$ is the vector of the boundary unknowns (among W, $\partial W/\partial n$, M_n and V_n) and $\{W_s\}$ is the vector of the deflections in the surface V. The system of Eqs. 41 can be recast in the classical eigenvalue form

$$[Z]\,\{W_s\} = (1/\beta^4)\,\{W_s\}$$

$$[Z] = [G_s][G_c]^{-1}[J_s] - [J_s^*] \qquad (42)$$

from which the eigenvalues and eigenvectors can be easily computed. Table 2 provides numerical values for the first 5 natural frequencies of a simply supported square plate as obtained by the BEM in Ref. (7) and clearly demonstrates the accuracy of the method. Presently, Providakis and Beskos (34) are working towards an improved version of this hybrid scheme by employing second order isoparametric boundary and finite elements, instead of the constant ones employed by Bézine (7). Preliminary free vibration results from (34) show a high accuracy with a small number of elements. In the same work Providakis and Beskos (34) were also able to extend this hybrid scheme idea to forced vibration problems with the aid of Laplace transform. Application of Laplace transform as defined by Eq. 14 on Eq. 21 under zero initial conditions yields

$$\nabla^4\bar{w} + \gamma^4\bar{w} = \bar{p}/D, \qquad \gamma^4 = mk^2/D \qquad (43)$$

indicating that in the transformed domain the equation of motion becomes of a static form with the load now being $\bar{p}/D - \gamma^4\bar{w}$. Thus, for example, one can write an integral equation analogous to Eq. 39 that has the form

$$\bar{c}\bar{w}(\xi,\eta) = \int_S G(\bar{p}/D)\,dS - \int_S G\gamma^4\bar{w}\,dS - \frac{1}{D}\int_C \Big[V_n(G)\bar{w} - M_n(G)\frac{\partial\bar{w}}{\partial n} + \frac{\partial G}{\partial n}M_n(\bar{w}) - GV_n(\bar{w})\Big]dC \qquad (44)$$

and finally cast the whole problem in the matrix form (34)

$$[\bar{G}_c]\,\{\bar{I}\} + \gamma^4[\bar{J}_s]\,\{\bar{w}_s\} = \{\bar{P}_1\}$$

$$[\bar{G}_s]\,\{\bar{I}\} + \gamma^4[\bar{J}_s^*]\,\{\bar{w}_s\} = \{\bar{P}_2\} \qquad (45)$$

Solution of the system of Eqs 45 can provide the transformed solution $\{\bar{I}\}$ and $\{\bar{w}_s\}$ for the unknown boundary values of $\bar{w}$, $\partial\bar{w}/\partial n$, $\bar{M}_n$ and $\bar{V}_n$ and the interior values of the deflection $\bar{w}$. A numerical inversion of the transformed solution, as described in the previous section, can provide the time domain response of the plate.

The first study on forced vibrations of plates by the BEM was that of Bézine and Gamby (8) which is based on a time domain direct approach. It starts with the generalized Green's identity in space and time, which

	β_1	β_2	β_3	β_4	β_5	β_6	
Leissa (23)	3.196	4.611	5.906	6.306	7.144	7.799	clamped circular
BEM (28)	3.2	4.6	5.9	6.3	7.2	7.9	
error %	0.13	0.29	0.10	0.10	0.78	1.30	
Leissa (23)	4.443	7.025	8.886	9.935	11.327	12.953	simply supported square
BEM (28)	4.4	7.0	8.9	9.9	11.3	13.0	
error %	0.97	0.36	0.16	0.35	0.24	0.36	

Table 1. First 6 natural frequencies of a clamped circular plate and a simply supported square plate; N = 28 boundary elements; Poisson's ratio $\nu = 0.3$

	β_1^2	β_2^2	β_3^2	β_4^2	β_5^2
Leissa (23)	19.74	49.35	78.96	98.70	128.30
BEM (7) 4x4 internal	20.252	52.491	86.565	107.904	142.15
error %	2.6	6.4	9.6	9.4	10.8
BEM (7) 8x8 internal	19.866	50.145	80.971	101.864	133.681
error %	0.6	1.6	2.5	3.2	4.2

Table 2. First 5 natural frequencies of a simply supported square plate; N = 48 boundary elements; Poisson's ratio $\nu = 0.3$.

for the case of no corners has the form

$$\int_S (G* \nabla^4 w - w * \nabla^4 G)\, dS = -\frac{1}{D}\int_C \Big[V_n(w) * G - M_n(w) * \frac{\partial G}{\partial n} + \frac{\partial w}{\partial n} * M_n(G) - w * V_n(G)\Big]\, dC \tag{46}$$

where the star symbol denotes convolution defined as

$$G * w = \int_0^t G(x,y,t-\tau)\, w(x,y,\tau)\, d\tau , \quad t > 0$$
$$G * w = 0, \quad t < 0 \tag{47}$$

The lateral deflection w naturally satisfies Eq. 21, while the fundamental solution G satisfies the equation

$$\nabla^4 G + K^2 \frac{\partial G}{\partial t} = \delta(x,y)\, H(t), \quad K^2 = m/D \tag{48}$$

where H(t) is the Heaviside function. The solution of Eq. 48 is given for two points P(x,y) and $Q(\xi,\eta)$ as

$$G(P,Q,t) = \frac{At}{2K}\, F\left(\frac{Kr^2}{4t}\right) \tag{49}$$

where

$$F(\varphi) = \frac{\pi}{2} - \int_0^{\varphi} \frac{\sin\zeta}{\zeta}\, d\zeta - \varphi\int_{\varphi}^{\infty} \frac{\cos\zeta}{\zeta}\, d\zeta - \sin\varphi \tag{50}$$

$r = |P-Q|$ and A is a constant that takes the value of 1/2. In view of Eqs 21 and 48 and under the assumption of zero initial conditions, Eq. 46 becomes for two points P and Q on C the boundary integral equation

$$\frac{1}{2} H(t)*w(P,t) = \int_S G(P,Q,t)*p(Q,t)\,dS + \frac{1}{D}\int_C \Big[V_n(w(Q,t))*G(P,Q,t) - M_n(w(Q,t))*\frac{\partial G}{\partial n}(p,Q,t) + \frac{\partial w}{\partial n}(Q,t)*M_n(G(P,Q,t)) - w(Q,t)* * V_n(G(P,Q,t))\Big]\, dC \tag{51}$$

Directional differentiation of Eq. 51 provides the second boundary integral equation

$$\frac{1}{2} H(t) * \frac{\partial w}{\partial v} (P,t) = \int_S \frac{\partial G}{\partial v} (P,Q,t) * p(Q,t) dS + \frac{1}{D} \int_C \Big[V_n(w(Q,t)) * \frac{\partial G}{\partial v} (P,Q,t)$$

$$- M_n(w(Q,t)) * \frac{\partial^2 G}{\partial n \partial v} (P,Q,t) + \frac{\partial w}{\partial n} (Q,t) * \frac{\partial M_n}{\partial v} (G(P,Q,t)) - w(Q,t) *$$

$$* \frac{\partial M_n}{\partial v} (G(P,Q,t)) \Big] dC \tag{52}$$

For a point P inside S, on the other hand, one has

$$w(P,t) = \int_S \frac{\partial G}{\partial t} (P,Q,t) * \frac{p(Q,t)}{D} dS + \frac{1}{D} \int_C \Big[V_n(w(Q,t)) * \frac{\partial G}{\partial t} (P,Q,t)$$

$$- M_n(w(Q,t)) * \frac{\partial 2 G}{\partial t \partial n} (P,Q,t) + \frac{\partial w}{\partial n} (Q,t) * \frac{\partial M_n}{\partial t} (G(P,Q,t)) -$$

$$- w(Q,t) * \frac{\partial V_n}{\partial t} (G(P,Q,t)) \Big] dC \tag{53}$$

The system of Eqs 51 and 52 is solved numerically by both a time and space discretization. The total time interval of interest is divided into M segments of length Δt, while the perimeter C of the plate is divided into N straight segments or boundary elements over which displacements and forces are assumed to be constant. It is also assumed that displacements and forces remain constant inside every time segment. The computation of the integral over S, that is characterized by a known integrand, is generally done numerically by the creation of domain cells and presents no difficulties. The solution for the unknown boundary values of w, $\partial w/\partial n$, M_n and V_n is obtained by a step-by-step marching scheme in time, where a solution of Eqs 51 and 52 in space is obtained for every time step as in a static BEM. Once all the boundary values are known one can use Eq. 53 to determine values of the deflection at any interior point. Figure 2, which is taken from Ref (8), provides the central deflection of a simply supported circular plate loaded on its center by a rectangular load pulse of magnitude F_o and time duration T, as obtained by the time domain BEM (8) and the analytical solution of Sneddon (39).

For reasons of completeness, in this section, the case of in-plane vibrations of plates which is actually a plane stress elastodynamic problem that can be treated as described in the seventh chapter of this book, is also very briefly discussed. Free vibration analysis of plates under conditions of plane stress by the BEM together with numerical results can be found in Niwa, Kobayashi and Kitahara (29, 31) and Kitahara (21), while forced vibration analysis results for the case of harmonic in-plane forces are available in Dominguez and Alarcón (13). An interesting BEM approach to the free vibration problem of plates under conditions of plane stress has been recently proposed by Shaw (37). The idea is here to consider the plate as a three-dimensional structure and use the BEM in conjunction with the displacement potentials approach to reduce the problem to a two-dimensional one thereby accomplishing the goal of the plate theory, i.e., the dimensionality reduction of the governing

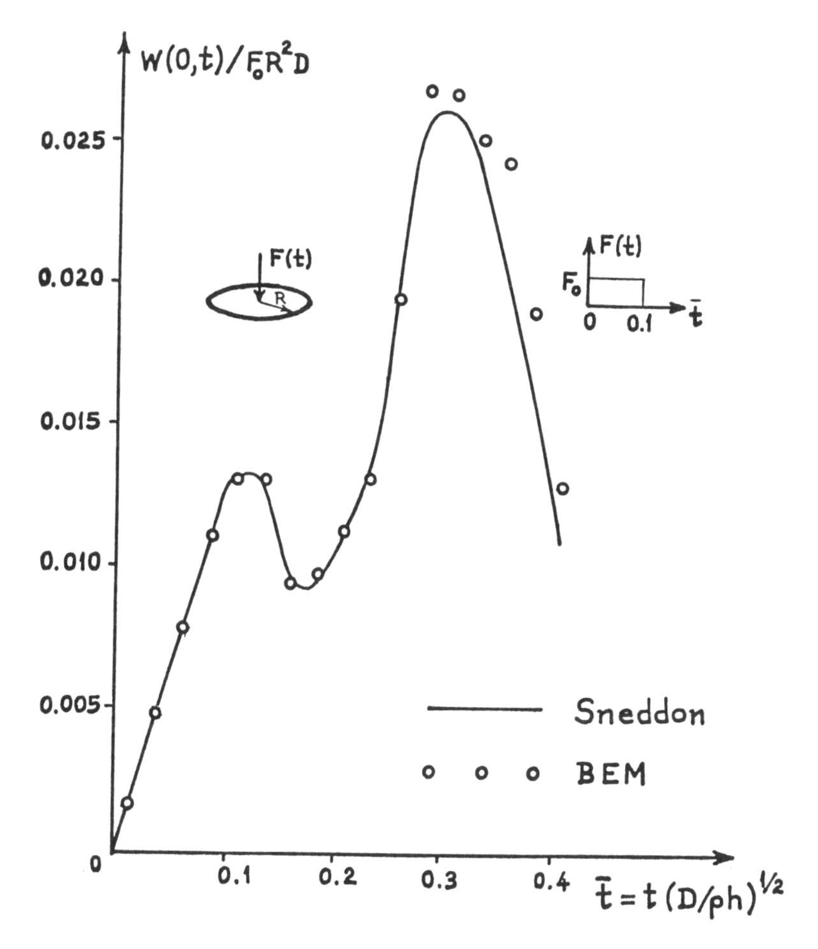

Figure 2. History of central deflection of a simply supported circular plate due to a rectangular load pulse applied at its center; Ref. (8).

equations, without paying the corresponding price of a physical approximation.

Finally, two general boundary methods as applied to free vibrations of flexural plates are also discussed here because of their similarities with the conventional BEM. Both of these methods assume that the solution W of Eq. 22 can be written as a linear combination of a set of functions which simply satisfy Eq. 22, i.e.,

$$W = \sum_{n=1}^{N} A_n \Phi_n(x,y,\omega) \tag{54}$$

where

$$\nabla^4\Phi_n - \beta^4\Phi_n = 0, \quad \beta^4 = m\omega^2/D \tag{55}$$

According to the first method, which has recently been equipped by a sound mathematical basis developed by Herrera (17) for potential and some elasticity problems, the problem is here the construction of a complete set of functions Φ_n. There are no as yet theoretical results available to ensure that the various sets of Φ_n that are used in practical plate problems are indeed complete, eventhough these sets are used with considerable success. According to this Generalized Boundary Method (GBM) the boundary conditions are satisfied approximately by various techniques, such as the point collocation method or the line element method, thereby providing the necessary number of linear equations for the determination of the unknown coefficients A_n. In free vibration problems, of course, the resulting system is of the form

$$[K(\omega)]\ \{A_n\} = \{0\} \tag{56}$$

which is an eigenvalue problem to be solved for ω and A_n. The point collocation or point matching method is a rather old technique that has been applied for plate vibration problems in the past by Conway and Farnham (11, 12) and most recently by Akkari and Hutchinson (1). The line element method developed by Akkari and Hutchinson (1) is a weighted residual approach, whereby the boundary conditions are approximated by first dividing the boundary into a finite number of line (straight or curved) elements and then satisfying the boundary conditions along each segment approximately by setting the weighted line integrals of the boundary condition residuals equal to zero. Both of these approaches have been successfully used in Ref. (1) for the free vibration analysis of various simple plates.

According to the second boundary method, better known as the Edge Function Method (EFM) and originally developed by Quinlan (35) for elastostatic problems, the functions Φ_n are not only functions that simply satisfy Eq. 22 but also functions that in addition have the property of rapidly decaying with increasing distance from the edges of the plate boundary which is approximated by straight segments. The difficulty here lies in the construction of these special asymptotic functions but once these have been constructed the problem can be easily solved with fewer functions than in the previous method. The EFM has been successfully applied for the free vibration analysis of simple isotropic and orthotropic plates by Nash et al (27) and O'Callaghan and Studdert (32), respectively.

Dynamic Analysis of Membranes

Membranes are very thin and flexible structures whose free lateral vibrations are governed by

$$T\nabla^2 w - \rho \frac{\partial^2 w}{\partial t^2} = 0 \tag{57}$$

where $w = w(x,y,t)$ is the lateral deflection, T is the constant surface tension and ρ is the constant surface density. Assuming a solution of the form

$$w = W(x,y) \sin\omega t \tag{58}$$

Eq. 57 becomes the Helmholtz equation

$$\nabla^2 W + \lambda^2 W = 0, \quad \lambda^2 = \omega^2(\rho/T) \tag{59}$$

which is the basic equation in linear acoustics that has been solved by the BEM for its eigenvalues λ and eigenvectors W by various researchers, e.g., Tai and Shaw (41) and Shaw (36, 38). A brief description of the direct BEM as used for the solution of Eq. 59 is given, however, in the following for reasons of completeness.

The Green's function G of the problem satisfies the equation

$$\nabla^2 G + \lambda^2 G = \delta \tag{60}$$

and has the explicit form

$$G(x,y;\xi,\eta) = \frac{i}{4} H_o^{(1)}(\lambda r) \tag{61}$$

where $H_o^{(1)}(\cdot)$ is the Hankel function of the first kind and zero order and $r^2 \triangleq (x-\xi)^2 + (y-\eta)^2$.

Multiplication of Eqs 59 and 60 by G and W, respectively, subtraction of the resulting expressions and integration of the final result over the area S yields

$$W(\xi,\eta) = \int_S (W\nabla^2 G - G\nabla^2 W)\ dS \tag{62}$$

On account of Green's second identity

$$\int_S (W\nabla^2 G - G\nabla^2 W)\ dS = \int_C (W \frac{dG}{dn} - G \frac{dW}{dn})\ dC \tag{63}$$

where C indicates the perimeter (boundary) of the membrane and n is the unit outward normal at C, Eq. 62 becomes for a point $P(\xi,\eta)$ the integral equation

$$cW(\xi,\eta) = \int_C (W \frac{dG}{dn} - G \frac{dW}{dn})\ dC \tag{64}$$

where $c = 1$ for $P \varepsilon S$ and $c = 1/2$ for $P \varepsilon C$

The solution of Eq. 64 with $c = 1/2$ is accomplished numerically by simply dividing the boundary C into a number N of boundary elements. Assuming for simplicity straight constant boundary elements one receives the discretized form of Eq. 64 with $c = 1/2$ as

$$[G_n]\ \{W\} - [G]\ \{W_n\} = 0 \tag{65}$$

where $[G]$ and $[G_n]$ are influence matrices consisting of integrals over boundary elements of G and dG/dn, respectively and $\{W\}$ and $\{W_n\}$ are the vectors of the boundary deflections and slopes of the membrane. Application of the boundary conditions and rearrangement reduces Eq. 65 to

$$[A(\lambda)]\ \{a\} = \{0\} \tag{66}$$

where the vector $\{a\}$ contains unknown boundary values of W and dW/dn. Solution of the eigenvalue equation 66 provides the eigenvalues λ and eigenvectors $\{a\}$ from which the natural frequencies ω and modal shapes $\{W\}$ can be obtained.

Hutchinson (18) used the direct BEM in a special form and with $G = Y_o(\lambda r)$ to successfully study free vibrations of membranes with various shapes. This approach which uses just the real part of G has the advantage that $[A(\lambda)]$ is a real valued matrix, but it provides fictitious values of λ along with the correct ones, which, however, can be sorted out by looking at the corresponding modal shapes (18, 20). A boundary method in conjunction with direct point collocation has recently also been successfully used by Hutchinson (20) for free vibrations of membranes.

Dynamic Analysis of Shells

To the author's best knowledge there is no any work available in the literature on free or forced vibrations of shells by the conventional BEM. A brief discussion on the application of the EFM to free vibrations of shallow spherical shells is given in Nash et al (27) together with some numerical results. The equations of free vibration of a shallow spherical shell are of the form

$$D\nabla^4 W + (1/R)\nabla^2\Phi - \rho h\omega^2 W = 0 \tag{67}$$

$$(1/Eh)\nabla^4\Phi - (1/R)\nabla^2 W = 0 \tag{68}$$

where W is the amplitude of the normal deflection, Φ is the amplitude of the stress function, E is the modulus of elasticity, D is the flexural rigidity, R is the radius, h is the thickness and ρ is the density of the shell, while ω is the circular frequency of vibration. Uncoupling of Eqs 67 and 68 results in the equations

$$\nabla^6 W - p^4\nabla^2 W = 0 \tag{69}$$

$$\nabla^8\Phi - p^4\nabla^4\Phi = 0 \tag{70}$$

where

$$p^4 = (h/D)\left[\rho\omega^2 - (E/R^2)\right] \tag{71}$$

Setting

$$\psi = \nabla^4 W - p^4 W \tag{72}$$

one has from Eq. 69 that

$$\nabla^2\psi = 0 \tag{73}$$

with a solution in terms of edge functions of the form

$$\psi = \sum_{j=1}^{J}\sum_{m_j}\left[A_m^j \sin(m_j x_j + \alpha_m) + B_m^j \cos(m_j x_j + \beta_m)\right] e^{-m_j y_j} \tag{74}$$

where J is the total number of boundary sides or edges. An analogous expression can be written down for Φ and finally the unknown coefficients can be determined by enforcing satisfaction of the boundary conditions as described in Ref. (27). From Ref. (27) is apparent that, even for this special case, the EFM is rather complicated.

Conclusions

This chapter critically reviewed the applications of the BEM in dynamic problems of beams, plates, membranes and shells. The main conclusions that can be drawn on the basis of the discussion in the previous sections are the following:

1) The BEM is at best as efficient as the FEM for free and forced vibrations of beams, because for these one-dimensional structures the method looses its major advantage over the FEM, which is the dimensionality reduction. Presentation of the dynamic analysis of beams was included in this chapter for reasons of completeness and tutorial purposes.

2) The BEM is ideally suited for free and forced vibrations of plates, where indeed there is a dimensionality reduction of the problem. Thus, the BEM appears to be more efficient than either the FEM or the FDM for this class of problems. More work in this area is, however, required in order to go in depth and refine and generalize both the frequency and time domain BEM algorithms.

3) No definite answer can be given at this moment about the advantages and disadvantages of the BEM as applied to thin shell dynamic problems because there is no as yet work available on this subject in the literature.

4) The BEM is ideally suited for membrane vibration problems and there is a considerable amount of work in linear acoustics, which is characterized by the same governing equation as vibrating membranes, that can be profitably adopted here.

5) Two boundary methods, the GBM and the EFM, similar in idea with the BEM in that they require only a boundary discretization but different in their construction, have been successfully applied for the solution of certain classes of dynamic problems of membranes, plates ans shells and show a significant future promise.

Acknowledgements

The author would like to express his thanks to Ms. H. Alexandridis for her conscientious typing of the manuscript.

Appendix.- References

1. Akkari, M.M. and Hutchinson, J.R., "An Alternative BEM Formulation Applied to Plate Vibrations", in "Boundary Elements VII", C.A. Brebbia and G. Maier, Editors, Springer-Verlag, Berlin, 1985, pp. 6.111-6.126.

2. Banerjee, P.K. and Butterfield, R., "Boundary Element Methods in Engineering Science", McGraw-Hill Book Company, London, 1981.

3. Bathe, K.J., "Finite Element Procedures in Engineering Analysis", Prentice-Hall, Englewood Cliffs, New Jersey, 1982.

4. Beskos, D.E. and Narayanan, G.V., "Dynamic Response of Frameworks by Numerical Laplace Transform", Computer Methods in Applied Mechanics and Engineering, Vol. 37, 1983, pp. 289-307.

5. Beskos, D.E., Editor, "Boundary Element Methods in Mechanics", North-Holland Publishing Co., Amsterdam, 1987.

6. Beskos, D.E., "Boundary Element Methods in Dynamic Analysis", Applied Mechanics Reviews, Vol. 40, 1987, pp. 1-23.

7. Bézine, G., "A Mixed Boundary Integral - Finite Element Approach to Plate Vibration Problems", Mechanics Research Communications, Vol. 7, 1980, pp. 141-150.

8. Bézine, G. and Gamby, D., "Étude des Mouvements Transitoires de Flexion d'une Plaque par la Méthode des Équations Intégrales de Frontière", Journal de Mécanique Théorique et Appliquée, Vol. 1, 1982, pp. 451-466.

9. Brebbia, C.A., Telles, J.C.F. and Wrobel, L.C., "Boundary Element Techniques", Springer-Verlag, Berlin, 1984.

10. Clough, R.W. and Penzien, J., "Dynamics of Structures", McGraw-Hill Book Co., New York, 1975.

11. Conway, H.D., "The Bending, Buckling and Flexural Vibrations of Simply Supported Polygonal Plates by Point-Matching", Journal of Applied Mechanics, Vol. 28, 1961, pp. 288-291.

12. Conway, H.D. and Farnham, K.A., "The Free Flexural Vibration of Triangular, Rhombic and Parallelogram Plates and Some Analogies", International Journal of Mechanical Sciences, Vol. 7, 1965, pp. 811-816.

13. Dominguez, J. and Alarcón, E., "Elastodynamics", in "Progress in Boundary Element Methods", Vol. 1, C.A. Brebbia, Editor, Halsted Press, New York, 1981, pp. 213-257.

14. Durbin, F., "Numerical Inversion of Laplace Transforms: An Efficient Improvement to Dubner and Abate's Method", Computer Journal, Vol. 17, 1974, pp. 371-376.

15. Ghali, A. and Neville, A.M., "Structural Analysis", Intext Educational Publishers, Scranton, Pennsylvania, 1972.

16. Graff, K.F., "Wave Motion in Elastic Solids", Ohio State University Press, Columbus, 1975.

17. Herrera, I., "Boundary Methods: An Algebraic Theory", Pitman, Boston, 1984.

18. Hutchinson, J.R., "Determination of Membrane Vibrational Characteristics by the Boundary-Integral Equation Method", in "Recent Advances in Boundary Element Methods", C.A. Brebbia, Editor, Pentech Press, London, 1978, pp. 301-316.

19. Hutchinson, J.R. and Wong, G.K.K., "The Boundary Element Method for Plate Vibrations", Proceedings of the ASCE 7th Conference on Electronic Computation, St. Louis, Missouri, 1979, pp. 297-311.

20. Hutchinson, J.R., "An Alternative BEM Formulation Applied to Membrane Vibrations", in "Boundary Elements VII", C.A. Brebbia and G. Maier, Editors, Springer-Verlag, Berlin, 1985, pp. 6.13-6.25.

21. Kitahara, M., "Boundary Integral Equation Methods in Eigenvalue Problems of Elastodynamics and Thin Plates", Elsevier, Amsterdam, 1985.

22. Kraus,H., "Thin Elastic Shells", John Wiley and Sons, New York, 1967.

23. Leissa, A.W., "Vibration of Plates", NASA SP-160, NASA, Washington, D.C., 1969.

24. Leissa, A.W., "Vibration of Shells", NASA SP-288, NASA, Washington, D.C., 1973.

25. Love, A.E.H., "A Treatise on the Mathematical Theory of Elasticity", 4th Edition, Dover Publication, New York, 1944.

26. Narayanan, G.V. and Beskos, D.E., "Numerical Operational Methods for Time-Dependent Linear Problems", International Journal for Numerical Methods in Engineering, Vol. 18, 1982, pp. 1829-1854.

27. Nash, W.A., Tai, I.H., O'Callaghan, M.J.A. and Quinlan, P.M., "Statics and Dynamics of Elastic Bodies - A New Approach", Proceedings of the International Symposium on Innovative Numerical Analysis in Applied Engineering Science, Versailles, France, T.A. Cruse et al., Editors, CETIM, Cenlins, France, 1977, pp. 8.3-8.8.

28. Niwa, Y., Kobayashi, S. and Kitahara, M., "Eigenfrequency Analysis of a Plate by the Integral Equation Method", Theoretical and Applied Mechanics, Vol. 29, University of Tokyo Press, Tokyo, 1981, pp. 287-307.

29. Niwa, Y., Kobayashi, S. and Kitahara, M., "Analysis of the Eigenvalue Problems of Elasticity by the Boundary Integral Equation Method", Theoretical and Applied Mechanics, Vol. 30, University of Tokyo Press, Tokyo, 1981, pp. 335-356.

30. Niwa, Y., Kobayashi, S. and Kitahara, M., "Determination of Eigenvalues by Boundary Element Methods", in "Developments in Boundary Element Methods - 2 ", P.K. Banerjee and R.P. Shaw, Editors, Applied Science, London, 1982, pp. 143-176.

31. Niwa, Y., Kobayashi, S. and Kitahara, M., "Applications of the Boundary Integral Equation Method to Eigenvalue Problems of Elastodynamics", in "Boundary Element Methods in Engineering", C.A. Brebbia, Editor, Springer-Verlag, Berlin, 1982, pp. 297-311.

32. O'Callaghan, M.J.A. and Studdert, R.P., "The Edge-Function Method for the Free Vibrations of Thin Orthotropic Plates", in "Boundary Elements VII", C.A. Brebbia and G. Maier, Editors, Springer-Verlag, Berlin, 1985, pp. 6.37-6.52.

33. Providakis, C.P. and Beskos, D.E., "Dynamic Analysis of Beams by the Boundary Element Method", Computers and Structures, Vol. 22, 1986, pp. 957-964.

34. Providakis, C.P. and Beskos, D.E., "Free and Forced Plate Vibrations by the Boundary Element Method", in preparation.

35. Quinlan, P.M., "The Torsion of an Irregular Polygon", Proceedings of the Royal Society of London, Vol. 282 (A), 1964, pp. 208-227.

36. Shaw, R.P., "Boundary Integral Equation Methods Applied to Wave Problems", in Developments in Boundary Element Methods - 1", P.K. Banerjee and R. Butterfield, Editors, Applied Science, London, 1979, pp. 121-153.

37. Shaw, R.P., "Elastic Plate Vibrations by Boundary Integral Equations Part 1: Infinite Plates", Res Mechanica, Vol. 4, 1982, pp. 83-88.

38. Shaw, R.P., "Acoustics" in "Boundary Element Methods in Mechanics", D.E. Beskos, Editor, North-Holland Publishing Co., Amsterdam, 1986, Chapter 9.

39. Sneddon, I.N., "The Symmetrical Vibrations of a Thin Elastic Plate", Proceedings of the Cambridge Philosophical Society, Vol. 41, 1945, pp. 27-43.

40. Stern, M., "A General Boundary Integral Formulation for the Numerical Solution of Plate Bending Problems", International Journal of Solids and Structures, Vol. 15, 1979, pp. 769-782.

41. Tai, G.R.C. and Shaw, R.P., "Helmholtz Equation Eigenvalues and Eigenmodes for Arbitrary Domains", Journal of the Acoustical Society of America, Vol. 56, 1974, pp. 796-804.

42. Vivoli, J., "Vibrations de Plaques et Potentiels de Couches", Acustica, Vol. 26, 1972, pp. 305-314.

43. Vivoli, J. and Filippi, P., "Eigenfrequencies of Thin Plates and Layer Potentials", Journal of the Acoustical Society of America, Vol. 55, 1974, pp. 562-567.

44. Wong, G.K.K. and Hutchinson, J.R., "An Improved Boundary Element Method for Plate Vibrations", in "Boundary Element Methods", C.A. Brebbia, Editor, Springer-Verlag, Berlin, 1981, pp. 272-289.

Elastodynamics

Marijan Dravinski*

An outline of direct and indirect boundary integral equation formulations for transient and steady state problems of elastodynamics is presented together with an approximate indirect boundary integral equation approach. The numerical solution of the integral equations is discussed for time domain and transform domain formulations. The emphasis is placed upon the problems of diffraction of elastic waves by an obstacle embedded within an elastic full-space or half-space which are of particular interest in structural dynamics and earthquake engineering. Some problems in the application of these techniques to realistic models are identified.

Presented numerical results are chosen to illustrate the applicability of the integral equation approach for a wide class of problems in two- and three-dimensional elastodynamics.

Introduction

The aim of this paper is to present an outline of the boundary integral equation formulations in elastodynamics for a class of problems of interest in structural dynamics and earthquake engineering. The same methodology can also be applied to problems in geophysics and nondestructive testing of materials. A vast number of publications on boundary integral equation methods in elastodynamics which have appeared in technical literature in the past several years are beyond the scope of this paper. For an exhaustive list of references on this topic the reader is referred to Refs. (18,33).

Problems of diffraction of elastic waves by an obstacle embedded within an elastic medium are usually solved by two methods: i) numerical methods, and ii) analytical methods. Each of these has limitations. The analytical solutions apply mainly to linear, isotropic, and homogeneous materials, and simple geometries. The numerical solutions, on the other hand, are often inapplicable to problems of interest in earthquake engineering and structural dynamics. Namely, the most commonly used numerical methods, finite elements and finite differences, require construction of a computational grid which fills the solution domain, in space and time (of the problem under consideration). This reduces the effectiveness of these methods for geotechnical problems which involve large dimensions.

---

*Associate Professor, Department of Mechanical Engineering, University of Southern California, Los Angeles, CA 90089-1453

For many problems, it is possible to construct an integral representation of the solution. Corresponding integral equations involve only the boundary and the initial values (and possibly the interior sources). The boundary value problems are thus formulated in terms of boundary values and the solution at interior points need not be considered in order to solve the integral equation, Refs.(10,14). Once the integral equation is solved, the solution at any interior point can be determined through the original integral representation.

The main advantages of the boundary integral equation methods are i) only the boundary of the body is being discretized, thus reducing the number of unknown variables significantly in comparison to the finite elements and finite difference methods, and ii) the methods are specifically suitable for exterior problems, since the boundary integral equation methods can model infinite domains directly, Ref.(33).

Boundary integral equation formulation of a problem may be stated in a time domain or in a transform domain. The latter one usually makesuse of the Fourier or the Laplace integral transforms. In addition, these formulations may be of a direct or an indirect type.

Although the formulation of the integral equation methods dates back to works of Betti(6), Somigliana(49), Lauricella(36), and Fredholm(26), their numerical solutions are rather recent. Among the first numerical solutions were those of Friedman and Shaw(27), in acoustics, and Banaugh and Goldsmith(3), in steady state elastodynamics. Recent developments in the time domain formulations can be found in the papers by Cole et al.(10), Niwa et al.(41), Manolis(38), Nardini and Brebbia(40), Beskos and Spyrakos(5), and Karabalis and Beskos(31). For recent developments on the Fourier transform domain formulation the reader is referred to the papers by Kobayashi(33) and Rizzo et al.(44). Advances on the Laplace transform formulations in boundary integral equation methods were reported by Cruse and Rizzo(15), Cruse(14), and Manolis and Beskos(39).

A detailed review of the literature pertinent to the evolution of the boundary integral equation methods is given by Brebbia(8). Theory and applications of the boundary integral equation methods in engineering and science are discussed in great detail by Shaw(48), Brebbia et al.(9), Crouch and Starfield(13), and Banerjee and Butterfield(4).

Solutions of the soil-structure-interaction problems using the boundary integral equation approach in time domain are presented in Refs.(5,31) while the corresponding frequency domain problems are addressed in Refs.(16,17).

The approximate boundary integral equation method considered in this paper originates in the works of Kupradze(34), Copley(11), and Oshaki(43). Extension of the method to the wave propagation problems in geophysics and earthquake engineering can be found in Refs.(2,19,20,21,46,52).

Section two of this paper deals with the boundary integral formulation of elastodynamics. Direct and indirect boundary integral equation approach is presented for transient and steady state problems. Corresponding integral equations are stated and modified into a form convenient for studies of diffraction of elastic waves by an obstacle in an elastic full-space or a half-space.

Section three describes an approximate indirect boundary integral equation method which has been successfully used by a number of researchers in the area of geophysics and earthquake engineering.

Section four deals with numerical solution of the boundary integral equations. Discretization of the boundary and the volume integrals and removal of the Cauchy principal value integrals are outlined for the transform domain formulations. A time-step approach for the time domain formulation is illustrated.

Section five discusses the applications of the boundary integral equation approach to problems of two-dimensional and three-dimensional elastodynamics.

Section six contains the summary and conclusions, while section seven incorporates the references.

Boundary Element Formulations In Elastodynamics

Transient Case.-Let us consider a linearly elastic, homogeneous body occupying volume V within a smooth surface S (see Fig. 1). The body is subjected to surface forces for $\underline{x}\varepsilon S$ and to body forces for $\underline{x}\varepsilon V$. For the sake of simplification it is assumed that the body was of a quiescent past (i.e., that all disturbances start at time t = 0). Equation of motion is given by

$$\sigma_{ij,j}(\underline{x},t) + b_i(\underline{x},t) = \ddot{u}_i(\underline{x},t) \qquad i,j = 1,2,3. \qquad \underline{x}\varepsilon V. \tag{1}$$

σ_{ij}, b_i, and u_i are components of stress tensor, body force and displacement vector, respectively and ρ is the mass density. Throughout the derivations, summation over repeated indices is understood, a comma denotes the partial diferentiation with respect to a spatial coordinate x, and $(\dot{\ }) \equiv \partial()/\partial t$. The boundary conditions on S are, in general

$$u_i(\underline{x},t) = f_i(\underline{x},t) \ ; \ \underline{x}\varepsilon S_1 \tag{2a}$$

$$t_i(\underline{x},t) = \sigma_{ij}(\underline{x},t)n_j(\underline{x}) = g_i(\underline{x},t) \ ; \ \underline{x}\varepsilon S_2 \tag{2b}$$

where $S_1 + S_2 = S$, t_i denotes the components of the traction, n_i represents the components of the unit normal on S (pointing away from V), and $\underline{f}$ and $\underline{g}$ are assumed to be known. The stresses are given by

$$\sigma_{ij} = c_{ijkl}u_{k,l}, \tag{3a}$$

$$c_{ijkl} = \lambda\delta_{ij}\delta_{kl} + \mu(\delta_{ik}\delta_{jl} + \delta_{il}\delta_{jk}) \tag{3b}$$

where λ and μ are the Lamé constants, Ref. (29).

At this point, we introduce the displacement field from a simple unit impulse force within the elastic body which is called the elastodynamic Green function, Ref. (1), or the singular solution, Ref. (4). If the unit impulse is applied at $\underline{x} = \underline{\xi}$ and at $t = 0$ and in the $\underline{n}$ direction, then we denote the i-th component of displacement at general $(\underline{x},t)$ by $G_{in}(\underline{x},t;\underline{\xi},0)$. To specify $\underline{G}$ uniquely under zero initial conditions, it remains to state the boundary conditions on S and this may vary from problem to problem (e.g., full-space versus half-space problems, etc.). Therefore, for an impulsive body force of unit magnitude $e_i(\underline{\xi})$ at $\underline{\xi}$ in direction of the x_i-axis we have

$$b_i(\underline{x},t) = \delta(t)\delta(\underline{x}-\underline{\xi})e_i(\underline{\xi}) \tag{4}$$

and the displacement at a point $\underline{x}$ at time t is given by, Ref. (4),

$$u_i(\underline{x},t) = G_{ij}(\underline{x},t;\underline{\xi},0)e_j(\underline{\xi}). \tag{5}$$

Corresponding stresses and tractions are specified by

$$\sigma_{ij}(\underline{x},t) = T_{ijk}(\underline{x},t;\underline{\xi},0)e_k(\underline{\xi}) \tag{6a}$$

$$t_i(\underline{x},t) = F_{ik}(\underline{x},t;\underline{\xi},0)e_k(\underline{\xi}), \tag{6b}$$

where

$$T_{ijk}(\underline{x},t;\underline{\xi},0) = c_{ijml}G_{mk,l} \tag{7a}$$

$$F_{ik}(\underline{x},t;\underline{\xi},0) = c_{ijml}G_{mk,l}n_j. \tag{7b}$$

Direct Boundary Element Formulation.-The dynamic reciprocal theorem provides a basis from which the boundary integral equations for elastodynamics can be constructed. The theorem (4,51) can be stated as follows. If there exist two unrelated quiescent elastodynamic states $(\underline{b},\underline{t},\underline{u})$ and $(\underline{b}^*,\underline{t}^*,\underline{u}^*)$ defined in the same region V bounded by a surface S, then for every $t \geq 0$

$$\int_S t_i * u_i^* dS(\underline{x}) + \int_V b_i * u_i^* dV(\underline{x})$$

$$= \int_S t_i^* * u_i dS(\underline{x}) + \int_V b_i^* * u_i dV(\underline{x}), \tag{8}$$

where

$$t_i * u_i^* = \int_{-\infty}^{\infty} t_i(\underline{x},t-\tau)u_i^*(\underline{x},\tau)d\tau. \tag{9}$$

Let us assume that the body of domain V within a boundary S (see Fig. 1a), characterized by an elastodynamic state $(\underline{b},\underline{t},\underline{u})$, is embedded within another domain V^* within S^* (which may be at infinity, see Fig. 1b) characterized by an elastodynamic state $(\underline{b}^*,\underline{t}^*,\underline{u}^*)$. This new region is assumed to be in equilibrium as well as the original body within S. It can be shown, Ref. (9), that the reciprocal relation 8 for these two states is still valid. By choosing the elastodynamic state $(\underline{b}^*,\underline{t}^*,\underline{u}^*)$ to correspond to the elastodynamic Green function 5-7 we obtain from 8, for an interior point $\underline{\xi}$,

$$u_j(\underline{\xi},t) = \int_S \{G_{ij} * t_i - F_{ij} * u_i\}dS(\underline{x}) + \int_V G_{ij} * b_i dV(\underline{x}), \tag{10}$$

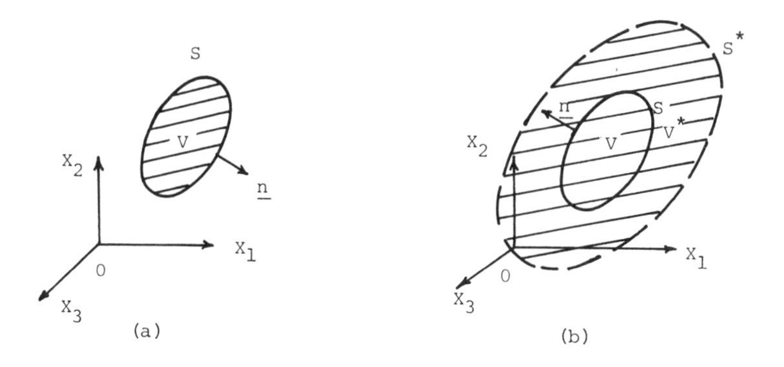

Figure 1. – a) Interior Problem Model, b) Fictitious Problem Model

where the convolution integrals are defined by

$$G_{ij} * t_i \equiv \int_{-\infty}^{\infty} G_{ij}(\underline{x},t-\tau;\underline{\xi},0)t_i(\underline{x},\tau)d\tau \tag{11}$$

$$F_{ij} * u_i \equiv \int_{-\infty}^{\infty} G_{ij}(\underline{x},t-\tau;\underline{\xi},0)u_i(\underline{x},\tau)d\tau. \tag{12}$$

Thus for a general point $\underline{\xi}_o$ on the boundary ($\underline{\xi} \rightarrow \underline{\xi}_o$ from inside of S) we obtain

$$C_{ij}(\underline{\xi}_o)u_i(\underline{\xi}_o,t) = \int_S (G_{ij}*t_i - F_{ij}*u_i)dS(\underline{x}) + \int_V G_{ij} * b_i dV(\underline{x}), \tag{13}$$

where $C_{ij} = \frac{1}{2}\delta_{ij}$ if the boundary at ξ_o is smooth (i.e., it has a unique tangent plane). The term C_{ij} arises from the treatment of the improper integral involving F_{ij} and the integral over S is the Cauchy principal value integral, Ref. (4). Equation 13 is the required integral equation obtained through direct formulation. For prescribed tractions on S (Neumann problem) the integral equation is solved for the unknown surface displacement. Displacement field throughout the elastic medium is then calculated by using the integral representation 10. For prescribed displacement on S (Dirichlet problem) Eq. 13 is solved for the surface traction and displacement field within S is evaluated as before through the integral representation 10. A similar procedure applies for the case of a mixed boundary value problem with boundary conditions specified by Eqs. 2a,b. Once the displacement field is known, corresponding stress and traction fields can be evaluated using Eqs. 3a,b,2b.

Indirect Boundary Element Foumulation.–As before, we consider a quiescent, isotropic, linear, homogeneous elastic region V enclosed within a surface S. It is customary to assume, Refs. (4,9), that the body of volume V and surface S is embedded within a domain V^* with surface S^*, which may be at infinity. Domain V^* is in equilibrium. Now, we introduce fictitious sources of initially unknown density $\phi(\underline{\xi})$ on S, Ref. (4). Known body forces $\underline{b}$ are specified throughout the volume. The displacement field $\underline{u}$ at any interior point $\underline{\xi}$ can be obtained by integrating the unit solutions related to $\underline{\phi}$ and $\underline{b}$ over S and V, respectively to obtain

$$u_j(\underline{\xi},t) = \int_S G_{ij} * \phi_i dS(\underline{x}) + \int_V G_{ij} * b_i dV(\underline{x}), \tag{14}$$

where the convolution integrals are defined in a manner analogous to that of Eq. 11. Corresponding traction on a surface through the interior point $\underline{\xi}$ with a normal $\underline{n}(\underline{\xi})$ is given by

$$t_i(\underline{\xi},t) = \int_S F_{ij} * \phi_i dS(\underline{x}) + \int_V F_{ij} * b_i dV(\underline{x}). \tag{15}$$

By letting $\underline{\xi}$ approach a point $\underline{\xi}_o$ on the boundary from inside S, Eqs. 14,15 provide the following discrete indirect boundary element equations

$$u_j(\underline{\xi}_o,t) = \int_{-\infty}^{\infty} d\tau \int_S G_{ij}(\underline{x},t-\tau;\underline{\xi}_o,0)\phi_i(\underline{x},\tau)dS(\underline{x}) + \int_{-\infty}^{\infty} d\tau \int_V G_{ij}(\underline{x},t-\tau;\underline{\xi}_o,0)b_i(\underline{x},\tau)dV(\underline{x}) \tag{16}$$

$$t_i(\underline{\xi}_o,t) = -\frac{1}{2}\phi_i(\underline{\xi}_o,t) + \int_{-\infty}^{\infty} d\tau \int_S F_{ij}(\underline{x},t-\tau;\underline{\xi}_o,0)\phi_i(\underline{x},\tau)dS(\underline{x}) + \int_{-\infty}^{\infty} d\tau \int_V F_{ij}(\underline{x},t-\tau;\underline{\xi}_o,0)b_i(\underline{x},\tau)dV(\underline{x}) \tag{17}$$

provided the point $\underline{\xi}_o$ is not located at any edge or a corner (i.e., there must be a unique tangent plane at $\underline{\xi}_o$) and the surface integral in Eq. 17 must be understood as a Cauchy principal value integral. Equations 16,17 are the two boundary integral equations for the elastodynamic problem obtained by the indirect boundary element method. If, for example, the displacement field is specified on S then the solution of Eq. 16 will provide the surface loading $\phi_j(\underline{\xi}_o)$. If the tractions are specified on S, then Eq. 17 needs to be solved for $\phi_j(\underline{\xi}_o)$. For a mixed boundary value problem Eq. 16 may be used for that portion of the boundary where displacements are specified and Eq. 17 can be used for the portion on which the surface forces are specified. The resulting equations are then combined and solved together, Ref. (4).

Steady State Elastodynamics.-For steady state motion of an elastic medium we are assuming the following time dependence for the elements of an elastodynamic state

$$u_i(\underline{x},t) = u_i(\underline{x},\omega)e^{i\omega t} \tag{18a}$$

$$b_i(\underline{x},t) = b_i(\underline{x},\omega)e^{i\omega t} \tag{18b}$$

$$t_i(\underline{x},t) = t_i(\underline{x},\omega)e^{i\omega t}, \tag{18c}$$

where ω denotes the circular frequency and $i = \sqrt{-1}$. In this case the equation of equilibrium becomes

$$\mu u_{i,jj} + (\lambda+\mu)u_{j,ij} + b_i + \rho\omega^2 u_i = 0. \tag{19}$$

The Green function must now satisfy the following equation

$$c_{ijkl}G_{km,lj}(\underline{x},\underline{\xi},\omega) + \delta_{im}\delta(\underline{x}-\underline{\xi}) + \rho\omega^2 G_{im}(\underline{x},\underline{\xi},\omega) = 0 \tag{20}$$

with corresponding displacement, stress and traction field specified by

$$\begin{aligned} u_i(\underline{x},\delta) &= G_{ij}(\underline{x},\underline{\xi},\omega)e_j(\underline{\xi}) \\ \sigma_{ij}(\underline{x},\omega) &= T_{ijk}(\underline{x},\underline{\xi},\omega)e_k(\underline{\xi}) \\ t_i(\underline{x},\omega) &= F_{ik}(\underline{x},\underline{\xi},\omega)e_k(\underline{\xi}), \end{aligned} \tag{21}$$

where the tensors $\underline{T}$ and $\underline{F}$ are defined by Eqs. 7a,b.

Equation 20 can be used to derive the reciprocal identity for steady state elasto-dynamics, Ref. 4. If $(\underline{u},\underline{b})$ and $(\underline{u}^*,\underline{b}^*)$ are two solutions of equation of motion 19, then the reciprocal identity is specified by

$$\begin{aligned} &\int_S t_i(\underline{x},\omega)u_i^*(\underline{x},\omega)dS(\underline{x}) + \int_V b_i(\underline{x},\omega)u_i^*(\underline{x},\omega)dV(\underline{x}) \\ &= \int_S t_i^*(\underline{x},\omega)u_i(\underline{x},\omega)dS(\underline{x}) + \int_V b_i^*(\underline{x},\omega)u_i(\underline{x},\omega)dV(\underline{x}) \end{aligned} \tag{22}$$

Using the Green function, defined by Eq. 20, for the elastodynamic state $(\underline{u}^*,\underline{b}^*)$ the reciprocal identity 22 provides an integral representation for direct boundary element formulation at an interior point $\underline{\xi}$

$$u_j(\underline{\xi},\omega) = \int_S [t_i(\underline{x},\omega)G_{ij}(\underline{x},\underline{\xi},\omega) - F_{ij}(\underline{x},\underline{\xi},\omega)u_i(\underline{x},\omega)]dS(\underline{x})$$

$$+ \int_V b_i(\underline{x},\omega)G_{ij}(\underline{x},\underline{\xi},\omega)dV(\underline{x}). \tag{23}$$

Letting $\underline{\xi}$ to approach $\underline{\xi}_o$ on boundary S, Eq. 24 provides the following integral equation

$$\frac{1}{2} u_j(\underline{\xi}_o,\omega) = \int_S [t_i(\underline{x},\omega)G_{ij}(\underline{x},\underline{\xi}_o,\omega) - F_{ij}(\underline{x},\underline{\xi}_o,\omega)u_i(\underline{x},\omega)]dS(\underline{x})$$

$$+ \int_V b_i(\underline{x},\omega)G_{ij}(\underline{x},\underline{\xi}_o,\omega)dV(\underline{x}), \tag{24}$$

where S is assumed to be smooth at $\underline{\xi}_o$ and integral over S is a Cauchy principal value integral. The integral equation is solved for surface displacement if surface tractions are specified, or for surface traction if surface displacement is specified. Displacement throughout the elastic medium is then determined by the integral representation 23. This concludes the direct boundary element formulation for steady state elastodynamics. Indirect formulation is analogous to that of the transient case and therefore it will be omitted here.

Steady State Wave Propagation Problems.–For many problems of diffraction of elastic waves the input disturbance is present in the form of an incident wave. When the incident wave strikes an obstacle or an interface between two different materials, a scattered wave field is created. Therefore, outside of the obstacle, the total displacement field consists of incident and scattered wave fields

$$\underline{u} = \underline{u}^{inc} + \underline{u}^s, \tag{25}$$

where superscripts inc and s denote the incident and scattered wave field, respectively. Therefore, the direct boundary element integral equation 23 for steady state waves becomes, Ref. (4),

$$C_{ij}(\underline{\xi})u_i(\underline{\xi},\omega)=\int_S[G_{ij}(\underline{x},\underline{\xi},\omega)t_i(\underline{x},\omega)-F_{ij}(\underline{x},\underline{\xi},\omega)u_i(\underline{x},\omega)]dS(\underline{x})+u_j^{inc}(\underline{\xi},\omega), \tag{26}$$

where zero body force is assumed in V and C_{ij} is defined to be

$$C_{ij}(\underline{\xi}) = \begin{cases} \frac{1}{2}\,\delta_{ij}, & \text{for } \underline{\xi}\,\varepsilon\, S \end{cases} \tag{27}$$

δ_{ij}, for $\xi \varepsilon V$

where integral over S is to be interpreted as a Cauchy principal value integral. Equation 26 is valid for three or two dimensions. It provides a relationship that must be satisfied between surface displacements, surface tractions and the incident field.

Infinite and Semi-Infinite Regions.-In our analysis so far, only bounded bodies were considered. This interior formulation describes the elastodynamic fields in V within S. We are interested in extending this to exterior problems in an infinite regular region, Ref. (9), as shown by Fig. 2a. The extension to infinite region requires certain hypotheses on the behavior of the elastodynamic state at infinity, Ref. (51).

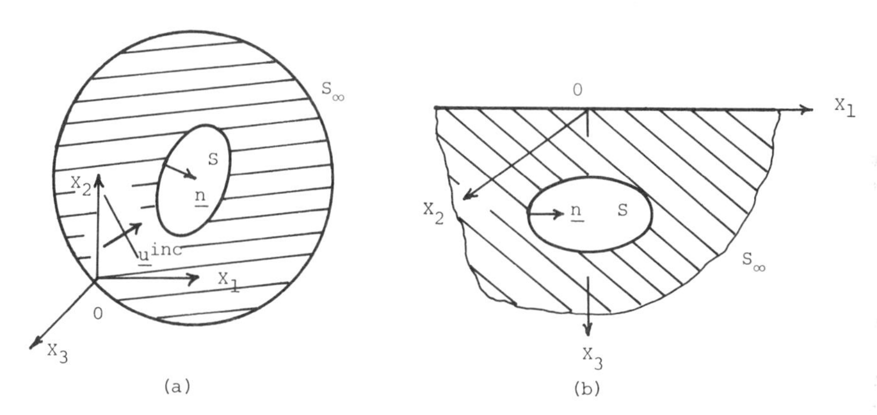

Figure 2. - An Exterior Problem a) In a Full-Space b) In a Half-Space

Let a sphere S_∞ enclose a cavity S (Fig. 2a), Ref. (9). The area of integration in Eq. 26 becomes S + S_∞ for the exterior problem. Due to attenuation of the elastodynamic field at infinity, only the integrals over S will remain. Consequently, the problem of diffraction of elastic waves by a cavity embedded in an elastic full-space is formally stated by Eq. 26 which was derived for an interior problem. The same integral equation is valid for problems of diffraction of elastic waves by a cavity embedded in an elastic half-space with stress free surface. The only difference in application of Eq. 26 to these two exterior problems is that the unit normal on S is now pointing into the cavity, Figs. (2a,b).

It should be emphasized that for the half-space (full-space) problems corresponding singular solution is that for a unit load within a half-space (full-space). As a result, Green functions for the half space problems are considerably more complicated than the Green functions for the full-space problem, Ref. (35). Problems in numerical evaluation of elastodynamic Green functions is discussed in detail in Refs. (2,37).

An Approximate Method

Statement of Problem.–Diffraction of harmonic waves by an obstacle of arbitrary shape embedded into an elastic half-space is characterized by great complexity. Consequently, approximate methods were proposed in order to simplify the numerical calculations of the results. One discussed here originates in the works of Kupradze (34) and Oshaki (43). This method has been successfully used in a number of investigations of interest in geophysics and earthquake engineering, Refs. (2,19,20,21,52). For the sake of illustration of the method, we consider a case of a plane strain model for diffraction of a plane harmonic P and SV wave by a cavity of arbitrary shape partially embedded in a weakly anelastic half-space. An incident plane harmonic (P or SV) wave strikes the cavity thus causing a scattered wave field to be created. Since the model is of the plane strain type this implies that the medium extends to infinity perpendicularly to the plane of drawing and the motion takes place in the xy-plane only, i.e., $\underline{u} = (u,v)$, where u and v denote the displacement components along the x and y axis, respectively. The displacement field is related to the displacement potentials through

$$u = \frac{\partial \phi}{\partial x} + \frac{\partial \psi}{\partial y} \tag{28a}$$

$$v = \frac{\partial \phi}{\partial y} - \frac{\partial \psi}{\partial y} \tag{28b}$$

which in turn satisfy the following equations

$$(\nabla^2 + h^{*2})\phi(x,y,\omega) = 0 \tag{29a}$$

$$(\nabla^2 + k^{*2})\psi(x,y,\omega) = 0. \tag{29b}$$

In the last equation h^* and k^* represent the wave numbers associated with P and SV waves, respectively, and the factor $\exp(i\omega t)$ is understood. Since the medium is assumed to be weakly anelastic, one distinguishes two types of body wave velocities (1): i) the intrinsic elastic velocities in the absence of dissipation (α for P and β for SV waves) and ii) the velocities of the waves in the presence of dissipation

$$\frac{1}{\alpha^*} = \frac{1}{\alpha}\left(1 - \frac{i}{2Q_\alpha}\right) \; ; \; i=\sqrt{-1} \tag{30a}$$

$$\frac{1}{\beta^*} = \frac{1}{\beta}\left(1 - \frac{i}{2Q_\beta}\right), \tag{30b}$$

where Q_α and Q_β are taken to be frequency independent. Introduction of attenuation can be thought of as replacing the real wavenumbers (h,k) by complex-valued wavenumbers (h^*,k^*), Ref. (1),

$$h^* = h\left(1 - \frac{i}{2Q_\alpha}\right) \tag{31a}$$

$$k^* = k\left(1 - \frac{i}{2Q_\beta}\right). \tag{31b}$$

Zero tractions on the cavity implies

$$\sigma_{ij}(\underline{x},\omega)n_j(\underline{x}) = 0 \ , \ i,j=1,2 \ ; \quad \underline{x}\varepsilon S. \tag{32}$$

In addition, the field at infinity should contain only outgoing waves.

Solution of Problem.-The total displacement field in the half-space is specified by

$$\underline{u}(\underline{x},\omega) = \underline{u}^{ff}(\underline{x},\omega) + \underline{u}^{s}(\underline{x},\omega), \tag{33}$$

where the superscripts ff and s denote the free and scattered wave field, respectively. The free field is the one which would exist in the half-space in the absence of the cavity when subjected to the incident wave field. By assuming the displacement potentials in terms of single layer potentials, Ref. (50), we have

$$\phi^s(\underline{x},\omega) = \int_S q(\underline{\xi})\phi(\underline{x},\underline{\xi},\omega)dS(\underline{\xi}) \tag{34a}$$

$$\psi^s(\underline{x},\omega) = \int_S \underline{q}(\underline{\xi})\psi(\underline{x},\underline{\xi},\omega)dS(\underline{\xi}), \tag{34b}$$

where q and $\underline{q}$ are the unknown density functions. The functions ϕ and ψ are the Green functions corresponding to P and SV line loads in a half-space, respectively and they satisfy the equations

$$(\nabla^2 + h^{*2})\phi(\underline{x},\underline{\xi},\omega) = -\delta(\underline{x}-\underline{\xi}) \tag{35a}$$

$$(\nabla^2 + k^{*2})\psi(\underline{x},\underline{\xi},\omega) = -\delta(\underline{x}-\underline{\xi}). \tag{35b}$$

with appropriate boundary conditions

$$\sigma_{xy}(\underline{x},\underline{\xi},\omega) = \sigma_{yy}(\underline{x},\underline{\xi},\omega) = 0 \quad \text{on} \quad y = 0. \tag{36}$$

Introduction of the auxiliary surface S^- in the integral representation for the scattered wave field is the fundamental characteristic of this approximate method. The surface S^- is assumed to be defined inside of the surface S, Ref. (34). This simplifies the numerical procedure considerably since the resulting integral equations are of the regular type as opposed to the singular integral equations which resulted from the exact formulations. However, the method is not without difficulties of its own, Ref. (28), since the location of the auxiliary surface must be chosen carefully in order to obtain accurate results, Refs. (2,19,20,21,47,52).

If the scattered wave field is expressed in terms of discrete line sources, it follows then from Eqs. 34a,b

$$\phi^s(\underline{x},\omega) = A_m \phi(\underline{x},\underline{\xi}_m,\omega) \quad ; \quad \underline{x}_m \varepsilon S^- \ , \quad m=1,2,\ldots,M. \tag{37a}$$

$$\underline{\psi}^s(\underline{x},\omega) = \underline{\underline{A}}_m \underline{\psi}(\underline{x},\underline{\xi}_m,\omega), \tag{37b}$$

where M represents the order of the approximation. The unknown source intensities are determined through Eq. 32 which specifies zero tractions along the cavity. By choosing, say, N collocation points along S to impose the zero traction condition 32, the source intensities can be determined in the least square sense by

$$(A_1,A_2,\ldots,A_M,\underline{\underline{A}}_1,\underline{\underline{A}}_2,\ldots,\underline{\underline{A}}_M)^T = (\underline{G}^*,\underline{G})^{-1}\underline{G}^*\underline{f}, \tag{38}$$

where matrix $\underline{G}$ and vector $\underline{f}$ are known and $\underline{G}^*$ denotes the transpose complex conjugate of $\underline{G}$. Matrix $\underline{G}$ contains information associated with the Green functions and vector $\underline{f}$ incorporates the free field. Once the source intensities are determined, the scattered field throughout the elastic medium can be evaluated using Eqs. 38 and 39.

Numerical Solutions

Integral Eqs. 13,16,17,25 are the exact formulations of the solution of of an elastodynamic problem. Numerical errors in the solution will arise only due to discretization of the integrals and the subsequent solution of the algebraic equations. For the sake of illustration only the simplest possible algorithm for a two-dimensional problem is discussed here. The algorithm utilizes linear boundary elements and triangular internal cells, Ref. (4). Figure 3 depicts the situation when a region is divided into M triangular cells and the boundary of the region into N line segments. It is assumed that the unknown displacement and the traction fields, together with the body forces, vary linearly over the boundary elements or triangular cells. Although the interior cells have the appearance of elements in a finite element discretization scheme they are merely a convenient way of calculating the effects of the body forces (sources) distributed throughout the volume,Ref. (4). The number of linear algebraic equations generated by a discretization scheme in a boundary integral equation approach depends only upon the number of boundary elements used and is not related to the number of cells.

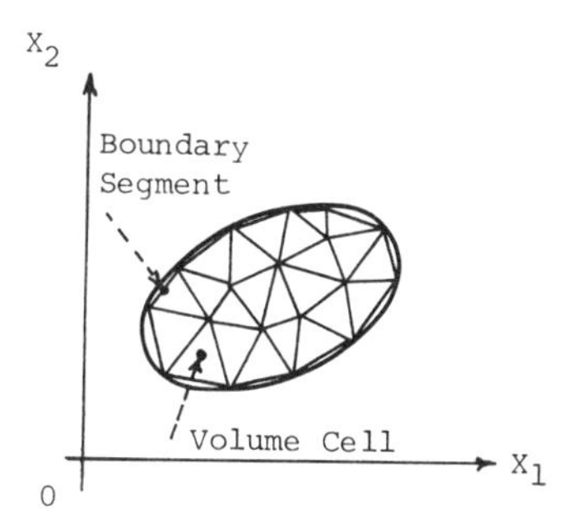

Figure 3. - Discretization of the Surface and Volume Integrals

Fourier Transform Formulation: A Direct Method.- If the displacements, tractions, and the body forces are distributed uniformly over each element, we obtain from Eq. 24 that the displacement vector of a representative nodal point on the pth boundary element is given by

$$1/2u_j(\xi_o^p,\omega) = \sum_{q=1}^{N}[t_i(\underline{x}^q)\int_{\Delta S_q} G_{ij}(\underline{x}^q,\xi_o^p,\omega)dS(\underline{x}^q) - u_i(\underline{x}^q)\int_{\Delta S_q} F_{ij}(\underline{x}^q,\xi_o^p)dS(\underline{x}^q)] + \sum_{m=1}^{M} b_i(\underline{x}^m)\int_{\Delta Y_m} G_{ij}(\underline{x}^m,\xi_o^p)dV(\underline{x}^m) \quad (39)$$

where ξ_o^p are the coordinates of a representative field point on the pth boundary element, such as at the center of the element, ΔS_q is the length of the qth boundary element and ΔV_m is the area of the mth body cell, Ref. (4). In matrix notation this can be written in the form

$$1/2\ \underline{I}\ \underline{u}^p = \sum_{q=1}^{N}[(\int_{\Delta S_q}\underline{G}^{pq}dS)\underline{t}^q - (\int_{\Delta S_q}\underline{F}^{pq}dS)\underline{u}^q] + \sum_{m=1}^{M}(\int_{\Delta V_m}\underline{G}^{pm}dV)\underline{b}^m. \quad (40)$$

If the tractions, displacements, and body forces vary linearly over each element, then the quantities $\underline{t}^q$, $\underline{u}^q$, and $\underline{b}^m$ cannot be taken outside the integrals as was done in Eq. 40. In this case, it is possible to express $\underline{t}^q$ and $\underline{u}^q$ as functions of the nodal values by using the shape function $\underline{N}^q$ for the qth boundary element with nodes r and s. Similarly, $\underline{b}^m$ can be defined as a function of the nodal values using the shape function $\underline{M}^m$ for the mth cell which has nodes r, s, and t as it is shown by Fig. 4.

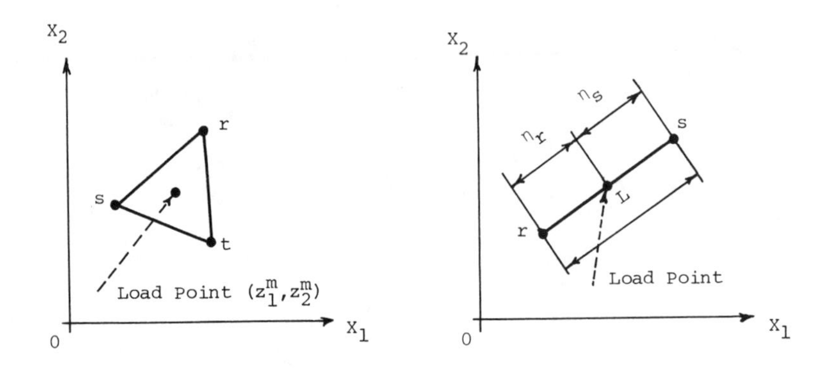

Figure 4. - Boundary Element and Body Cell Element

Therefore, we can write

$$\underline{t}^q = \underline{N}^q t_n \tag{41a}$$

$$\underline{u}^q = \underline{N}^q \underline{u}_n \tag{41b}$$

$$\underline{b}^m = \underline{N}^m \underline{b}_n, \tag{41c}$$

where nodal displacement vector $\underline{u}_n$ is defined in terms of displacements at nodes r and s according to

$$\underline{u}_n = (\underline{u}_r, \underline{u}_s) \tag{42a}$$

and the shape function $\underline{N}$ is specified through

$$\underline{N}^q = \begin{bmatrix} N_1 & 0 & N_2 & 0 \\ \\ 0 & N_1 & 0 & N_2 \end{bmatrix} \tag{42b}$$

$$N_1 = 1 - \eta_r/L \tag{42c}$$

$$N_2 = 1 - \eta_s/L \tag{42d}$$

Coordinates (η_r, η_s) in Eqs. 42c,d describe the load point. Nodal body force vector $\underline{b}_n$ is defined by the body forces at the nodes r, s, and t according to

$$\underline{b}_n = (\underline{b}_r, \underline{b}_s, \underline{b}_t) \tag{43a}$$

and the shape function $\underline{M}^m$ is specified through

$$\underline{M}^m = 1/(2\Delta) \begin{bmatrix} M_1 & 0 & M_2 & 0 & M_3 & 0 \\ & & & & & \\ 0 & M_1 & 0 & M_2 & 0 & M_3 \end{bmatrix} \tag{43b}$$

where Δ is the area of the mth cell and M_i, i = 1, 2, 3 are known functions of the coordinates of the nodes r, s, and t as well as the coordinates of a typical load point (z_1^m, z_2^m). Therefore, for linear boundary elements and triangular internal cells, Eq. 40 becomes

$$\begin{aligned} 1/2 \; \underline{I} \; \underline{u}^p = & \sum_{q=1}^{N} [(\int_{\Delta S_q} \underline{G}^{pq}\underline{N}^q dS)\underline{t}_n - (\int_{\Delta S_q} \underline{F}^{pq}\underline{N}^q dS)\underline{u}_n] \\ & + \sum_{m=1}^{M} (\int_{\Delta V_m} \underline{G}^{pm}\underline{M}^m dV)\underline{b}_n \end{aligned} \tag{44}$$

when a representative point on the pth boundary element is not located on a corner. If the problems has corners, Eq. 44 has to be modified for a field point at a corner node. In general case, with or without the corners,it follows that

$$\begin{aligned} \underline{C} \; \underline{u}^p = & \sum_{q=1}^{N} [(\int_{\Delta S_q} \underline{G}^{pq}\underline{N}^q dS)\underline{t}_n - (\int_{\Delta S_q} \underline{F}^{pq}\underline{N}^q dS)\underline{u}_n] \\ & + \sum_{m=1}^{M} (\int_{\Delta V_m} \underline{G}^{pm}\underline{M}^m dV)\underline{b}_n \end{aligned} \tag{45}$$

where $\underline{C}$ is a matrix (2X2 for a two-dimensional problems) whose coefficients are functions of the solid angle substended at the corner node.

By taking the field points successively at the nodal points on the boundary and absorbing the $\underline{C}$-matrix with the $\underline{u}_n$-term on the RHS of Eq. 45, we obtain a set of linear algebraic equations of the form

$$\underline{G} \; \underline{t} - \underline{F} \; \underline{u} + \underline{G} \; \underline{b} = \underline{0}, \tag{46}$$

which can be solved for the unknowns displacements and/or tractions, depending upon the problem under consideration.

For cases of distribution of the displacement, traction, and the body-force fields over the elements with quadratic, cubic, and higher order variations, the reader is referred to chapter eight of Ref. (4).

Removal of Cauchy Principal Value Integrals.–Since some of the integrals in Eqs. 44 and 45 are of the Cauchy principal value type, special care must be taken to evaluate these integrals numerically. Recently, Rizzo et al. (44) proposed a technique which removes all the Cauchy principal value integrals thus allowing a standard Gaussian quadrature rule to be used. This procedure is outlined here within the framework of direct boundary integral formulation in the Fourier transform domain. An appropriate integral equation is defined by Eq. 13 for a body which occupies the interior V of the surface S (see Fig. 1a). For the same interior problem, with loading of static nature, a corresponding integral equation is given by, Ref. (44),

$$C^*_{ij}(\underline{\xi})u_i(\underline{\xi}) = \int_S [G^*_{ij}(\underline{x},\underline{\xi})t_i(\underline{x}) - F^*_{ij}(\underline{x},\underline{\xi})u_i(\underline{x})]dS(\underline{x}), \tag{47}$$

where the superscript * denotes the elastostatic counterparts of the tensors $\underline{C}$, $\underline{G}$, and $\underline{F}$. Since a rigid body motion of the volume V is a solution of the equations of elastostatics, Eq. 47 provides a way to define tensor $\underline{C}^*$ in a form

$$C^*_{ij}(\underline{\xi}) = -\int_S F^*_{ij}(\underline{x},\underline{\xi})dS(\underline{x}). \tag{48}$$

Rizzo et al.(44) showed that this tensor is related to the corresponding elastodynamic one in the following manner

$$C^*_{ij}(\underline{\xi}) + C_{ij}(\underline{\xi}) = \delta_{ij}. \tag{49}$$

Combining Eqs. 48 and 49 it follows that

$$C_{ij}(\underline{\xi}) = \delta_{ij} - \int_S F^*_{ij}(\underline{x},\underline{\xi})dS(\underline{x}). \tag{50}$$

Substituting Eq. 50 into the original integral equation, Eq. 14, leads to the following result

$$\begin{aligned}
&\{\delta_{ij} - \int_{S_c} F^*_{ij}(\underline{x},\underline{\xi})dS(\underline{x}) + \int_{\Delta S} [F_{ij}(\underline{x},\underline{\xi},\omega) - F^*_{ij}(\underline{x},\underline{\xi})]dS(\underline{x})\}u_i(\underline{\xi},\omega) \\
&+ \int_{\Delta S} F_{ij}(\underline{x},\underline{\xi},\omega)[u_i(\underline{x},\omega) - u_i(\underline{\xi},\omega)]dS(\underline{x}) \\
&+ \int_{S_c} F_{ij}(\underline{x},\underline{\xi},\omega)u_i(\underline{x},\omega)dS(\underline{x}) \\
&= \int_S G_{ij}(\underline{x},\underline{\xi},\omega)t_i(\underline{x},\omega)dS(\underline{x}) + u^{inc}_j(\underline{\xi},\omega),
\end{aligned} \tag{51}$$

where the area S is split into two parts: i) ΔS a portion which contains $\underline{\xi}$, and ii) S_c is the remaining part so that the total area S is the sum of the two. As $\underline{x}$ approaches $\underline{\xi}$ the orders of singularities of $\underline{F}$ and $\underline{F}^*$ are the same, Ref. (44). This implies that the first integral over ΔS in Eq. 51 is not a Cauchy principal value integral. In fact, this integral is not singular at all so that a standard Gaussian quadrature may be used throughout the numerical computations.

Time Domain Formulation: A Direct Method.–For the sake of completeness a brief illustration of the numerical procedure for a boundary integral equation formulation of a transient elastodynamic problem is presented next. This particular procedure was applied by Cristensen(12) and Noble(42) in solving different problems of viscoelasticity. The method is discussed in detail in Ref. (4).

For an elastodynamic process which started at time, say, $t = 0$, it is necessary first to discretize the time variable. Then for a time t, which lies between the two time successive nodes $(r - 1, r)$, the displacements and tractions can be approximated by the following implicit time discretization scheme

$$u_i(\underline{x},t) = 1/\Delta t[(t_r - t)u_i(\underline{x},t_{r-1}) + (t-t_{r-1})u_i(\underline{x},t_r)] \tag{52a}$$

$$t_i(\underline{x},t) = 1/\Delta t[(t_r - t)t_i(\underline{x},t_{r-1}) + (t-t_{r-1})t_i(\underline{x},t_r)] \tag{52b}$$

$$\Delta t = t_r - t_{r-1}. \tag{52c}$$

It follows then from Eq. 13, with no body forces, that for a time interval $(m-1,m)$ the displacement field at node m can be written in terms of displacement field at the time node $m-1$ as follows

$$C_{ij}u_i(\underline{\xi},t_m) = C_{ij}u_i(\underline{\xi},t_{m-1}) + \int_{\Delta S}\int_{t_{m-1}}^{t_m} [G_{ij}(\underline{x},\underline{\xi},t_m-\tau)t_i(\underline{x},\tau) - F_{ij}(\underline{x},\underline{\xi},t_m-\tau)u_i(\underline{x},\tau)]d\tau dS. \tag{53}$$

Substituting the displacements and tractions given by Eqs. 52a,b into Eq. 53 the following result is derived

$$\begin{aligned} C_{ij}u_i(\underline{\xi},t_m) = {} & \sum_{r=1}^{m-1} \int_S\int_{t_{r-1}}^{t_r}[t_i(\underline{x},\tau)G_{ij}(\underline{x},\underline{\xi},t_m-\tau) - u_i(\underline{x},\tau)F_{ij}(\underline{x},\underline{\xi},t_m-\tau)]d\tau dS(\underline{x}) \\ & + \int_S t_i(\underline{x},t_m)dS(\underline{x})1/\Delta t\int_{t_{m-1}}^{t_m} G_{ij}(\underline{x},\underline{\xi},t_m-\tau)(\tau-t_{m-1})d\tau \\ & - \int_S u_i(\underline{x},t_m)dS(\underline{x})1/\Delta t\int_{t_{m-1}}^{t_m} F_{ij}(\underline{x},\underline{\xi},t_m-\tau)(\tau-t_{m-1})d\tau \\ & + \int_S t_i(\underline{x},t_{m-1})dS(\underline{x})1/\Delta t\int_{t_{m-1}}^{t_m} G_{ij}(\underline{x},\underline{\xi},t_m-\tau)(t_m-\tau)d\tau \\ & - \int_S u_i(\underline{x},t_{m-1})dS(\underline{x})1/\Delta t\int_{t_{m-1}}^{t_m} F_{ij}(\underline{x},\underline{\xi},t_m-\tau)(t_m-\tau)d\tau, \end{aligned} \tag{54}$$

where the first term on the RHS of Eq. 54 can be identified as the past history from time $t = 0$ up to time $t = t_{m-1}$. The last result illustrates very clearly the complexities in the numerical evaluation of the solution of an elastodynamic problem using boundary integral equation formulation in the time domain. Equation 54 can be now reduced to a set of algebraic equations in the usual manner.

This concludes the discussion on the numerical solutions of the integral equations which arise in the boundary integral equation formulations of elastodynamics. Application of these numerical procedures to various problems of interest in dynamics of structures and earthquake engineering is considered next.

Applications

In this section we present some recent results from the application of the boundary integral equation methods to the problems of diffraction of elastic waves by an obstacle embedded within an elastic full-space or half-space.

Manolis and Beskos(39) used the Laplace transform formulation, together with numerical inversion algorithm of Durbin(24) to solve the transient plane strain problem of diffraction of elastic plane P wave by a cavity of arbitrary shape within an elastic full-space. We present their results for a case of a square hole of side $a = 20$ in. placed at the origin of the coordinate system so that the direction of the incident wave propagation is perpendicular to one of its sides. The incident wave is assumed to be of the form

$$\sigma_{xx} = s_o H(t-t_n) \tag{55a}$$

$$\sigma_{yy} = s_o \nu/(1-\nu) H(t-t_n) \tag{55b}$$

$$\sigma_{xy} = 0, \tag{55c}$$

where ν is the Poisson's ratio, H( ) is the Heaviside step function, s_o is the prescribed amplitude, and t_n is the time required for the wave traveling with dilatational speed α along the x-direction with starting point ($x = a$, $y = 0$) to reach a point n on the boundary of the hole. The hole is assumed to have the rounded-off corners with the round-off radius r_o and the surrounding medium is chosen to be steel. The stress concentration factor is defined to be the ratio σ_{yy}/s_o, for the vertical side of the first quadrant, and the ratio σ_{xx}/s_o on the horizontal side of the square hole. For a point on the side of the square defined by a polar angle $\theta = 50.2°$ transient stress concentration history is depicted by Fig. 5. Since the characteristic time of the problem is $T = 2a/\alpha = 17.34 \times 10^{-5}$ sec., where α is dilatational wave speed, it is evident that the result depicted by Fig. 5 is calculated for a rather long time interval. The maximum stress concentration factor is about 13 per cent larger than the static one. The discretization scheme consists of twelve equal segments along the boundary of the half-structure, and the round-off radius is chosen to be r_o = 0.083a. If the point of interest is closer to the corner (or the angle of 45°) more refined discretization and /or use of singular elements have to be used to achieve acceptable accuracy of the results.

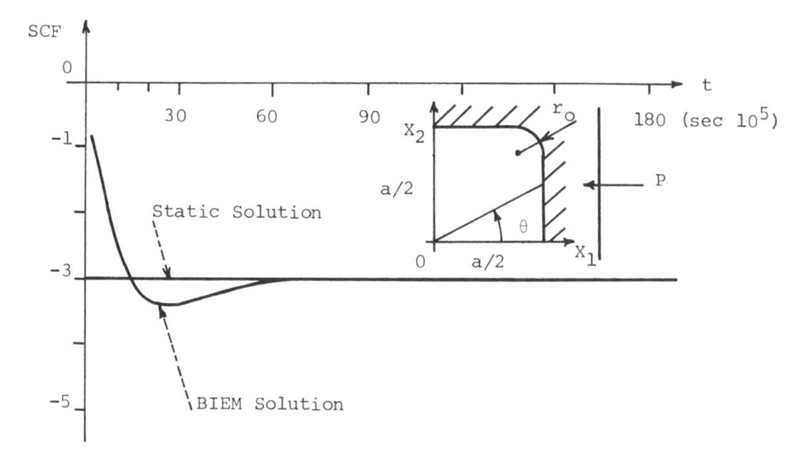

Figure 5. - Transient Stress Concentration History of the Square Hole at θ = 50.5° for Incident P-Wave. Plane Strain Model. After Ref. (39)

We are considering next the transient plane strain problem of diffraction of a plane P wave by a circular hole embedded in an elastic full-space. Using the Fourier transform formulation, a solution of this problem was given by Kobayashi(33). The incident stress wave is assumed to be in the form of a Heaviside step function of amplitude τ_o. Isoparametric boundary elements were used in order to discretize the boundary integral equation in the frequency domain and obtain a set of the algebraic equations which are then solved. For the transient analysis the Fast Fourier transform algorithm is used to evaluate numerically the Fourier transform. Twentyfour quadratic isoparametric elements are used and frequency contributions of up to ak = 7 (a and k being the radius and the shear wave number,respectively) are taken into account. Durations are chosen sufficiently long compared to the characteristic time a/β (about 120 times a/β, where β is the shear wave speed).

Fig. 6 shows the stresses on the boundary as a function of time measured from the arrival time of the incident wave at the center of the hole. The results by boundary integral equation method show very good agreement with the analytical solutions.

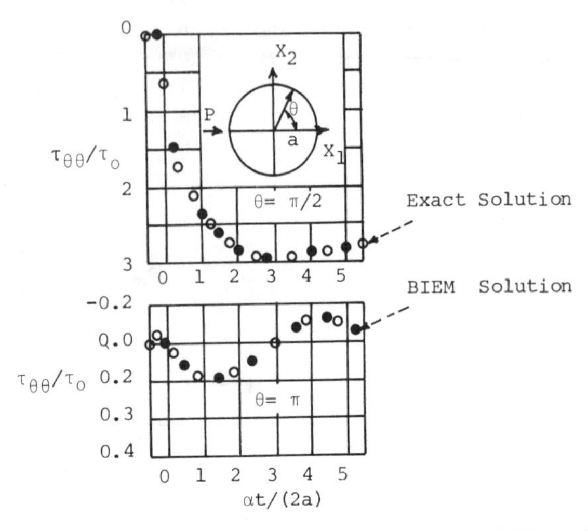

Figure 6. -Stress on the Boundary of a Circular Hole Subjected to an Incident P-Wave. Plane Strain Model. After Ref. (33)

For illustration of the boundary integral equation methods to the problems of three-dimensional elastodynamics we present a result by Rizzo et al.(44) for diffraction of a plane harmonic P wave by a spherical cavity of radius a embedded at the origin of an elastic full-space. The incident wave is assumed to be of the form

$$\underline{u}^{inc}(\underline{x},\omega) = u_o \underline{x}_3 \; e^{-i(hx_3 + \omega t)}, \tag{56}$$

where u_o, h, and $\underline{x}_3$ are the amplitude, the P wave number, and the unit vector along the x_3-axis, respectively. The stress concentration factor is defined by

$$SCF = \sigma_{\theta\theta}/\sigma_{rr} \;, \quad \text{at } r = a \tag{57}$$

where σ_{rr} is the radial stress at r = a in the absence of the cavity. Due to the symmetry of the problem the results are evaluated for a plane with the azimuthal angle being zero as a function of the so called polar angle which is being measured from the x_3-axis. It is sufficient to consider the polar angle in the range 0 - 180 degrees.

The problem is solved in frequency domain (Fourier transform domain formulation) through the application of a direct boundary integral equation approach. All the Cauchy principal value integrals were eliminated according to the procedure discussed here earlier and the subsequent algebraic equations were obtained utilizing standard Gaussian quadrature. Using 24 curvilinear triangular boundary elements with 50 nodes the stress concentration factor is evaluated for two frequencies as shown by Fig. 7

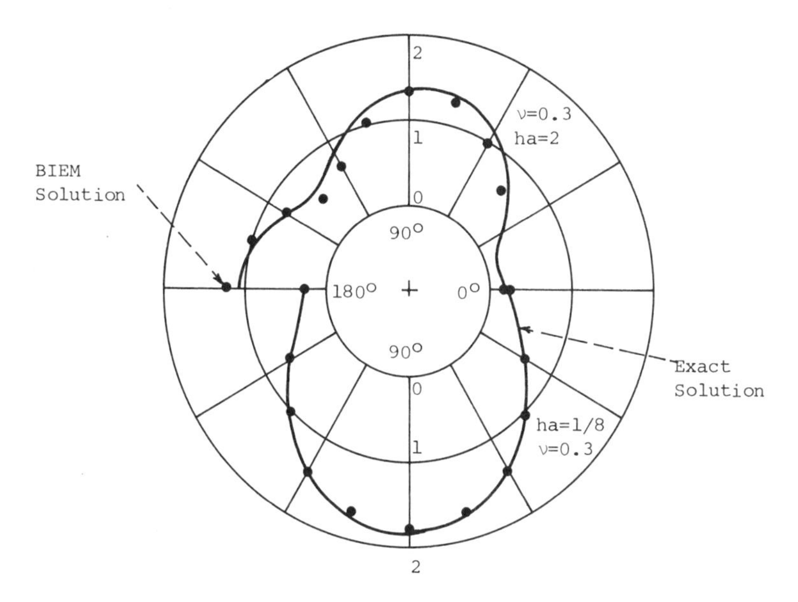

Figure 7. - Stress Concentration Versus Polar Angle for Incident P-Wave Upon a Spherical Cavity of Radius a. After Ref. (44)

Problems presented so far deal with a cavity within an elastic full-space. For most applications of practical interest it is necessary to replace the full-space with a half-space. Therefore, the last two examples deal with the half-space problems.

First we present the results which correspond to diffraction of a harmonic plane SV wave by a semicircular cavity embedded in a viscoelastic half-space. A solution of this problem was derived by Dravinski and Mossessian(23) using the approximate boundary integral equation method discussed in this chapter. For convenience, a dimensionless frequency is defined as a ratio of the total width of the cavity and the wavelength of the incident wave. Dimensionless frequency is chosen to be one half. Shear modulus and intrinsic shear wave velocity are taken to be equal unity, intrinsic dilatational wave speed is chosen to be two, and the attenuation factors for P and SV waves are assumed to be 50 and 100, respectively (see the approximate method formulation).

Horizontal and vertical displacement amplitudes for incident SV wave and three angles of incidence are presented by Figs. 8a and 8b, respectively. For those particular calculations very small attenuation allows the comparison with the results of Wong(52) and Sánchez-Sesma(46) obtained for purely elastic material. For the results shown by Figs. 8a,8b solid lines,open circles, and solid circles correspond to the results of Refs. (23,52, and 46), respectively.

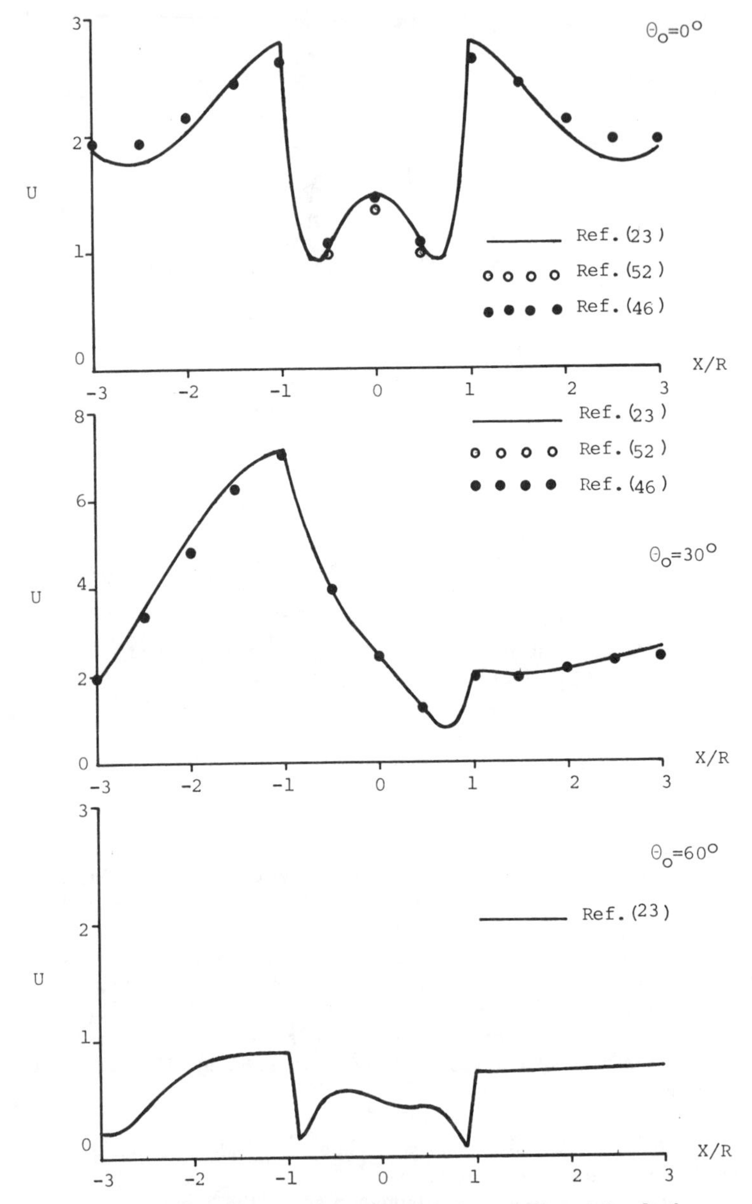

Figure. 8a. – Amplitude of the Horizontal Component of the Surface Displacement Spectra for Incident SV-Wave upon a Semicircular Cavity in a Half-Space. After Ref. (23)

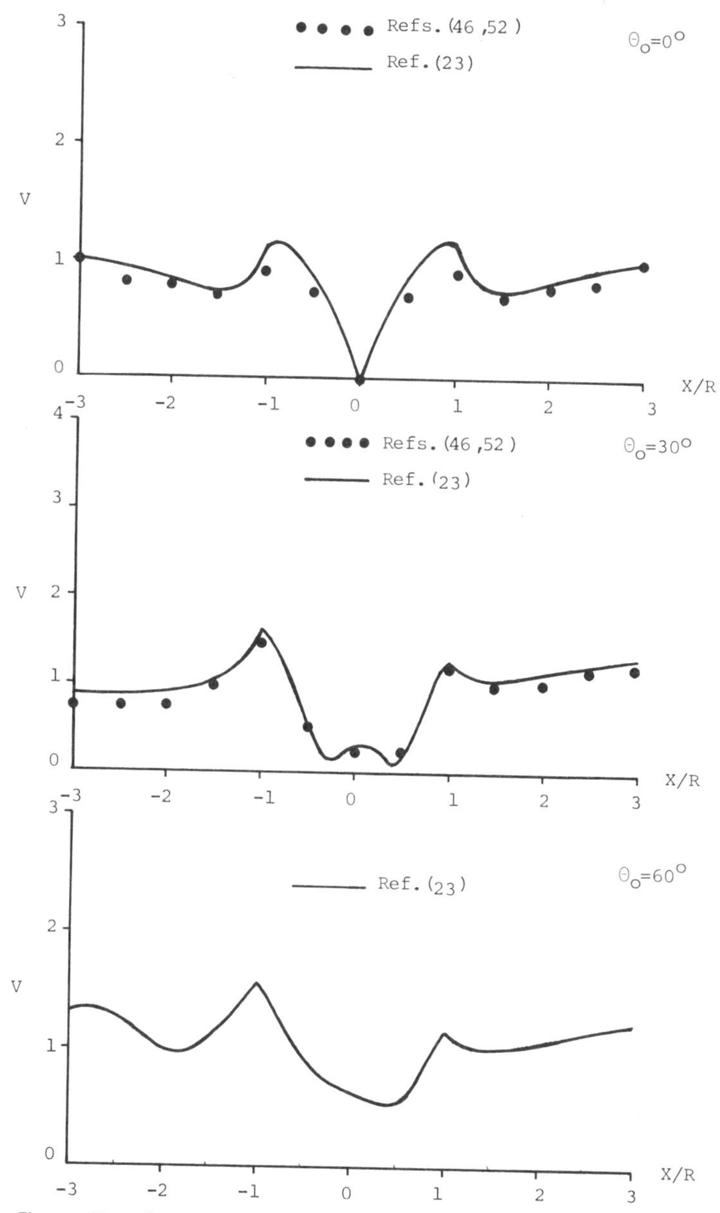

Figure. 8b. - Amplitude of the Vertical Component of the Surface Displacement Spectra for Incident SV-Wave upon a Semicircular Cavity in a Half-Space. After Ref. (23)

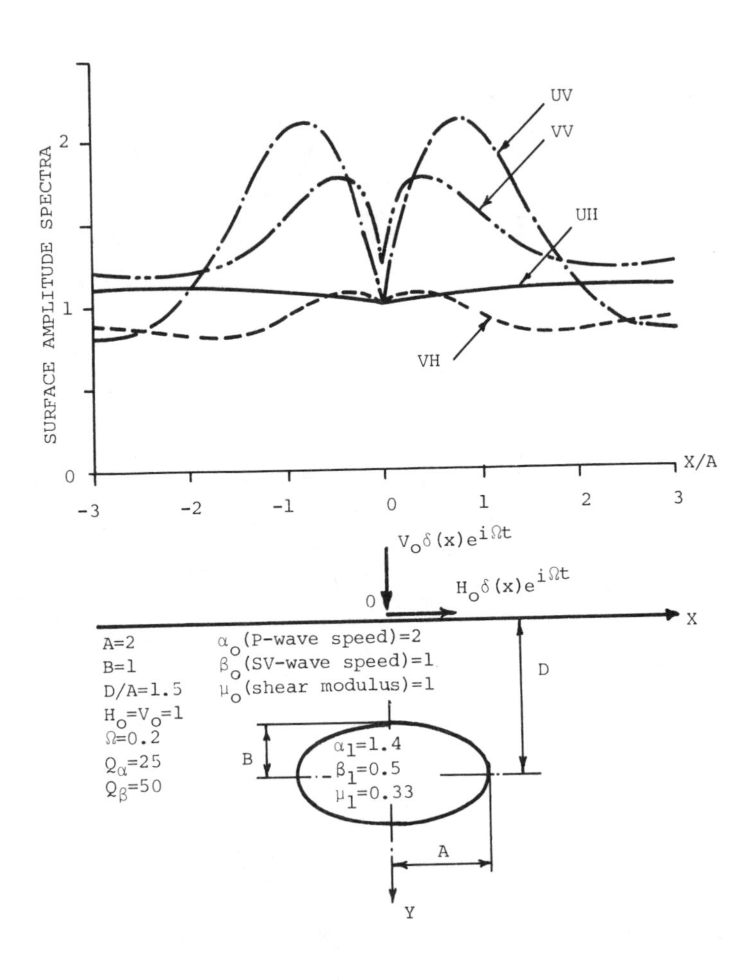

Figure. 9. - Surface Displacement Amplitude Spectra for Harmonic Line Loads on the Surface of a Half-Space which Contains an Anelastic Inclusion of Elliptical Shape. After Ref. (22)

The results depicted by Figs. 8a,8b show very good agreement for the three independent calculations of the same problem. For the numerical results presented here, Ref. (23), sixteen sources are placed along a semicircular auxiliary surface of the radius one half that of the cavity radius. Along the surface of the canyon, thirty-four collocation points are chosen. Details on the choice of the auxiliary surfaces can be found in Refs. (21, 22,52).

As the final numerical example we present a plane strain model of the following problem: An anelastic half-space which contains an anelastic inclusion of arbitrary shape is subjected to horizontal or vertical harmonic surface line loads. Using the approximate boundary integral equation method discussed here earlier, Dravinski and Mossessian(22) solved numerically the displacement field of the problem. Here we present the results which correspond to an elliptical inclusion. The amplitude of the surface displacement spectra for dimensionless frequency (ratio of the maximum width of the inclusion and the wavelength of the incident wave) of one half is shown by Fig. 9. Here, H and V denote the horizontal and vertical surface line load, respectively while U (V) denotes the displacement amplitude of the horizontal (vertical) component of the surface displacement field. Results of the surface displacement filed are presented normalized with respect to the corresponding surface free field motion. Similarly, all the distances are normalized with respect to the characteristic length in the problem, i. e., the major axis of the elliptical inclusion.

It is evident from the presented results that the presence of a subsurface and/or surface irregularity may cause locally very large amplification effects of the surface motion when compared with the free-field motion which would be present in the absence of the irregularity. Consequently, the analysis of these types of problems are of great interest in strong ground motion seismology and earthquake engineering. However, realistic problems of interest are characterized by a great degree of irregularities in geometry and the material properties which should be incorporated in the model. In addition, the results are needed for a wide range of frequencies or large time duration (see Eq. 54). A problem which must be resolved is how to incorporate all these requirements into the boundary integral equation modeling and still remain within an acceptable amount of required computation time. Once the problem is solved we can successfully apply the boundary integral approach to problems of interest in engineering and seismology for which at the present time neither numerical nor analytical solutions are available. This is of particular importance for three-dimensional problems of elastodynamics which are often characterized by considerable numerical difficulties in the course of solving the problems, Refs. (44,45). For the boundary integral equation formulation in transform domain, the primary one arises from the numerical evaluation of the singular solutions or the Green functions. Therefore, in future research special emphasis has to be placed on development of numerical procedures which would facilitate fast evaluation of the required Green functions, Ref. (44).

An additional problem in the numerical evaluation of the results for diffraction of elastic waves by an embedded obstacle may arise at certain frequencies. Namely, there exists a nonuniqueness of the solution of the integral equations for certain frequencies, Refs. (44,50). For example, for an elastic field in an exterior of surface S, when the tractions are prescribed on S, corresponding boundary integral equation in frequency domain has no unique solution at certain frequencies. These frequencies correspond to eigenfrequencies of vibration of an (imaginary) elastic body occupying the interior of S under prescribed zero displacement on S, Ref. (44). It is possible, by the one of the several procedures, to alter the integral equation solution procedure to obtain an accurate result at or near the eigenfrequencies, Refs. (44,50).

Conclusions

A brief review of the boundary integral equation formulations for problems in elastodynamics is presented for transient and steady state problems. Basic difficulties in the application of this approach to problems of interest in geophysics and earthquake engineering are identified. Presented numerical results correspond to the problems of diffraction of elastic waves by an obstacle for two and three dimensional geometries. It is evident from these results that the boundary integral equation methods can be successfully applied to a variety of problems of interest in structural dynamics, earthquake engineering, and geophysics.

Acknowledgment

Completion of this work was made possible in part through a support by a contract from the United States Geological Survey. The author is grateful to D. E. Beskos for his comments and suggestions during the course of the preparation of this manuscript.

APPENDIX I.–References

1. Aki, K., and Richards, P. G., Quantitative Seismology, Theory and Methods, Vol. 1, W. H. Freeman and Co., San Francisco, California, 1980.

2. Apsel, R. J., "Dynamic Green's Functions for Layered Media and Application to Boundary Value Problems," thesis presented to the University of California, at San Diego, Ca., in 1979, in a partial fulfillment of the requirements for the degree of Doctor of Philosophy.

3. Banaugh, R. P., and Goldsmith, W., " Diffraction of Steady Elastic Waves by Surfaces of Arbitrary Shape," Journal of Applied Mechanics, Vol. 30, 1963, pp. 589 – 595.

4. Banerjee, P. K., and Butterfield, R., Boundary Element Methods in Engineering and Science, McGraw-Hill, New York, New York, 1981.

5. Beskos, D. E. and Spyrakos, C. C., "Dynamic Response of Strip Foundations by the Time Domain BEM - FEM Method," Report CEE - 8024725, Part B, National Science Foundation, Washington, D. C., Jan., 1984.

6. Betti, E., "Teori dell Elasticita," Il Nuovo Ciemento, Vols. 6 - 10, 1872.

7. Brebbia, C. A., "Introductory Remarks," Boundary Element Methods, C. A. Brebbia, ed., Proceedings of the Third International Seminar, Irvine, California, Springer - Verlag, New York, New York, 1981.

8. Brebbia, C. A., Futagami, T., and Tanaka, M., eds.,Boundary Elements, Proceedings of the Fifth International Conference, Hiroshima, Japan, Nov., Springer - Verlag, New York, New York, 1983.

9. Brebbia, C. A., Telles, J. C. F., and Wrobel, L. C., Boundary Element Techniques, Theory and Applications in Engineering, Springer - Verlag, New York, New York, 1984.

10. Cole, D. M., Kosloff, D. D., and Minster, J. B., "A Numerical Boundary Integral Equation Method for Elastodynamics," Bulletin of the Seismological Society of America, Vol. 68, No. 5, Oct., 1978, pp. 1331 - 1357.

11. Copley, L. A., "Integral Equation Method for Radiation from Vibrating Surfaces," Journal of Acoustical Society of America, Vol. 41, 1967, pp. 807 - 816.

12. Cristensen, R. M., Theory of Viscoelasticity, Academic Press, New York, New York, 1971.

13. Crouch, S. L., and Starfield, A. M., Boundary Element Methods in Solid Mechanics, George Allen and Unwin, London, UK, 1983.

14. Cruse, T. A., "A Direct Formulation and Numerical Solution of the General Transient Elastodynamic Problem. II," Journal of Mathematical Analysis and Applications, Vol. 22, 1968, pp. 341 - 355.

15. Cruse, T. A., and Rizzo, F. J., "A Direct Formulation and Numerical Solution of the General Transient Elastodynamics Problem. I," Journal of Mathematical Analysis and Applications, Vol. 22, 1968, pp. 244 - 259.

16. Dominguez, J., "Dynamic Stiffness of Rectangular Foundations," MIT Research Report R - 78 - 20, Civil Engineering Department, August, 1978.

17. Dominguez, J., "Response of Embedded Foundations to Traveling Waves," MIT Research Report R - 78 - 24, Civil Engineering Department, August, 1978.

18. Dominguez, J., and Alarcón, E., "Elastodynamics," in Progress in Boundary Element Methods, Vol. 1, C. A. Brebbia, ed., Halsted Press, New York, New York, 1981, pp. 213 - 257.

19. Dravinski, M., "Scattering of SH Waves by Subsurface Topography," Journal of Engineering Mechanics Division, ASCE, Vol. 108, No. EM1, Feb., 1982, pp. 1 - 17.

20. Dravinski, M., "Scattering of Elastic Waves by an Alluvial Valley," Journal of Engineering Mechanics Division, ASCE, Vol. 108, No. EM1, Feb., 1982, pp. 19 - 31.

21. Dravinski, M., "Scattering of Plane Harmonic SH Waves by Dipping Layers of Arbitrary Shape," Bulletin of the Seismological Society of America, Vol. 73, No. 5, Oct., 1983, pp. 1303 - 1319.

22. Dravinski, M. and T. K. Mossessian, "Amplification of Surface Ground Motion by an Inclusion of Arbitrary Shape for Harmonic Surface Line Loadings," submitted for publication.

23. Dravinski, M. and Mossessian, T. K., "Scattering of Harmonic P, SV, and Rayleigh Waves by Dipping Layers of Arbitrary Shape", in preparation.

24. Durbin, F., "Numerical Inversion of Laplace Transforms: An Efficient Improvement to Dubner and Abate's Method," Computer Journal, Vol. 17, 1974, pp. 371 - 376.

25. Eringen, A. C., and Suhubi, E. S., Elastodynamics, Vol. 2, Academic Press, New York, New York, 1975.

26. Fredholm, I., "Solution of Fundamental Problems in Theory of Elasticity," Ark. Mat. Astronom. Fysik, Vol. 2, No. 28, 1905, pp. 1 - 8.

27. Friedman, M. B., and Shaw, R. P., "Diffraction of a Plane Shock Wave by an Arbitrary Rigid Cylindrical Obstacle," Journal of Applied Mechanics, Vol. 29, No. 1, 1962, pp. 40 - 46.

28. Fairweather, G., and Johnston, R. L., " The Method of Fundamental Solutions for Problems in Potential Theory," in Treatment of Integral Equations by Numerical Methods, C. T. H. Baker and G. F. Miller, eds., Academic Press, New York, New York, 1982, pp. 349 - 360.

29. Fung, Y. C., Foundations of Solid Mechanics, Prentice Hall, Englewood Cliffs, New Jersey, 1965.

30. Johnson, L. R., "Green's Function for Lamb's Problem," Geophysical Journal of Royal Astronomical Society, Vol. 37, 1974, pp. 99 - 131.

31. Karabalis, D. L., and Beskos, D. E., "Dynamic Response of 3D Rigid Surface Foundations by Time Domain Boundary Element Method," Earthquake Engineering and Structural Dynamics, Vol. 12, 1984, pp. 73 - 93.

32. Kobayashi, S., "Elastodynamics," in Boundary Element Methods in Mechanics, D. E. Beskos, ed., North-Holland, Amsterdam, 1987.

33. Kobayashi, S., "Some Problems of the Boundary Integral Equation Method in Elastodynamics," in Boundary Elements, Proceedings of the Fifth International Conference, Hiroshima, Japan, C. A. Brebbia, T. Futagami, and M. Tanaka, eds., Springer-Verlag, New York, New York, 1983, pp. 775 - 784.

34. Kupradze, V. D., "Dynamical Problems in Elasticity," in Progress in Solid Mechanics, Vol. 3, I. N. Sneddon and R. Hill, eds., North Holland, Amsterdam, 1963.

35. Lamb, H., "On the Propagation of Tremors Over the Surface of an Elastic Solid," Philosophical Transaction of the Royal Society of London, Vol. A203, 1904, pp. 1 - 42.

36. Lauricella, C., "Atti della Reale Academia dei licei," Vol. 16, No. 1, 1906, pp. 426 - 432.

37. Luco, J. E., and Apsel, R. J., " On the Green's Functions for a Layered Half-Space," Bulletin of the Seismological Society of America, Vol. 73, 1983, pp. 909 - 929.

38. Manolis, G. D., "A Comparative Study on Three Boundary Element Method Approaches to Problems in Elastodynamics," International Journal for Numerical Methods in Engineering, Vol. 19, 1983, pp. 73 - 91.

39. Manolis, G. D., and Beskos, D. E., "Dynamic Stress Concentration Studies by Boundary Integrals and Laplace Transform," International Journal for Numerical Methods in Engineering, Vol. 17, 1981, pp. 573 - 599.

40. Nardini, D., and Brebbia, C. A., "Transient Dynamic Analysis by the Boundary Element Method," in Boundary Methods, Proceedings of the Fifth International Conference, Hiroshima, Japan, Nov., C. A. Brebbia, T. Futagami, and M. Tanaka, eds., Springer-Verlag, New York, New York, 1983.

41. Niwa, Y., Kobayashi, S., and Azuma, N., "Analysis of Transient Stresses Produced Around Cavities of Arbitrary Shape During the Passage of Traveling Waves,", Memoirs of Faculty of Engineering, Kyoto University, Vol. 37, pt. 2, 1975, pp. 28 -46.

42. Noble, B., "The Numerical Solution of Non-linear Integral Equations and Related Topics," in Nonlinear Integral Equations, P. M. Anselone, ed., University of Wisconsin Press, Madison, Wisconsin, 1964.

43. Oshaki, Y., "On Movements of a Rigid Body in Semiinfinite Elastic Medium,", Proceedings of Japanese Earthquake Symposium, Aug. - Sept., 1973, pp. 245 - 252.

44. Rizzo, F. J., Shippy, D. J., and Rezayat, M., " A Boundary Integral Method for Radiation and Scattering of Elastic Waves in Three Dimensions," International Journal for Numerical Methods in Engineering, Vol. 21, 1985, pp. 115 - 129.

45. Rizzo, F. J., Shippy, D. J., and Rezayat, M., "Boundary Integral Equation Analysis for a Class of Earth-Structure Interaction Problems," Report, Department of Engineering Mechanics, University of Kentucky, Lexington, Kentucky, 1985.

46. Sánchez-Sesma, F. J., Bravo, M. A., and Herrera, I., "Surface Motion of Topographical Irregularities for Incident P, SV and Rayleigh Waves," Bulletin of the Seismological Society of America, Vol. 75, 1985, pp. 263 - 270.

47. Sánchez-Sesma, F. J., and Rosenblueth, E., "Ground Motions at Canyons of Arbitrary Shape Under incident SH Waves," Earthquake Engineering and Structural Dynamics, Vol. 7, 1979, pp. 441 - 450.

48. Shaw, R. P., "Boundary Integral Equation Methods Applied to Wave Problems," in Developments in Boundary Element Methods, Vol. 1, P. K. Banerjee and R. Butterfield, eds.,, Applied Science Publishers, London, UK, 1979.

49. Somigliana, C., "Sopra l'equilibrio di un corpo elastico isotropo," Il Nuovo Ciemento, Vols. 17 - 19, 1886.

50. Ursell, F., "On the Exterior Problems of Acoustics," Proceedings of the Cambridge Philosophical Society, Vol. 74, 1973, pp. 117 - 125.

51. Wheeler, L. T., and Sternberg, E., "Some Theorems in Classical Elastodynamics," Archive for Rational Mechanics and Analysis, Vol. 31, 1968, pp. 51 - 90.

52. Wong, H. L., "Diffraction of P, SV, and Rayleigh Waves by Surface Topographies," Bulletin of the Seismological Society of America, Vol. 72, pp. 1167 - 1184.

Nonlinear Structural Analysis

Subrata Mukherjee[1]
Abhijit Chandra[2]
Bimal Poddar[3]

This chapter presents applications of the boundary element method (BEM) for nonlinear problems - those with material nonlinearities such as plasticity or viscoplasticity, as well as fully nonlinear problems involving large strains and rotations. The elastic strains are assumed to remain small in all cases, as is generally the case with metal deformation.

Sample numerical results are presented for various cases. These include small elastic-viscoplastic deformation of axisymmetric solids as well as small elastic deformations of thin axisymmetric shells. The latter investigation is a first attempt at analysing deformation of shells of arbitrary shape by integral equation methods. Some recently obtained results for planar metal extrusion are also included. The BEM results are compared with those obtained by the finite element method (FEM) in many cases.

Introduction

This chapter presents a discussion of the applications of the boundary element method (BEM) in nonlinear structural analysis. This includes problems that are materially nonlinear due to the presence of plastic or viscoplastic strains in addition to elastic ones, but are geometrically linear (i.e., problems with small strains, rotations and displacements); as well as fully nonlinear problems that involve large strains and rotations as well as the above mentioned material nonlinearities. The discussion of the fully nonlinear problems is limited, however, to cases where the elastic strains remain small even though the rotations and nonelastic strains can be large. Such is typically the case in problems involving deformation of metallic components or structures.

---

[1]Associate Professor, Department of Theoretical and Applied Mechanics, Cornell University, Ithaca, NY 14853

[2]Assistant Professor, Department of Aero and Mechanical Engineering, University of Arizona, Tucson, AZ 85721

[3]Graduate Student. Same address as 1 above.

The BEM is rooted in classical integral equation formulations of boundary value problems such as the work of famous mathematicians like Fredholm and Volterra. As has been described in the earlier chapters of this book, it is a "natural" approach for linear problems, in that many linear partial differential operators can be transformed into equivalent integral forms through the use of appropriate Green's functions. It is not surprising to observe, therefore, that the principal BEM applications in computational solid mechanics in the 1960's primarily dealt with linear problems such as torsion of elastic bars (Jaswon and Ponter (22), bending of elastic plates (Jaswon and Maiti (23)) or classical elastostatics (e.g. Rizzo (44) and Cruse (17)). The method has been carried forward during the 1970's to include problems with material nonlinearities such as plasticity or viscoplasticity. Some of the important contributions in plasticity have been Ponter (42), Swedlow and Cruse (47), Mendelson (29), Riccardella (43), Mukherjee (31), Banerjee and Mustoe (1), Banerjee and Cathie (2), Kobayashi and Nishimura (26), Benitez et al. (4), Brunet (7), Oliveira et al. (40), Tanaka and Tanaka (48), Doblaré et al. (20), Maier and Novati (28) and Nishimura and Kobayashi (39). In the area of BEM applications for elasto-viscoplastic analysis, the following important contributions should be mentioned - Kumar and Mukherjee (25), Chaudonneret (15), Mukherjee and Kumar (32), Morjaria and Mukherjee (30), Mukherjee and Morjaria (33), Mukherjee (34) and Telles and Brebbia (49). Very recently, the BEM has been applied to fully nonlinear problems including both material and geometrical nonlinearities (Chandra and Mukherjee (10 - 14)) and Mukherjee and Chandra (36-38)). Some recent books on the BEM, those by Mukherjee (35), Banerjee and Butterfield (3) and Brebbia, Telles and Wrobel (6) deserve special mention at this point.

This chapter has two main parts. The first is a discussion of BEM applications for problems that are geometrically linear but materially nonlinear - elastic plastic and elastic-viscoplastic problems. A discussion of three-dimensional problems is followed by a presentation and some numerical results for axisymmetric problems and some exciting ongoing work on BEM applications in shell theory. The work on shell theory, discussed here, is limited to elastic analysis, but extension to inelastic analysis is being pursued at present and this is indicated at the appropriate places in this chapter. Fully nonlinear problems including plastic or viscoplastic effects as well as large rotations and inelastic strains, is the subject of the second half of this chapter. Some recently obtained numerical results for plane strain extrusion are included here.

Material Nonlinearities

Governing differential equations. The governing differential equations in rate form, for small elastic-plastic or elastic-viscoplastic deformations of a solid body, in the absence of body forces, can be written as

Equilibrium:

$$\dot{\sigma}_{ji,j} = 0 \qquad \text{in } B \tag{1}$$

$$n_j \dot{\sigma}_{ji} = \dot{\tau}_i \qquad \text{on} \quad \partial B \tag{2}$$

Kinematic:

$$\dot{\epsilon}_{ij} = \dot{\epsilon}_{ij}^{(e)} + \dot{\epsilon}_{ij}^{(n)} = \frac{1}{2}(\dot{u}_{i,j} + \dot{u}_{j,i}) \tag{3}$$

Constitutive:

$$\dot{\sigma}_{ij} = \lambda \dot{\epsilon}_{kk}^{(e)} \delta_{ij} + 2G\dot{\epsilon}_{ij}^{(e)} \tag{4}$$

In the above equations, u_i and τ_i are the components of the displacement and traction vectors, and ϵ_{ij} and σ_{ij} are the components of the strain and stress tensors, respectively. Equation (3) assumes the additive decomposition of strain rates into elastic and nonelastic parts. Also, λ and G are Lamé constants and a superposed dot denotes a real time derivative for elastic-viscoplastic and pseudo-time derivative for elastic-plastic problems. Finally, δ_{ij} is the Kronecker delta. These equations are written for three-dimensional situations so that the range of indices is 1,3.

In order to get a complete description of the problem, it is necessary to prescribe a constitutive description for the nonelastic strain rate $\dot{\epsilon}_{ij}^{(n)}$ in terms of relevant variables. This topic will be discussed later in the section called constitutive models. The exact form of such a model is kept unspecified for the time being.

An integral formulation for the above problem can be obtained by using, as a reference problem, the infinitesimal deformations of an isotropic linear elastic solid of infinite extent, due to a point body force. Using a superscript R to denote the different fields in this reference problem, the governing equations for this case can be written as

Equilibrium:

$$\sigma_{ji,j}^{(R)} = -\rho F_i^{(R)} = -\Delta(p,q)\delta_{ij} e_j \tag{5}$$

$$\sigma_{ji}^{(R)} n_j = \tau_i^{(R)} \tag{6}$$

Kinematic:

$$\epsilon_{ij}^{(R)} = \frac{1}{2}(u_{i,j}^{(R)} + u_{j,i}^{(R)}) \tag{7}$$

Constitutive:

$$\sigma_{ij}^{(R)} = \lambda \epsilon_{kk}^{(R)} \delta_{ij} + 2G\epsilon_{ij}^{(R)} \tag{8}$$

The new symbols in the above are the body force components $F_i^{(R)}$, the Dirac delta function Δ, unit orthogonal cartesian base vectors e_j and density ρ. The solution of the above problem, due to Lord Kelvin (Thomson (50)), can be written as

$$u_i^{(R)} = U_{ij}e_j \tag{9}$$

$$\tau_i^{(R)} = T_{ij}e_j \tag{10}$$

where U_{ij} and T_{ij} are the usual two point kernels of linear elasticity. They are given in many references (e.g., Mukherjee (35)) and are repeated here for completeness

$$U_{ij} = \frac{1}{16\pi(1-\nu)Gr}\left[(3-4\nu)\delta_{ij} + r,_i r,_j\right] \tag{11}$$

$$T_{ij} = -\frac{1}{8\pi(1-\nu)r^2}\left[\left[(1-2\nu)\delta_{ij} + 3r,_i r,_j\right]\frac{\partial r}{\partial n} + (1-2\nu)(r,_i n_j - r,_j n_i)\right] \tag{12}$$

A comma in the above denotes a derivative with respect to a field point and ν is Poisson's ratio.

<u>Boundary element formulation for displacement rates</u>. An integral formulation for the problem described by equations (1-4) can be obtained from the identity, adapted from the reciprocal theorem of Betti (5)

$$\int_B \dot{\sigma}_{ij}\epsilon_{ij}^{(R)}dV = \int_B \sigma_{ij}^{(R)}\dot{\epsilon}_{ij}^{(e)}dV \tag{13}$$

The integrands on either side of equation (13) are identical in view of equations (4) and (8).

Using equations (7), (1), (9) and (2), together with the divergence theorem, one can show that

$$\int_B \dot{\sigma}_{ij}\epsilon_{ij}^{(R)}dV = e_j \int_{\partial B} U_{ij}\dot{\tau}_i dS \tag{14}$$

Similarly, using equations (3), (8), (5), (7), (6), (10) and (9), together with the divergence theorem, one gets

$$\int_B \sigma_{ij}^{(R)}\dot{\epsilon}_{ij}^{(e)}dV = e_j\left[\int_{\partial B} T_{ij}\dot{u}_i dS + \int_B \Delta(p,q)\delta_{ij}\dot{u}_i dV - \int_B \left[\lambda U_{ij,i}\dot{\epsilon}_{kk}^{(n)} + 2GU_{ij,k}\dot{\epsilon}_{ik}^{(n)}\right]dV\right] \tag{15}$$

By virtue of equation (13) the right hand sides of equations (14) and (15) are equal. Equating these quantities and using the relevant property of the Dirac delta function, one obtains the equation

$$\dot{u}_j(p) = \int_{\partial B}\left[U_{ij}(p,Q)\dot{\tau}_i(Q) - T_{ij}(p,Q)\dot{u}_i(Q)\right]dS_Q + \int_B \left[\lambda U_{ij,i}(p,q)\dot{\epsilon}^{(n)}_{kk}(q) + 2GU_{ij,k}(p,q)\dot{\epsilon}^{(n)}_{ik}(q)\right]dV_q \qquad (16)$$

which is a modified form of the Somigliana identity (Somigliana (46).

A boundary integral equation for the unknown components of the traction and displacement rates in terms of the prescribed ones can be obtained by taking the limit as p approaches a boundary point P. This leads to the equation (Mukherjee (35))

$$C_{ij}(P)\dot{u}_i(P) = \int_{\partial B}\left[U_{ij}(P,Q)\dot{\tau}_i(Q) - T_{ij}(P,Q)\dot{u}_i(Q)\right]dS_Q + \int_B \left[\lambda U_{ij,i}(P,q)\dot{\epsilon}^{(n)}_{kk}(q) + 2GU_{ij,k}(P,q)\dot{\epsilon}^{(n)}_{ik}(q)\right]dV_q \qquad (17)$$

The surface integrals in the above equations (16) and (17) must be interpreted in the sense of Cauchy principal values. The coefficients C_{ij} multiplying $\dot{u}_i$ in the free term arise from the integration of $T_{ij}\dot{u}_i$. If ∂B is locally smooth at P, $C_{ij} = (1/2)\delta_{ij}$. Otherwise, C_{ij} can be evaluated in closed form for two-dimensional problems (Mukherjee (35)) and indirectly for three-dimensional problems (Cruse (17), Lachat (27)).

Once the nonelastic strain rate components are prescribed by a suitable constitutive model, equation (17) can be solved, at any time, for the unspecified boundary components of the displacement and traction rates, in terms of the prescribed ones. This can only be done, in general, by numerical discretization of equation (17) and solving of the resultant algebraic system. Further details of this procedure are given later in this chapter. It is important to note, however, that if an internal variable type constitutive model is used, the nonelastic strain rate components are known at any time as functions of stress components and internal variables at that time, so that the unknowns in equation (17) only lie on the boundary ∂B. The size of coefficient matrices in resultant algebraic systems, therefore, depend only on the number of boundary nodes, rather than on the total number of nodes in the entire three-dimensional solid, as is typical in domain type methods such as the finite element method (FEM).

The next step is the use of equation (16) to obtain the velocity distribution inside the body B. As many internal points, as necessary, can now be sampled for this purpose. The right hand side of equation (16) only involves known quantities at this stage.

Internal stress rates. Strain rate components inside the body B must now be determined by differentiation of equation (16). This can be achieved by several methods, the easiest and least accurate of which is numerical differentiation of the velocity field by interpolation of $\dot{u}_j$ over internal cells. This method, in general, leads to discontinuities in velocity gradients across boundaries of internal cells, and, therefore, can lead to significant numerical errors in the differentiated quantities.

A much more elegant approach is the analytical differentiation of equation (16) at a source point P. This gives

$$\dot{u}_{j,\bar{\ell}}(p) = \int_{\partial B}\left[U_{ij,\bar{\ell}}(p,Q)\dot{\tau}_i(Q) - T_{ij,\bar{\ell}}(p,Q)\dot{u}_i(Q)\right]dS_Q + \frac{\partial}{\partial x_{\bar{\ell}}}\int_B\left[\lambda W_{iji}(p,q)\dot{\epsilon}^{(n)}_{kk}(q) + 2GW_{ijk}(p,q)\dot{\epsilon}^{(n)}_{kk}(q)\right]dV_q \quad (18)$$

where $W_{ijk} = U_{ij,k}$ and the letter $\bar{\ell}$ following a comma denotes differentiation with respect to a source point. The differentiated kernels are given in Mukherjee (35) and will not be repeated here.

An immediate difficulty is that the kernel W_{ijk} has a singularity of order r^{-2} and differentiation of the domain integral must be handled with great care. One possibility is to evaluate the volume integral analytically for an arbitrary source point p and then differentiate this integral at p. This typically requires interpolation of the nonelastic strain rate field within a volume cell, evaluation of the integral over the cell in which the integrand is singular, and then differentiation of this integral at a source point p. This has been carried out successfully for planar problems by Mukherjee and Kumar (32).

Another alternative (Bui (8)) is to carry out a convected derivative of the domain integral in equation (18) by dividing the domain B into $B-B_\eta(p)$ and $B_\eta(p)$, where $B_\eta(p)$ is a sphere of radius η centered at p. Taking the limit as $\eta \to 0$ yields the appropriate free term and a principal value integral over $B-B_\eta$. Numerical evaluation of this principal value integral, for the case of planar and axisymmetric problems, has been discussed in a paper and in a recent book by Brebbia et al. (6). Evaluation of this integral for three-dimensional problems can be carried out in analogous fashion. Once the total strain rate at a point is determined, it is a simple matter to determine the elastic strain rate from equation (3) and then the stress rate from Hooke's Law (equation (4)).

Boundary stress rates. The boundary stress rate components are best obtained by using the method described below rather than by trying to take the limit of equation (18) as $p \to P$. Also, for the purpose of this discussion, P is a point on ∂B where it is locally smooth.

This calculation involves six vectors and two tensors at P. The vectors are the traction rate $\dot{\underset{\sim}{\tau}}$, the tangential derivatives of the velocity $\frac{\partial \dot{\underset{\sim}{u}}}{\partial s_1}$ and $\frac{\partial \dot{\underset{\sim}{u}}}{\partial s_2}$ along two linearly independent directions s_1 and s_2 on ∂B at P, and the unit normal and tangent vectors $\underset{\sim}{n}$, $\underset{\sim}{t}^{(1)}$ and $\underset{\sim}{t}^{(2)}$ at P. The tensors are the stress rate $\dot{\underset{\sim}{\sigma}}$ and the velocity gradient $\frac{\partial \dot{\underset{\sim}{u}}}{\partial \underset{\sim}{x}}$. Once equation (17) is solved, the velocity gradients $\frac{\partial \dot{\underset{\sim}{u}}}{\partial s_1}$, $\frac{\partial \dot{\underset{\sim}{u}}}{\partial s_2}$ must be evaluated numerically by differentiation of the appropriate shape functions for $\dot{\underset{\sim}{u}}$ along s_1 and s_2. Now, these six vectors $\frac{\partial \dot{\underset{\sim}{u}}}{\partial s_1}$, $\frac{\partial \dot{\underset{\sim}{u}}}{\partial s_2}$, $\dot{\underset{\sim}{\tau}}$, $\underset{\sim}{n}$, $\underset{\sim}{t}^{(1)}$ and $\underset{\sim}{t}^{(2)}$ (eighteen scalar com-ponents) are known at P. The fifteen unknown tensor components $\dot{\sigma}_{ij}$ and $\dot{u}_{i,j}$ ($\dot{\sigma}_{ij}$ is symmetric) can now be obtained from the following set of four equations (fifteen scalar linear algebraic equations) at P.

$$\dot{\sigma}_{ij} = \lambda\dot{u}_{k,k}\delta_{ij} + G(\dot{u}_{i,j} + \dot{u}_{j,i}) - \lambda\dot{\epsilon}_{kk}^{(n)}\delta_{ij} - 2G\dot{\epsilon}_{ij}^{(n)} \quad (19)$$

$$\dot{\tau}_i = \dot{\sigma}_{ji}n_j \quad (20)$$

$$\frac{\partial \dot{u}_i}{\partial s_1} = \dot{u}_{i,j}t_j^{(1)} \quad , \quad \frac{\partial \dot{u}_i}{\partial s_2} = \dot{u}_{i,j}t_j^{(2)} \quad (21a,b)$$

This procedure must often be carried out since the surface tangential stress rate components at P, which, in general, cannot be obtained from the traction rates at that point, can be very important in design.

Axisymmetric problems. Elastic-viscoplastic problems for axisymmetric bodies with axisymmetric loading are of importance in many applications such as pressure vessels and piping. This section describes a strategy for solving such problems by the BEM. Further details of this procedure are available in Sarihan and Mukherjee (45) and Mukherjee (35).

It is fairly straightforward to extend a planar finite element formulation to the axisymmetric case. Such is not the case for the BEM. The primary reason for the need of considerable effort for the BEM solution is the fact that the axisymmetric kernels contain elliptic functions which cannot be integrated analytically even over boundary elements or internal cells of simple shape. Thus, suitable methods must be developed for the efficient and accurate numerical integration of these singular and sensitive kernels over discrete elements.

An axisymmetric body with axisymmetric loading is considered in this section. Using polar coordinates R, θ and Z, the nonzero components of displacements, stresses and strains are u_R, u_Z, ϵ_{RR}, ϵ_{ZZ}, $\epsilon_{\theta\theta}$, ϵ_{RZ} $(=\epsilon_{ZR})$, σ_{RR}, $\sigma_{\theta\theta}$, σ_{ZZ} and σ_{RZ} $(=\sigma_{ZR})$. All dependent variables are functions of R, Z and t. Some torsion problems can be independent of θ, but these are not included in this formulation.

The BEM formulation given below is based on equation (16) for the case of incompressible inelastic deformation, i.e., $\dot{\epsilon}_{kk}^{(n)} = 0$. Compressible inelastic deformation can be included without any difficulty. The coordinate system chosen is shown in Fig. 1. The source point is denoted by $(R,0,Z)$ and the field point by (ρ,θ,ζ). Since the problem is axisymmetric, it is sufficient to choose the source point in the x_1-x_3 plane. The source point coordinates in this section are denoted by capital letters. It can, of course, lie inside or on the surface of the body.

An axisymmetric version of the three-dimensional equation (16) can be obtained by integrating the kernels U_{ij}, T_{ij} etc. for the field point moving around a ring with the source point fixed. Integrating equation (16) for the field point coordinate θ between 0 and 2π results in the equation (j = 1 and 3, no sum over ρ or ζ)

$$\begin{aligned} u_j(p) = & \int_{\partial B} (U_{\rho j}(p,Q)\dot{\tau}_\rho(Q) + U_{\xi j}(p,Q)\dot{\tau}_\xi(Q) \\ & - T_{\rho j}(p,Q)\dot{u}_\rho(Q) - T_{\xi j}(p,Q)\dot{u}_\xi(Q))\rho_Q dc_Q \\ & + 2G\int_B \Big[U_{\rho j,\rho}(p,q)\dot{\epsilon}_{\rho\rho}^{(n)}(q) + U_{\rho j,\xi}(p,q)\dot{\epsilon}_{\rho\xi}^{(n)}(q) \\ & + U_{\xi j,\rho}(p,q)\dot{\epsilon}_{\xi\rho}^{(n)}(q) + U_{\xi j,\xi}(p,q)\dot{\epsilon}_{\xi\xi}^{(n)}(q) \\ & + \frac{U_{\rho j}(p,q)\dot{\epsilon}_{\theta\theta}^{(n)}(q)}{\rho_q}\Big]\rho_q d\rho_q d\xi_q \end{aligned} \tag{22}$$

where, because of axisymmetry, $\dot{u}_R(p) = \dot{u}_1(p)$, $\dot{u}_Z(p) = \dot{u}_3(p)$, and $dc = \sqrt{(d\rho^2+d\xi^2)}$ is an element on the boundary of the ρ-ξ plane. The boundary ∂B in this section is that of a R-Z plane and the domain B is a R-Z section of the axisymmetric solid. Thus, only line integrals and area integrals must be evaluated in equation (22). This reduces the three-dimensional problem to a two-dimensional one.

The kernels $U_{\rho 1}$, etc. are given by the equations (Cruse, Snow, Wilson (19)).

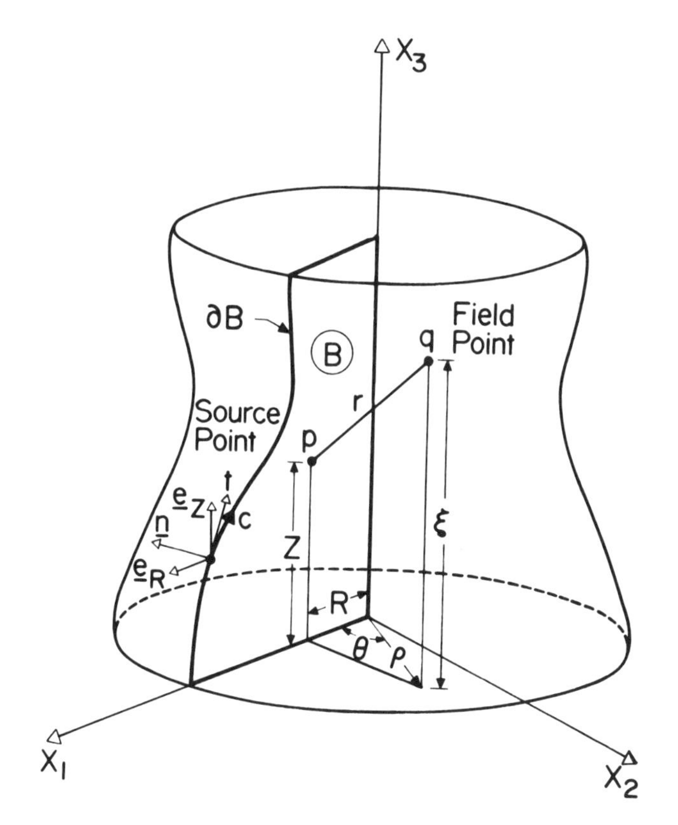

Figure 1. Geometry of the axisymmetric solid (from Mukherjee 1982b)

$$U_{\rho 1}(p,q) = \frac{1}{16\pi(1-\nu)G}\,\frac{k}{\sqrt{(R\rho)}}\left[\left[2(3-4\nu)\gamma+\frac{(Z-\xi)^2}{R\rho}\right]K(k) - \left[2(3-4\nu)(1+\gamma)+\left[\frac{\gamma}{1-\gamma}\right]\frac{(Z-\xi)^2}{R\rho}\right]E(k)\right]$$

$$U_{\rho 3}(p,q) = \frac{(\xi-Z)k}{16\pi(1-\nu)GR\rho\sqrt{(R\rho)}}\left[RK(k)+\frac{(\rho-R\gamma)}{\gamma-1}E(k)\right] \tag{23}$$

$$U_{\xi 1}(p,q) = \frac{(Z-\xi)k}{16\pi G(1-\nu)R\rho\sqrt{(R\rho)}}\left[\rho K(k)-\frac{(\rho\gamma-R)}{\gamma-1}E(k)\right]$$

$$U_{\xi 3}(p,q) = \frac{1}{8\pi(1-\nu)G}\,\frac{k}{\sqrt{(R\rho)}}\left[(3-4\nu)K(k)+\frac{(Z-\xi)^2}{2R\rho(\gamma-1)}E(k)\right]$$

where K(k) and E(k) are complete elliptic integrals of the first and second kind respectively and $k = \sqrt{[2/(1+\gamma)]}$. Further,

$$\gamma = 1 + \frac{((Z-\xi)^2+(R-\rho)^2)}{2R\rho}$$

The traction kernels are given in terms of derivatives of $U_{\rho j}$, etc. by the equations (j=1 and 3)

$$\frac{1}{2G}T_{\rho j}(p,Q) = \left[\left[\frac{1-\nu}{1-2\nu}\right]\frac{\partial U_{\rho j}}{\partial\rho} + \frac{\nu}{(1-2\nu)}\left[\frac{1}{\rho}U_{\rho j}+\frac{\partial U_{\xi j}}{\partial\xi}\right]\right]n_\rho + \frac{1}{2}\left[\frac{\partial U_{\rho j}}{\partial\xi}+\frac{\partial U_{\xi j}}{\partial\rho}\right]n_\xi$$
$$\frac{1}{2G}T_{\xi j}(p,Q) = \left[\left[\frac{1-\nu}{1-2\nu}\right]\frac{\partial U_{\xi j}}{\partial\xi} + \frac{\nu}{(1-2\nu)}\left[\frac{1}{\rho}U_{\rho j}+\frac{\partial U_{\rho j}}{\partial\rho}\right]\right]n_\xi + \frac{1}{2}\left[\frac{\partial U_{\rho j}}{\partial\xi}+\frac{\partial U_{\xi j}}{\partial\rho}\right]n_\rho \tag{24}$$

where n_ρ and n_ξ are the components of the outward unit normal at the field point on the boundary in the ρ and ξ directions.

The boundary integral equation when the source point p becomes a point P on the boundary ∂B of the body is (j=1 and 3, no sum over ρ or ξ)

$$\begin{aligned} c_{ij}(P)\dot{u}_i(P) = & \int_{\partial B}(U_{\rho j}(P,Q)\dot{\tau}_\rho(Q) + U_{\xi j}(P,Q)\dot{\tau}_\xi(Q) \\ & - T_{\rho j}(P,Q)\dot{u}_\rho(Q) - T_{\xi j}(P,Q)\dot{u}_\xi(Q))\rho_Q dc_Q \\ & + 2G\int_B \Big[U_{\rho j,\rho}(P,q)\dot{\epsilon}^{(n)}_{\rho\rho}(q) + U_{\rho j,\xi}(P,q)\dot{\epsilon}^{(n)}_{\rho\xi}(q) \\ & + U_{\xi j,\rho}(P,q)\dot{\epsilon}^{(n)}_{\xi\rho}(q) + U_{\xi j,\xi}(P,q)\dot{\epsilon}^{(n)}_{\xi\xi}(q) \\ & + \frac{U_{\rho j}(P,q)\dot{\epsilon}^{(n)}_{\theta\theta}(q)}{\rho_q}\Big]\rho_q d\rho_q d\xi_q \end{aligned} \tag{25}$$

The components of c_{ij} depend, as usual, on the geometry of the boundary at P and the location of P. These components are best determined indirectly (Mukherjee (35)).

Internal stress rates at a source point are most accurately obtained by analytical differentiation of the displacement rates at an internal source point. If equation (22) is used as a starting point for this purpose, second derivatives of the components of the kernel U_{ij} together with the associated free terms must be carefully dealt with (Bui (8)), Brebia, Telles, Wrobel (6)). An alternative approach, called the "strain rate gradient method" is described in Mukherjee (35)). This idea leads to the equation

$$
\begin{aligned}
u_{j,\bar{\ell}}(p) = & \int_{\partial B} (U_{\rho j,\bar{\ell}}(p,Q)\dot{\tau}_\rho(Q) + U_{\xi j,\bar{\ell}}(p,Q)\dot{\tau}_\xi(Q) \\
& - T_{\rho j,\bar{\ell}}(p,Q)\dot{u}_\rho(Q) - T_{\xi j,\bar{\ell}}(p,Q)\dot{u}_\xi(Q))\rho_Q dc_Q \\
& + 2G \int_{\partial B} \Big[U_{\rho j,\bar{\ell}}(p,Q)(\dot{\epsilon}^{(n)}_{\rho\rho}(Q)n_\rho(Q) + \dot{\epsilon}^{(n)}_{\rho\xi}(Q)n_\xi(Q)) \\
& + U_{\xi j,\bar{\ell}}(p,Q)(\dot{\epsilon}^{(n)}_{\xi\rho}(Q)n_\rho(Q) + \dot{\epsilon}^{(n)}_{\xi\xi}(Q)n_\xi(Q))\Big]\rho_Q dc_Q \\
& - 2G \int_B \Bigg[U_{\rho j,\bar{\ell}}(p,q)\left[\dot{\epsilon}^{(n)}_{\rho\rho,\rho}(q) + \frac{\dot{\epsilon}^{(n)}_{\rho\rho}(q) - \dot{\epsilon}^{(n)}_{\theta\theta}(q)}{\rho_q} + \dot{\epsilon}^{(n)}_{\rho\xi,\xi}(q)\right. \\
& + U_{\xi j,\bar{\ell}}(p,q)\left[\dot{\epsilon}^{(n)}_{\xi\rho,\rho}(q) + \frac{\dot{\epsilon}^{(n)}_{\xi\rho}(q)}{\rho_q} + \dot{\epsilon}^{(n)}_{\xi\xi,\xi}(q)\right]\Bigg]\rho_q d\rho_q d\xi_q \qquad (26)
\end{aligned}
$$

where $j = 1$ and 3, $\bar{\ell} = 1$ and 3 and there is no summmation over ρ or ξ. Differentiation with respect to $\bar{\ell}$ denotes a source point derivative. By virtue of axisymmetry (see Fig. 1),

$$
\begin{aligned}
u_{R,R} &= u_{1,1}\ , \quad u_{R,Z} = u_{1,3}, \\
u_{Z,R} &= u_{3,1}\ , \quad u_{Z,Z} = u_{3,3}.
\end{aligned}
$$

This method requires accurate values of $\dot{\epsilon}^{(n)}_{ij}$ on the boundary ∂B and these can be obtained from the boundary stresses. The boundary stresses are best obtained by using a method based on that described earlier for three-dimensional problems (see also Mukherjee, (35)). It should be noted that equation (26) has the drawback of requiring the divergence of the nonelastic strain rates over the domain of B. Numerical realization of these nonelastic strain rate derivatives requires piecewise interpolation of the nonelastic strain rates over internal cells. This might be less accurate than direct evaluation of $\dot{\epsilon}^{(n)}_{ij}$ at internal Gauss points.

Important details regarding the accurate evaluation of the integrals of U_{ij} and T_{ij} over boundary elements and evaluation of area integrals for terms like $U_{\rho j,\rho}$ are given in Sarihan and Mukherjee (45) and Mukherjee (35)).

Plate Analysis. The boundary element method has been used by Morjaria and Mukherjee (30) (see also Mukherjee (35)) to analyze the inelastic deformation of plates. The analysis is based on a direct BEM formulation for the rate of the transverse displacement of the plate, $\dot{w}$, in the form (Mukherjee (35))

$$8\pi\dot{w}(p) - \int_B (r^2 \ln r)_{pq} g(q)\,dA_q$$

$$= \int_{\partial B}\left[\frac{\partial}{\partial n_Q} \nabla^2 (r^2 \ln r)_{pQ} \dot{w}(Q) - \nabla^2 (r^2 \ln r)_{pQ} \frac{\partial \dot{w}(Q)}{\partial n_Q} + \frac{\partial}{\partial n_Q} (r^2 \ln r)_{pQ} \nabla^2 \dot{w}(Q) - (r^2 \ln r)_{pQ} \frac{\partial}{\partial n_Q} \{\nabla^2 \dot{w}(Q)\} \right] dc_Q \tag{27}$$

$$2\pi\nabla^2\dot{w}(p) - \int_B \ln r_{pq} g(q)\,dA_q$$

$$= \int_{\partial B}\left[\frac{\partial \ln r_{pQ}}{\partial n_Q} \nabla^2 \dot{w}(Q) - \ln r_{pQ} \frac{\partial}{\partial n_Q} \{\nabla^2 \dot{w}(Q)\} \right] dc_Q \tag{28}$$

where $g(q)$ involves the rates of the transverse loading and nonelastic strain rates. Also, B is the midsurface of the plate and ∂B is its boundary.

These two equations contain four quantities, $\dot{w}$, its normal derivative $\frac{\partial \dot{w}}{\partial n}$, $\nabla^2\dot{w}$ and $\frac{\partial}{\partial n}(\nabla^2\dot{w})$. Two equations relating these four quantites on the boundary are obtained from boundary conditions such as clamped edge: $\dot{w} = 0$, $\frac{\partial \dot{w}}{\partial n} = 0$ or simply supported edge: $\dot{w} = 0$, $\dot{M}_n = 0$.

A pair of coupled integral equations involving these four quantities are obtained by taking the limit as $p \to P$. If the plate boundary is locally smooth at P, the boundary integral equations are

$$4\pi\dot{w}(P) - \int_B (r^2 \ln r)_{Pq} g(q)\,dA_q$$

$$= \int_{\partial B}\left[\frac{\partial}{\partial n_Q} \nabla^2 (r^2 \ln r)_{PQ} \dot{w}(Q) - \nabla^2 (r^2 \ln r)_{PQ} \frac{\partial \dot{w}(Q)}{\partial n_Q} + \frac{\partial}{\partial n_Q} (r^2 \ln r)_{PQ} \nabla^2 \dot{w}(Q) - (r^2 \ln r)_{PQ} \frac{\partial}{\partial n_Q} \nabla^2 \dot{w}(Q) \right] dc_Q \tag{29}$$

$$\pi\dot{w}(P) - \int_B \ln r_{Pq} g(q)\,dA_q$$

$$= \int_{\partial B}\left[\frac{\partial}{\partial n_Q} (\ln r_{PQ}) \nabla^2 \dot{w}(Q) - \ln r_{PQ} \frac{\partial}{\partial n_Q} \{\nabla^2 \dot{w}(Q)\} \right] dc_Q \tag{30}$$

Equations (29) and (30), together with the appropriate boundary conditions, can be used to solve for the unspecified boundary rates. Next, equation (27) is used to solve for the displacement rates at desired locations inside the plate. Next, curvature rates are obtained by differentiating equation (27) at an internal point p. Stress rates and moment rates are obtained thereafter. Determination of curvature and moment rates at boundary points require special care (Mukherjee (35)).

Shell theory. Determination of accurate numerical solutions for the deformation of shells is a challenging problem. Many difficulties still exist with FEM analysis of such problems, in spite of the fact that many talented researchers have devoted many years to this research area. BEM applications to general shell theory are at their infancy, primarily because of the lack of free space Green's functions for shells of arbitrary shape. Some BEM solutions for shell deformation do exist in the literature (e.g. Tottenham (51)), but this work is restricted either to shallow shells or to shells of very simple shape such as cylinders or spheres.

A BEM shell formulation starting with three-dimensional elastic kernels has recently been completed by the authors of this chapter. An outline of this formulation for inelastic deformation of thin shells of arbitrary shape is presented below.

The starting point here is to take the 3D BEM equation (16) and transform the equation to the natural curvilinear coordinates of the shell. Equation (16) in general tensor notation has the form (with $\dot{\epsilon}^{(n)}_{kk} = 0$)

$$\dot{u}_j(p) = \int_{\partial B} [U^i_{\cdot j}(p,Q)\dot{\tau}_i(Q) - T^i_{\cdot j}(p,Q)\dot{u}_i(Q)]dS_Q + \int_B 2GW^i_{\cdot jk}(p,q)\dot{\epsilon}^{(n)k}_{\cdot i}(q)dv_q \tag{31}$$

where $W^i_{\cdot jk} = U^i_{\cdot j,k} = U_{ij,k}$

The next step requires transformation of the tensor components in equation (31) from the Cartesian basis $\underset{\sim}{e}_i$ to the shell midsurface basis $\underset{\sim}{a}_i$. The new basis is defined as

$$\underset{\sim}{a}_i = \frac{\partial \underset{\sim}{r}}{\partial \theta^i}, \qquad i = 1,2$$

$$\underset{\sim}{a}_3 = \frac{\underset{\sim}{a}_1 \times \underset{\sim}{a}_2}{|\underset{\sim}{a}_1 \times \underset{\sim}{a}_2|} \tag{32}$$

in terms of the radius vector

$$\underset{\sim}{r} = x^i \underset{\sim}{e}_i \tag{33}$$

to the shell midsurface and θ^1 and θ^2, the natural curvilinear coordinates for the shell midsurface. Also, a point in the shell space is written as

$$\underset{\sim}{r}^{*} = \underset{\sim}{r} + \xi \underset{\sim}{a}_3 \tag{34}$$

where ξ, the thickness coordinate, is measured from the shell midsurface in a direction normal to it.

The two sets of basis vectors are related by the transformation

$$\underset{\sim}{a}_i = \phi^{j}_{\cdot i} \underset{\sim}{e}_j \tag{35}$$

The two point kernels in equation (31) must be transformed carefully. The transformation takes the form

$$A^{i}_{\cdot j} \underset{\sim}{e}_i(q) \underset{\sim}{e}^j(p) = \bar{A}^{i}_{\cdot j} \underset{\sim}{a}_i(q) \underset{\sim}{a}^j(p)$$

where

$$\begin{aligned} \underset{\sim}{e}_i(q) &= (\phi^{-1}(q))^{j}_{\cdot i}\ \underset{\sim}{a}_j(q) \\ \underset{\sim}{e}^j(p) &= (\phi(p))^{j}_{\cdot i}\ \underset{\sim}{a}^i(p) \end{aligned} \tag{36}$$

so that a different ϕ must be used for the source and the field points.

The transformed equation (31) becomes

$$\begin{aligned} \dot{\bar{u}}_j(p) = {} & \int_{\partial B} \bar{U}^{i}_{\cdot j}(p,Q) \dot{\bar{\tau}}_i(Q) dS_Q - \int_{\partial B} \bar{T}^{i}_{\cdot j}(p,Q) \dot{\bar{u}}_i(Q) dS_Q \\ & + \int_{B} 2G\bar{W}^{i}_{\cdot jk}(p,q) \dot{\bar{\epsilon}}^{(n)k}_{\cdot i}(q) dv_q \end{aligned} \tag{37}$$

where

$$\begin{aligned} \bar{u}_i(Q) &= (\phi(Q))^{m}_{\cdot i} u_m(Q) \\ \bar{\tau}_i(Q) &= (\phi(Q))^{m}_{\cdot i} \tau_m(Q) \end{aligned}$$

and

$$\begin{aligned}\bar{U}^{i}_{\cdot j}(p,q) &= (\phi(p))^{\ell}_{\cdot j}(\phi^{-1}(q))^{i}_{\cdot k}U^{k}_{\cdot \ell}(p,q)\\ \bar{T}^{i}_{\cdot j}(p,q) &= (\phi(p))^{\ell}_{\cdot j}(\phi^{-1}(q))^{i}_{\cdot k}T^{k}_{\cdot \ell}(p,q)\\ \bar{W}^{i}_{\cdot jk}(p,q) &= (\phi(p))^{\ell}_{\cdot j}(\phi^{-1}(q))^{i}_{\cdot m}(\phi(q))^{n}_{\cdot k}W^{m}_{\cdot \ell n}(p,q)\end{aligned} \tag{38}$$

are the transformed components of the two point kernel tensors.

Finally, equation (37) takes the form

$$\begin{aligned}\dot{u}_{j}(p) &= (\phi(p))^{\ell}_{\cdot j}\int_{\partial B}(\phi^{-1}(Q))^{i}_{\cdot k}U^{k}_{\cdot \ell}(p,Q)\dot{\tau}_{i}(Q)dS_{Q}\\ &- (\phi(p))^{\ell}_{\cdot j}\int_{\partial B}(\phi^{-1}(Q))^{i}_{\cdot k}T^{k}_{\cdot \ell}(p,Q)\dot{u}_{i}(Q)dS_{Q}\\ &+ (\phi(p))^{\ell}_{\cdot j}\int_{\partial B}(\phi^{-1}(q))^{i}_{\cdot m}(\phi(q))^{s}_{\cdot k}W^{m}_{\cdot \ell s}\dot{\epsilon}^{(n)k}_{\cdot i}(q)dv_{q}\end{aligned} \tag{39}$$

The integrals must be evaluated over the entire boundary of the shell which includes the surfaces S^{+} and S^{-} which are parallel to the shell midsurface S, as well as the edges S_{e} of the shell. The idea here is to initially solve for the unknowns on the shell boundary by using equation (39) with $p \to P$ and treating the shell as a three-dimensional solid. Once this has been done, appropriate kinematic assumptions are made for the variation of displacements through the shell thickness and displacements, strains and stresses are determined throughout the shell. The stress resultants can then be obtained by integrating the stresses through the thickness of the shell. This idea is being pursued at present. The edges of the shell require special care. Attention is being limited in this section to an edgeless shell.

Attention is first being focussed on axisymmetric shells. The geometry for axisymmetric shells is shown in Fig. 2. Here, the natural curvilinear coordinates are $\theta^{1} = \theta$, $\theta^{2} = s$, and

$$\begin{aligned}x^{1} &= \rho(s)\cos\theta\\ x^{2} &= \rho(s)\sin\theta\\ x^{3} &= z(s)\end{aligned} \tag{40}$$

in terms of the angle θ between the x^{1}-x^{3} plane and a typical generator plane of the shell midsurface, and s, the distance measured along the intersection of this generator plane and the shell midsurface. Now ϕ in equation (35) has the matrix representation

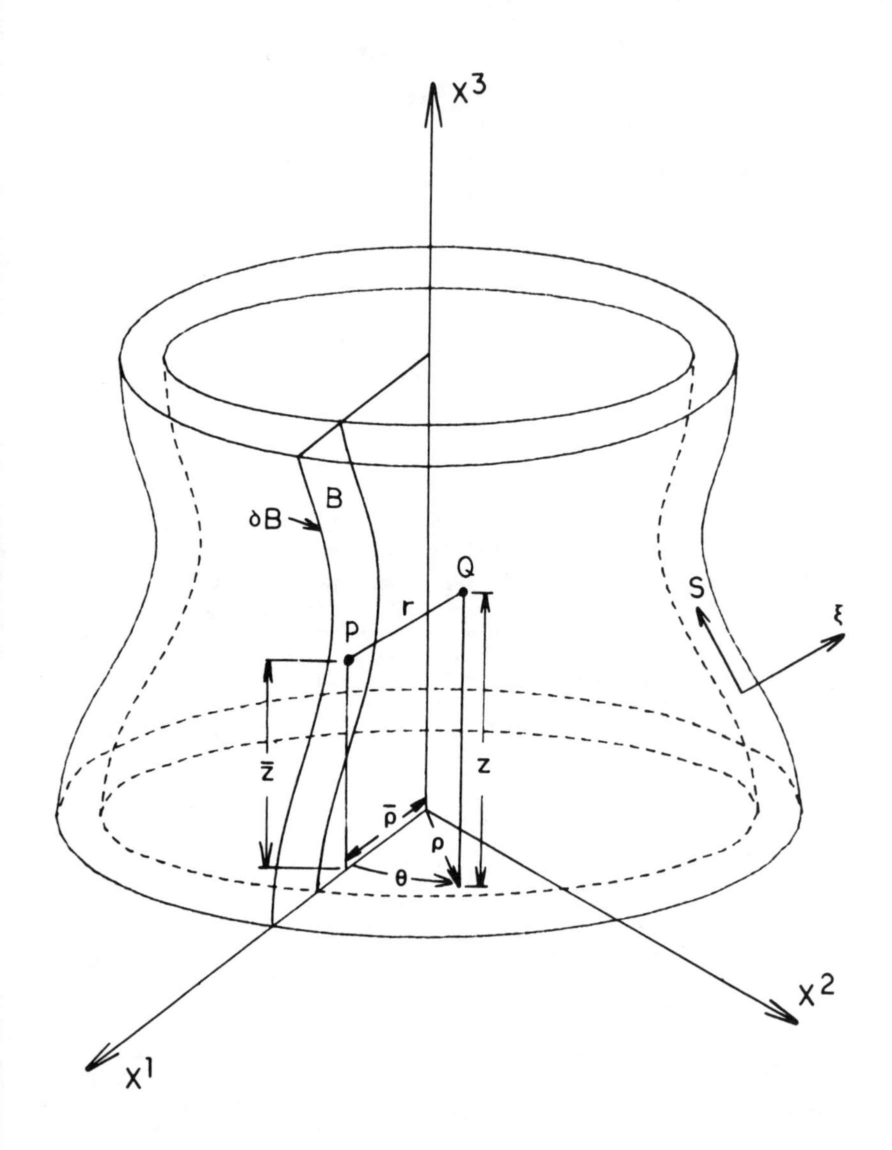

Figure 2. Geometry of the axisymmetric shell

$$(\phi)^{i}_{\cdot j} = \begin{bmatrix} -\rho(s)\sin\theta & \rho'(s)\cos\theta & z'(s)\cos\theta \\ \rho(s)\cos\theta & \rho'(s)\sin\theta & z'(s)\sin\theta \\ 0 & z'(s) & -\rho'(s) \end{bmatrix} \tag{41}$$

where a superposed prime denotes a derivative with respect to the curvilinear coordinate s.

Following the same idea as that for general axisymmetric problems, the kernels in equation (39) are integrated around a ring, so that

$$\dot{\bar{u}}_j(p) = \int_{\partial B_1} \hat{\bar{U}}^{i}_{\cdot j}(p,Q)\dot{\bar{\tau}}_i(Q)\rho(Q)dS_Q - \int_{\partial B_1} \hat{\bar{T}}^{i}_{\cdot j}(p,Q)\dot{\bar{u}}_i(Q)\rho(Q)dS_Q \tag{42}$$
$$+\int_{B_1} 2G\hat{\bar{W}}^{i}_{\cdot jk}(p,q)\dot{\bar{\epsilon}}^{(n)k}_{\cdot i}(q)\rho(q)dv_q$$

where the new kernels are

$$\hat{\bar{U}}^{i}_{\cdot j}(p,Q) = (\phi(p))^{\ell}_{\cdot j}\int_0^{2\pi}(\phi^{-1}(Q))^{i}_{\cdot k}U^{k}_{\cdot \ell}(p,Q)d\theta \tag{43}$$

and similarly for $\hat{\bar{T}}^{i}_{\cdot j}$ and $\hat{\bar{W}}^{i}_{\cdot jk}$, and ∂B_1 is the intersection of ∂B with the plane $\theta=0$.

<u>Constitutive models</u>. As mentioned before, it has been assumed that the strain rates can be additively decomposed into elastic and nonelastic parts and Hooke's law relates the rate of stress to that of the elastic strain rates (equations (3) and (4)). The nonelastic strain rate components must now be determined from an appropriate constitutive model. This can be a conventional rate independent plasticity model, in which case $\dot{\epsilon}^{(n)}_{ij}$ is the rate of plastic strain with respect to an appropriate loading variable (pseudo time). It can also be a combined creep-plasticity model with internal or state variables, in which case $\dot{\epsilon}^{(n)}_{ij}$ is the time rate of change of the nonelastic strain. This nonelastic strain, for the latter type of model, includes both irreversible as well as anelastic strains. The reader is referred to Mukherjee's book (35) for a further discussion of such constitutive models. A brief discussion of these models follows.

The mathematical structure of many of such constitutive models can be written in the form

$$\dot{\epsilon}^{(n)}_{ij} = f_{ij}(\sigma_{ij},\dot{\sigma}_{ij},q^{(k)}_{ij}) \tag{44}$$

and $$\dot{q}^{(k)}_{ij} = g^{(k)}_{ij}(\sigma_{ij},\dot{\sigma}_{ij},q^{(\ell)}_{ij}), \tag{45}$$

where $q_{ij}^{(k)}$ are the state variables.

Time-independent plasticity models generally prescribe the plastic strain as a rate function of stress as well as stress rate. The quantities $q_{ij}^{(k)}$ might denote, for example, the location of the center of the yield surface for a kinematic hardening model. A suitable equation for the yield surface must be added to equations (44) and (45) with the condition that $\dot{\epsilon}_{ij}^{(n)} = 0$ if the stress rate lies inside the yield surface. Examples of such models can be found in many plasticity texts as well as, for example, Mukherjee and Chandra (36, 37).

In case of viscoplasticity with state variables, equation (44) states that the history dependence of the nonelastic strain rate at any given time can be represented by the current values of the stress and suitably chosen state variables q_{ij}. The stress rate $\dot{\sigma}_{ij}$ is absent from equations (44) and (45). The non-elastic strain rate contains both slowly recoverable as well as irrecoverable components and thus includes both plastic and creep effects. Permanent as well as recoverable strains are considered to be time dependent. There is no explicit yield surface and $\dot{\epsilon}_{ij}^{(n)}$ is present at all stages of the deformation process. Its value, however, must be very small at low values of stress.

The state variables $q_{ij}^{(k)}$ (usually k equals 1 or 2) must be chosen carefully. They can be scalars or tensors and their choices vary in different models. They can physically represent quantities like average dislocation density, back stress, plastic work, etc. They are akin to quantities like yield stress and location of the centre of the yield surface in classical plasticity. The important assumption here is that the current values of the state variables completely characterize the present deformation state of the material.

The model due to Hart (21) has been used in some of the numerical applications presented later in this chapter. This model has two state variables, a scalar and a tensor. This model has been discussed in several previous publications such as Hart (21), Mukherjee (35) or Mukherjee and Chandra (36,37).

<u>Numerical implementation for 3-D problems</u>. Boundary value problems with irregular geometry must, in general, be solved by numerical methods. Some remarks on the numerical implementation of the integral equations are made in this section. The three-dimensional equations are considered here. Of course, similar procedures would apply to other problems.

The first step is the discretization of the three-dimensional body into surface elements and internal cells. As mentioned before, the internal discretization is necessary for the evaluation of volume integrals with known integrands. A discretized version of the boundary integral equation (17) for the displacement rate can be written as

$$
\begin{aligned}
C_{ij}(P_M)\dot{u}_i(P_M) &= \sum_{N_s} \int_{\Delta s_N} U_{ij}(P_M,Q)\dot{\tau}_i(Q)dS_Q \\
&- \sum_{N_s} \int_{\Delta s_N} T_{ij}(P_M,Q)\dot{u}_i(Q)dS_Q \\
&+ \sum_{N_s} \int_{\Delta V_N} 2GW_{ijk}(P_M,Q)\dot{\epsilon}_{ik}^{(n)}(q)dV_q \\
&+ \sum_{N_s} \int_{\Delta V_N} \lambda W_{iji}(P_M,Q)\dot{\epsilon}_{kk}^{(n)}(q)dV_q
\end{aligned} \tag{46}
$$

where the surface of the body ∂B has been divided into N_s boundary elements and the interior into n_i internal cells and $\dot{U}_i(P_M)$ are the components of the displacement rates at a point P which coincided with node M. In the above, $W_{ijk} = U_{ij,k}$.

Suitable shape functions must now be chosen for the variations of traction and displacement rates on the surface elements ΔS_N and the variation of the nonelastic strain rates over an internal cell ΔV_n. Special care must be taken for integration over an element in which a kernel is singular, i.e., over an element which contains the source point P_M. If the elements have simple geometry and the shape functions have simple form, it might be possible to carry out integrals of kernels analytically. Cruse (17) for example, has solved three-dimensional elasticity problems with flat triangular surface elements and piecewise uniform tractions and displacements on these elements. Each surface element is denoted by its centroidal point so that $C_{ij} = (\frac{1}{2})\delta_{ij}$. In this special case, it is possible to evaluate the integrals

$$
\Delta U_{ij} = \int_{\Delta s_N} U_{ij}(P_M,Q)dS_Q
$$

$$
\Delta T_{ij} = \int_{\Delta s_N} T_{ij}(P_M,Q)dS_Q
$$

exactly, knowing the size, orientation and location of the surface element ΔS_N and the point P_M.

Analytical integration of this type is not possible in general. Integration of these kernels over elements which are far from the source point can generally be carried out conveniently and very accurately by Gaussian integration. Integration in singular and near-singular situations, however, is another matter and special methods such as use of rigid body displacements (Mukherjee (35)) are useful in such cases.

Another comment pertains to the numerical modelling of possible jumps in normals or prescribed tractions across boundaries of surface elements. A simple way around this difficulty is to place source points inside rather than on the boundaries of surface elements. This is easily done in the boundary element method since the source points are really sampling points (or boundary collocation points). These points, for example, need not lie on the vertices of a triangular surface element over which traction and displacement rates are assumed to vary in a linear fashion. An alternative is to specify, if necessary, a 'zero area' element between coincident source points and assign different values of physical quantities on these points. This latter approach has been used for the solution of planar and axisymmetric viscoplastic problems which are described, for example, in Mukherjee (35).

Accurate evaluation of the domain integral in equation (46) presents a significant challenge. First, it should be noted that if a viscoplastic model with state variables is employed, the integrands in the domain integrals are known at any time and it is then a matter of accurate evaluation of known integrals. Use of a plasticity type model, on the other hand, introduces the unknown velocity field inside the domain integrals. In this case, only the boundary velocities can be treated as primary unknowns and an iterative procedure at each time step can be used. Alternatively, the velocities inside can also be treated as primary unknowns. The size of coefficient matrices in the latter case would then reflect the total number of boundary as well as internal nodes.

Even for viscoplasticity problems with known domain integrands, the accurate evaluation of the domain integrals is by no means trivial, since the kernel W_{ijk} becomes singular when the source point lies inside an internal cell over which the integral must be evaluated. The nonsingular portions of the integrals, of course, can be accurately obtained by numerical quadrature. One way to determine the singular portions of the integrals accurately is to use simple shape functions for $\dot{\epsilon}_{ij}^{(n)}$ over internal cells and evaluate the integrals analytically. Combined analytical/numerical methods are more versatile in that one is then not limited to internal cells of simple shape or to simple shape functions for the nonelastic strain rates. Such methods are discussed in books by Mukherjee (35) or Brebbia et al. (6).

Equation (46) is eventually reduced to an algebraic system of the type

$$[A]\{\dot{u}\} + [B]\{\dot{\tau}\} = \{\dot{b}\} \tag{47}$$

where the coefficient matrices [A] and [B] contain integrals of the kernels and the shape functions and the vector $\{\dot{b}\}$ contains the nonelastic strain rates and the kernel W_{ijk}. The vector $\{\dot{b}\}$, for viscoplasticity problems is known at any time through the constitutive equations, and the size of the matrices [A] and [B] depend only on the boundary discretization. A simple switching of the columns of [A] and [B] is carried out in order to determine the unknown components of the traction and displacement rates in terms of the prescribed ones. The displacement rate field throughout the body can be obtained from equation (16) discretized in an analogous manner.

The stress rates on the boundary and inside the body must now be determined from the velocities. The boundary stress rates can be easily determined by using the algorithm set forth earlier in this section. Determination of internal stress rates requires discretization of equation (18) over boundary elements and internal cells as before. There are no more unknowns at this stage but kernels with very strong singularities must be treated carefully. One possible approach, for three-dimensional problems, is the use of convected derivatives as suggested by Bui (8). Numerical realization of this approach is discussed by Brebbia et al. (6). Another possibility is the use of the strain rate gradient method as discussed by Mukherjee (see also (35)). A third approach is analytical integration over internal cells and then differentiation at a source point. Mukherjee and his group have successfully used the third approach for planar and the second approach for axisymmetric problems (35). Brebbia and his group, on the other hand, have used the Bui approach successfully for planar problems (Brebbia et al. (6)).

Solution strategy. The solution strategy for three-dimensional problems is described below. The initial values of the nonelastic strain components are taken to be zero and the initial values of the state variables, if any, are prescribed. The initial values of the stress and strain components are obtained from the solution for an identical body with boundary conditions same as those prescribed for the original nonelastic problem at zero time. This elastic solution is obtained by the boundary element method by solving the appropriate integral equations.

The rates of the nonelastic strains at zero time are next obtained from the constitutive equations (44). These rates are used in equation (47) which is solved for the unspecified components of the boundary traction and displacement rates at $t=0$. The displacement rates throughout the body are next obtained from a discretized version of equation (16). Displacement rate gradients are then determined from a differentiated version of the displacement rate equation (equation (18)), or by direct numerical differentiation of the displacement rate field. The stress rate components throughout the body follow from the components of the displacement rate gradients through Hooke's law. Finally, the rates of the state variables, if any, are determined from equation (45).

Since the rate equations often give rise to a stiff system of differential equations, it is very important, for these problems, to select an efficient time integration strategy with a suitable algorithm for time step control. Many such strategies - both implicit and explicit - exist in the literature and some of them have been reviewed in Mukherjee (35). By and large, implicit methods are unconditionally stable but are more difficult to implement than explicit ones. An example of recent research using the implicit scheme, for Hart's constitutive model, can be found in Cordts and Kollman (16).

Sample numerical results. This section presents a few sample numerical results for inelastic axisymmetric and plate bending problems, from Mukherjee (35). Hart's constitutive model is used to describe material behavior in these examples

Fig. 3 shows a comparison of direct, BEM and FEM results for internal pressurization of a uniform cylinder subjected to increasing pressure. The discretizations used for the various methods are given in Mukherjee (35). The "mixed" methods refer to a BEM solution for displacements and velocities followed by a piecewise quadratic interpolation of these quantities inside and on the boundary of the body. The stresses and stress rates inside and on the body are then obtained by differentiation of these quadratic interpolation functions. In this case, the FEM results, with piecewise quadratic interpolations of displacements and velocities over ring elements of triangular cross-section, are more accurate than the BEM results, which only use piecewise linear descriptions of displacements, tractions, and their rates on straight boundary elements. Errors in the piecewise linear representation of velocities are aggravated in the BEM which uses the boundary stress algorithm and therefore require numerical differentiation of velocities on the boundary ∂B. It is felt that a piecewise quadratic or spline approximation for boundary velocities would considerably improve the accuracy of the BEM from the present 7% error (compared to the direct solution) at a simulated strain of around 2%. Although one of the mixed methods does better than the BEM in this case, it is recommended that, in general, such methods be avoided and the BEM be used to determine the strain and stress rates inside a body.

The redistribution of stresses for the same problem, obtained from the BEM algorithm, are shown in Fig. 4. The results show the expected transition from an elastic to an elastic-plastic and finally a plastic stress distribution. The crosses in this figure refer to internal points and the circles to boundary points. The boundary stresses, especially at the inner radius, become progressively less accurate with time. This is attributed to the errors from the boundary stress algorithm as mentioned above. It should be realized that errors in boundary stress rates affect boundary stresses, which, in turn, cause inaccuracies in stress rates and stresses throughout the cylinder as integration proceeds in time.

Typical computer (cpu) times for the above calculations were of the order of 42 seconds for the BEM and 25 seconds for the FEM, respectively, on an IBM 370/168 computer. The mixed methods were an order of magnitude slower ($\simeq$ 600 seconds) and should be avoided.

In conclusion, it is felt that the BEM accuracy for these problems can be further improved either with the use of higher order shape functions for boundary displacements (and their rates), or by a smoothing operation on displacements obtained from a piecewise linear boundary interpolation, prior to differentiation to obtain boundary stresses. The reader is referred to Mukherjee (35) for further details.

A typical numerical result for the time-dependent inelastic bending of a clamped square plate is shown in Figure 5. The plate is of side 2 inches and the plate thickness is 2% of a. The loading is uniform pressure increasing at a constant rate of 0.5 psi/sec. The boundary and internal mesh used is given in the figure.

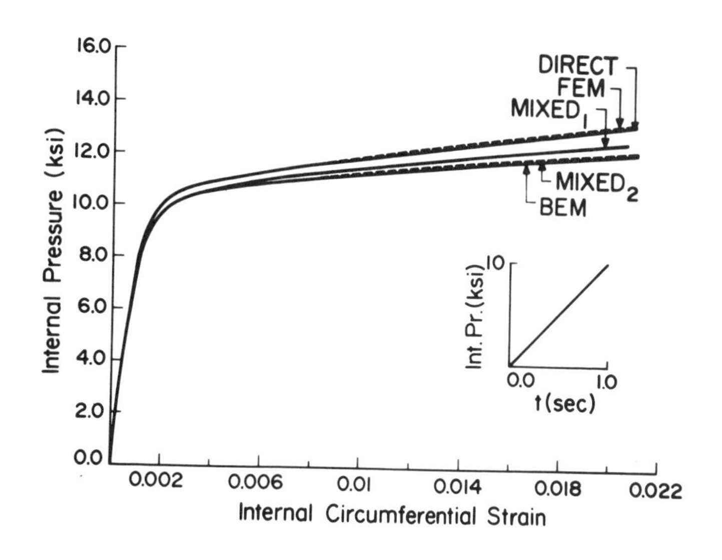

Figure 3. Results for internal pressure on a uniform cylinder. $\dot{p}$ = 10 ksi/sec (from Mukherjee 1982b)

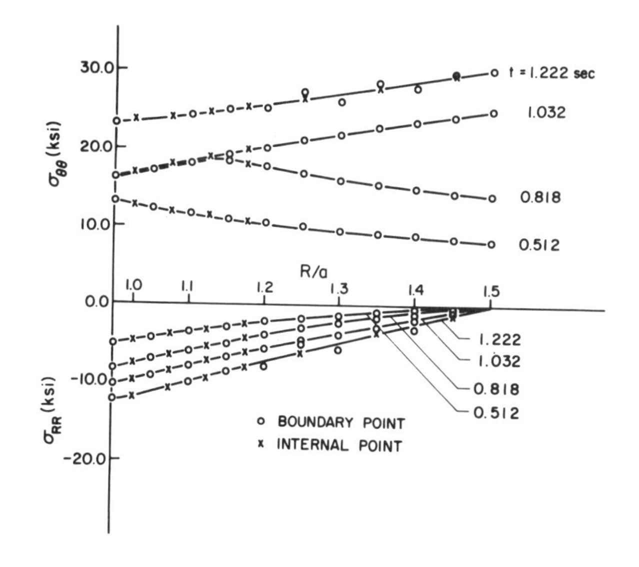

Figure 4. Redistribution of stresses for the cylinder under internal pressure. BEM solution (from Mukherjee 1982b)

The pressure as a function of central displacement of the plate is shown in Figure 5. Other numerical results are given in Mukherjee (35). This method works well for clamped or simply supported edges but leads to some problems for other boundary conditions such as plates with free edges. An improved BEM formulation, suitable for plates with more general boundary conditions, is currently under development.

As mentioned before, inelastic analysis of shells by the BEM is currently in progress. Comparisons of BEM results against exact elasticity solutions have been carried out for a toroid as well as for a circular cylinder, subjected to different types of loadings. The conparisons are uniformly excellent and this BEM approach looks extremely promising. A paper with details of the numerical implementation and numerical results for the elastic problem has been submitted for publication (Poddar and Mukherjee (41)).

Material and Geometrical Nonlinearities

This section of this chapter is devoted to a discussion of solutions of large strain-large rotation problems, in the presence of plasticity or viscoplasticity, by the boundary element method. It is assumed throughout that the elastic strains are infinitesimal while the nonelastic strains, as well as the rotations, can be arbitrary. The formulation presented here is based on an updated Lagrangian approach in which the configuration of a body of time t is used as reference for the deformation between t and $t + \Delta t$. This necessitates updating of the geometry during computer simulation of deformation but provides great simplifications in the BEM formulation. Numerical results, for a sample problem of plane extension of a metallic workpiece, are presented at the end of the section.

Kinematics. A three-dimensional body is considered in this section. Referring to a set of spatially fixed rectangular cartesian coordinates, a material particle in the body in a reference configuration is assumed to have coordinates X_i. The same particle has coordinates x_i in the current configuration. Unless otherwise indicated, the range of indices in this Section is 1,2,3.

The displacement vector u_i is defined by the equation

$$x_i = X_i + u_i \tag{48}$$

The velocity of this material point during deformation is denoted by v_i. The deformation rate d_{ij} is the symmetric part of the velocity gradient.

$$d_{ij} = (1/2)\left[\frac{\partial v_i}{\partial x_j} + \frac{\partial v_j}{\partial x_i}\right] \tag{49}$$

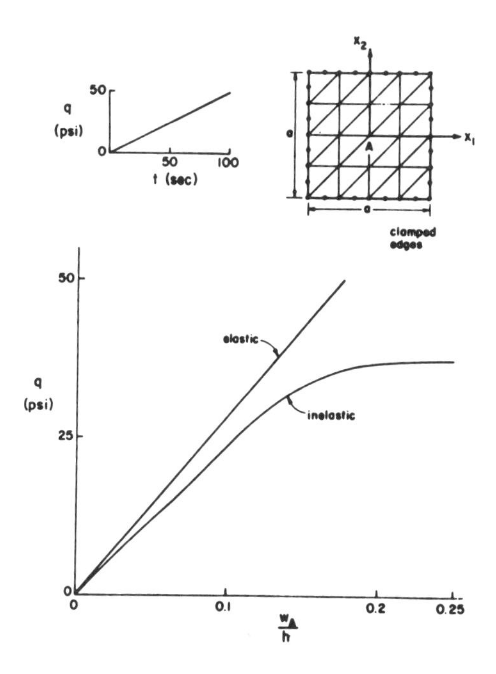

Figure 5. Load as a function of central displacement for a clamped square plate subjected to a uniform load increasing at a constant rate

whereas the spin ω_{ij} is the anti-symmetric part

$$\omega_{ij} = (1/2)\left[\frac{\partial v_i}{\partial x_j} - \frac{\partial v_j}{\partial x_i}\right] \tag{50}$$

<u>Consitutive assumptions</u>. The boundary element formulation presented next is valid for a wide range of elastic-viscoplastic consitutive models. While some details about these models are given later in this section, the key assumptions used in the BEM formulation are presented here.

The first assumption is that the deformation rate gradient tensor can be linearly decomposed into an elastic and a nonelastic part

$$d_{ij} = d_{ij}^{(e)} + d_{ij}^{(n)} \tag{51}$$

The next step relating $d_{ij}^{(e)}$ to the stress rate is very important. A hypoelastic material is one in which the components of a (proper) stress rate are homogenoeus linear functions of the components of a (proper) rate of deformation. One form of such a law, under conditions of material isotropy (of the virgin specimen) can be written as

$$\overset{*}{\sigma}_{ij} = \lambda d_{kk}^{(e)} \delta_{ij} + 2G d_{ij}^{(e)} \tag{52}$$

where $\overset{*}{\sigma}_{ij}$ is the Jaumann rate of the Cauchy stress.

Equation (52) is one of the simplest hypoelastic laws imaginable and considerable controversy exists in the mechanics community regarding the proper form for such a law one should use as well as the appropriate choice of the rate of effective stress. A spirited debate on these issues ensued recently at an ASME meeting and the interested reader is referred to a recent ASME publication (Willam (52)) containing papers presented there. The chosen stress rate must, of course, be objective with respect to rigid body rotations, but, unfortunately, many such rates exist. The general feeling among the researchers currently is that, unless very large shearing strains are involved, equation (52) is probably sufficient, and further experimentation and analysis, both at the micro and macro levels, is necessary before the verdict with regard to the best constitutive assumption is out (Chandra and Mukherjee, (14)). Equation (52) is adopted in this work.

The Jaumann (sometimes called the Zaremba-Jaumann-Noll rigid body rate or the corotational rate) (Jaumann (24)) of a second rank tensor is related to its material rate by the formula

$$\overset{*}{\underset{\sim}{T}} = \dot{\underset{\sim}{T}} + \underset{\sim}{T}\cdot\underset{\sim}{\omega} - \underset{\sim}{\omega}\cdot\underset{\sim}{T} \tag{53}$$

or, in component form,

$$\overset{*}{t}_{ij} = \dot{t}_{ij} + t_{ik}\omega_{kj} - \omega_{ik}t_{kj} \tag{54}$$

where the material derivative is defined as

$$\dot{t}_{ij} = \frac{\partial t_{ij}}{\partial t} + \frac{\partial t_{ij}}{\partial x_k} v_k \tag{55}$$

It is clear that unlike the case of small deformations discussed in the previous section, the choice of tensor rates is no longer unambiguous and must be given very careful attention. Physically, the Jaumann rate is the rate of change of a tensor as evidenced by an observer which translates with a material point as well as rotates, in some sense, with the differential material element around that point.

The above constitutive assumptions are sufficient for the purpose of developing a BEM formulation for this class of problems. Further assumptions regarding the specifications of $d_{ij}^{(n)}$ are, of course, necessary for the solution of the complete problem, but this discussion is being postponed till later in this chapter.

BEM formulation for velocities. The key to the BEM formulation is an appropriate form of Betti's (5) reciprocal theorem. Analogous to equation (13), the applicable version for the large strain problem is

$$\int_B \overset{*}{\sigma}_{ij} \epsilon_{ij}^{(R)} dV = \int_B \sigma_{ij}^{(R)} d_{ij}^{(e)} dV \tag{56}$$

where the reference solution is exactly the same as the one used for small deformations (equations 5-10). Again, the integrands on either side of the above equation are identical by virtue of equations (8) and (52). Also, it should be noted that the domain of integration of equation (56) is the current volume

Equation (56) must next be written in terms of the known reference region B^o. Following previous publications (Chandra and Mukherjee (9,10), Mukherjee and Chandra (36,37,38))

$$\overset{*}{\sigma}_{ij} = \overset{*}{\tau}_{ij} - v_{k,k} \sigma_{ij} \tag{57}$$

where τ_{ij} are the components of the Kirchhoff stress. It is more convenient to write the above equation in the form

$$\overset{*}{\sigma}_{ij} = \overset{*}{\tau}_{ij} + C_{ijk\ell} v_{k,\ell} \tag{58}$$

where $C_{ijk\ell} = -\sigma_{ij}\delta_{k\ell}$ in terms of the Cauchy stress and the Kronecker delta.

The rate of equilibrium equation in terms of the material rate of the nonsymmetric Lagrange stress is given as

$$\dot{s}_{ji,j} = 0 \tag{59}$$

The next step requires the relationship between the Jaumann rate of the Kirchhoff stress and the material rate of the Lagrange stress. This is, in general, rather complicated (Mukherjee and Chandra (37)) but, in the updated Lagrangian description, simplifies to

$$\dot{s}_{ji} = \overset{*}{\tau}_{ji} - \sigma_{jk}\omega_{ki} - d_{jk}\sigma_{ki} \tag{60}$$

Again, it is convenient to write this equation in the form

$$\overset{*}{\tau}_{ji} = \dot{s}_{ji} + G_{jik\ell} v_{k,\ell} \tag{61}$$

where

$$2G_{ijk\ell} = \sigma_{ik}\delta_{j\ell} + \sigma_{j\ell}\delta_{ik} + \sigma_{jk}\delta_{\ell i} - \sigma_{\ell i}\delta_{jk}$$

The tensor $G_{ijk\ell}$ is, therefore, a function only of the components of the Cauchy stress σ_{ij}. Using the convention

$$\{v_{k,\ell}\}^T = \left[v_{1,1} v_{2,1} v_{3,1} v_{1,2} v_{2,2} v_{3,2} v_{1,3} v_{2,3} v_{3,3}\right]$$

the matrix corresponding to $G_{ijk\ell}$ is given in Mukherjee and Chandra (37).

The form of equation (56) in the reference configuration B^o is now

$$\int_{B^o} \left[\dot{s}_{ji} + G_{jik\ell} v_{k,\ell} + C_{ijk\ell} v_{k,\ell}\right] \epsilon_{ij}^{(R)} dV = \int_{B^o} \sigma_{ij}^{(R)} \dot{E}_{ij}^{(e)} dV \qquad (62)$$

where, in the updated Lagrangian frame,

$$d_{ij} = \dot{E}_{ij} \qquad (63)$$

with E_{ij} the Green-Saint Venant strain.

The rest of the derivation parallels that of the small deformation situation. Using equations (7), (59), (9) and the divergence theorem, the left hand side of equation (62) becomes

$$\begin{aligned} \text{L.H.S.} = e_j \Big[& \int_{\partial B^o} U_{ij} \dot{s}_{ki} n_k^o dS^o \\ & + \int_{B^o} U_{ij,m} G_{mik\ell} v_{k,\ell} dV^o \\ & + \int_{B^o} U_{ij,m} C_{mik\ell} v_{k,\ell} dV^o \Big] \end{aligned} \qquad (64)$$

Similarly using equations (63), (51), (49), (8), (5), (7), (6), (10) and (9) and the divergence theorem, the right hand side of equation (62) becomes

$$\begin{aligned} \text{R.H.S.} = e_j \Big[& \int_{\partial B^o} T_{ij} v_i dS^o + \int_{B^o} \Lambda(p,q) \delta_{ij} v_i dV^o \\ & - \int_{B^o} \left[\lambda U_{ij,i} d_{kk}^{(n)} + 2G U_{ij,k} d_{ik}^{(n)}\right] dV^o \Big] \end{aligned} \qquad (65)$$

It is interesting to compare equations (64) and (65) with equations (14) and (15) for small deformation.

Finally, equating the expressions in equations (64) and (65) and using the relevant property of the Dirac delta function $\Delta(p,q)$, one obtains the equation

$$\begin{aligned} v_j(p) = & \int_{\partial B^o} \left[U_{ij}(p,Q)\dot{\tau}_i(Q) - T_{ij}(p,Q)v_i(Q)\right]dS_Q^o \\ & + \int_{B^o} \left[\lambda U_{ij,i}(p,q)d_{kk}^{(n)}(q) + 2GU_{ij,k}(p,q)d_{ik}^{(n)}(q)\right]dV_q^o \\ & + \int_{B^o} U_{ij,m}(p,q)G_{mik\ell}(q)v_{k,\ell}(q)dV_q^o \\ & + \int_{B^o} U_{ij,m}(p,q)C_{mik\ell}(q)v_{k,\ell}(q)dV_q^o \end{aligned} \tag{66}$$

where $\dot{\tau}_i = n_j^o \dot{s}_{ji}$.

Comparing equation (66) with the corresponding equation (16) for small strains, it is seen that the first two integrals on the right hand side of the above equation are analagous to the small strain formulation. The traction rate term $\dot{\tau}_i$ requires special care and is discussed a little later. The third term is the so called "geometric correction" term which arises due to finite deformations and rotations inside the body. The last integral is a consequence of bulk compressibility of the material. If incompressible metal deformation is being studied, this last integral vanishes altogether (since $d_{kk}^{(n)} = 0$ and one usually assumes $d_{kk} = 0$ in such cases) and so does the term containing $d_{kk}^{(n)}$ in the second integral (Mukherjee and Chandra (36)).

Using equation (60), the traction rate $\dot{\tau}_i$ in equation (66) can be written as

$$\dot{\tau}_i = n_j\dot{s}_{ji} = \dot{t}_i - n_j\sigma_{jk}\omega_{ki} - n_jd_{jk}\sigma_{ki} = \dot{t}_i - n_jG_{jik\ell}v_{k,\ell} \tag{67}$$

where

$$\dot{t}_i = n_j\overset{*}{\tau}_{ji} = n_j\overset{*}{\sigma}_{ji} - n_jC_{jik\ell}v_{k,\ell}$$

The presence of the G tensor in the boundary traction expression is sometimes referred to as "load correction" and is a consequence, during deformation, of the change in the area of a surface element and the rotation of the unit normal at a point on it. The rate $\dot{t}_i$ can be interpreted as a component of the rate of the prescribed follower force, per unit deformed surface area, on the deforming boundary.

It can be seen that, unlike elastic-viscoplastic problems with small strains and rotations, the unknown velocity gradient now occurs in the surface as well as some of the domain integrals in equation (66). Thus, iterations now become necessary at each time step. This has been carried out in order to obtain the numerical results presented later in this chapter.

As before, the limit $p \to P$ is taken so that equation (66) becomes

$$\begin{aligned} C_{ij}(P)v_i(P) = & \int_{\partial B^o} \left[U_{ij}(P,Q)\dot{\tau}_i(Q) - T_{ij}(P,Q)v_i(Q)\right]dS_Q^o \\ & + \int_{B^o} \left[\lambda U_{ij,i}(P,q)d_{kk}^{(n)}(q) + 2GU_{ij,k}(P,q)d_{ik}^{(n)}(q)\right]dV_q^o \\ & + \int_{B^o} U_{ij,m}(P,q)G_{mik\ell}(q)v_{k,\ell}(q)dV_q^o \\ & + \int_{B^o} U_{ij,m}(P,q)C_{mik\ell}(q)v_{k,\ell}(q)dV_q^o \end{aligned} \tag{68}$$

The coefficient C_{ij} is the same as in the case of small strain-small rotation problems.

Internal stress rates. As mentioned before, analytical differentiation of equation (66) at a source point is the best way to obtain velocity rates at internal points. To this end, a differentiated version of equation (66) is

$$\begin{aligned} v_{j,\bar{\ell}}(p) = & \int_{\partial B^o} \left[U_{ij,\bar{\ell}}(p,Q)\dot{\tau}_i(Q) - T_{ij,\bar{\ell}}(p,Q)v_i(Q)\right]dS_Q^o \\ & + \frac{\partial}{\partial x_{\bar{\ell}}}\int_{B^o} \left[\lambda U_{ij,i}(p,q)d_{kk}^{(n)}(q) + 2GU_{ij,k}(p,q)d_{ik}^{(n)}(q)\right]dV_q^o \\ & + \frac{\partial}{\partial x_{\bar{\ell}}}\int_{B^o} U_{ij,m}(p,q)G_{mikn}(q)v_{k,n}(q)dV_q^o \\ & + \frac{\partial}{\partial x_{\bar{\ell}}}\int_{B^o} U_{ij,m}(p,q)C_{mikn}(q)v_{k,n}(q)dV_q^o \end{aligned} \tag{69}$$

As before, one must be careful with differentiation of kernels like $U_{ij,m}$. The extra integrals in the large strain formulation, however, present no added difficulties relative to the small strain problem. Thus, all the earlier discussion regarding numerical evaluation of the domain integrals applies here also. So far, numerical results from these equations have only been obtained for planar problems. In such cases, it has been possible to evaluate each of the volume integrals analytically (for a specific simple choice of shape functions for the unknowns in these domain integrals) and then differentiate these integrals at an arbitrary source point p.

Once the velocity gradients and then d_{ij} have been determined, the nonelastic deformation rate $d_{ij}^{(n)}$ is subtracted from it to get the elastic deformation rate $d_{ij}^{(e)}$. The components of the Jaumann rates of the Cauchy stress are now obtained from the hypoelasticity relationship (equation (52)).

The corresponding equations for plane strain and plane stress are given in Mukherjee and Chandra (37).

Constitutive models. As mentioned before, additive decomposition of d_{ij} (equation (51)) and the hypoelastic law (equation (52)) have been assumed previously. The nonelastic velocity gradient tensor $d_{ij}^{(n)}$ must be determined from an appropriate plastic or viscoplastic constitutive law.

A general form of such constitutive models can be written as

$$d_{ij}^{(n)} = f_{ij}(\sigma_{ij}, \overset{*}{\sigma}_{ij}, q_{ij}^{(k)}) \tag{70}$$

$$\overset{*}{q}{}_{ij}^{(k)} = g_{ij}^{(k)}(\sigma_{ij}, \overset{*}{\sigma}_{ij}, q_{ij}^{(\ell)}) \tag{71}$$

The viscoplastic constitutive model due to Hart (21) (Fig. 6) generalized to the case of large strains, has been employed to obtain the numerical results that are presented later in this section. This model has been discussed in several previous publications (e.g. Mukherjee and Chandra (36,37,38)).

Numerical implementation. Most of the earlier discussion regarding numerical implementation of the small deformation BEM equations apply here to the case of large deformations also. A discretized version of equation (68) for the velocity is now

$$\begin{aligned}
C_{ij}(P_M)v_i(P_M) &= \sum_{N_S}\int_{\Delta S_N}\left[U_{ij}(P_M,Q)\dot{\tau}_i(Q) - T_{ij}(P_M,Q)v_i(Q)\right]dS_Q^o \\
&+ \sum_{n_i}\int_{\Delta A_n}\left[\lambda U_{ij,i}(P_M,q)d_{kk}^{(n)}(q) + 2GU_{ij,k}(P_M,q)d_{ik}^{(n)}(q)\right]dV_q^o \\
&+ \sum_{n_i}\int_{\Delta A_n} U_{ij,m}(P_M,q)G_{mik\ell}(q)v_{k,\ell}(q)dV_Q^o \\
&+ \sum_{n_i}\int_{\Delta A_n} U_{ij,m}(P_M,q)C_{mik\ell}(q)v_{k,\ell}(q)dV_q^o
\end{aligned} \tag{72}$$

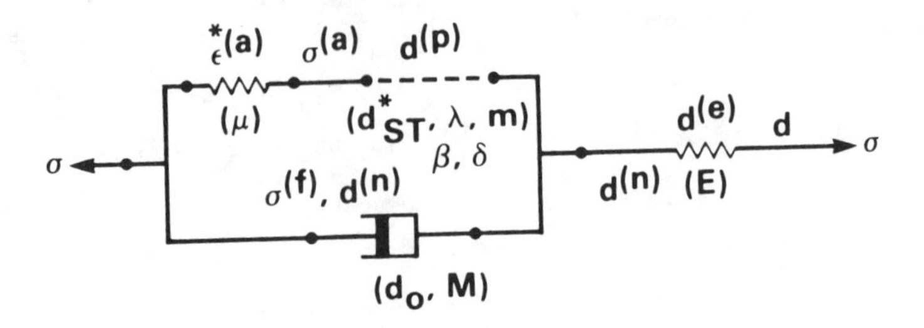

Figure 6. Hart's model in tension for large strain

where the boundary of the body in the reference configuration ∂B^o is divided into N_s boundary segments and the interior into n_i internal cells and $v_i(P_M)$ are the components of velocities at a point P which coincides with node M.

Suitable shape functions must now be chosen for the variation of tractions and velocities on the surface element ΔS_N and the variation of the nonelastic strain rates and velocities over an internal cell ΔA_n.

Following essentially the same procedure as for small strain problems, one obtains the algebraic system

$$[A]\{v\} + [B]\{\dot{\tau}\} = \{b\} \tag{73}$$

where the coefficient matrices [A] and [B] contain integrals of the kernels and the shape functions; and the vector {b} contains the contributions of various quantities from the three domain integrals.

Equations (66) and (69) for the velocity field and the velocity gradients at an internal point are discretized in similar fashion.

<u>Interface modelling for extrusion problems</u>. Interface conditions at the die-workpiece boundary are considered next for problems of plane strain extrusion (Mukherjee and Chandra (37)). The assumptions stated here are the same as those used earlier for FEM analysis of the same problem (Chandra and Mukherjee, (11)).

The assumptions are best explained in terms of a local coordinate system α,β,γ. The origin of this coordinate system is positioned on the die-workpiece interface. The α axis is tangential to the die surface and the β axis is the outward normal to the die surface.

A consequence of the assumption of plane strain is that

$$v_\gamma = 0 \;, \; \sigma_{\alpha\gamma} = \sigma_{\gamma\alpha} = \sigma_{\beta\gamma} = \sigma_{\gamma\beta} = 0$$

It is further assumed that the contact or lubricant layer adjacent to the die surface cannot provide any resistance to tensile or compressive deformation. Thus, the scheme is equivalent to having an interface or bond element with zero stiffness in the direction tangential to the contact surface. Therefore,

$$\sigma_{\alpha\alpha} = 0. \tag{74}$$

and only pressures and shear loadings get transferred across the interface and the stretch or compression of either the die or the workpiece is not transferred to the other.

Another assumption made here is that the material is incompressible so that

$$d_{\alpha\alpha} + d_{\beta\beta} = 0 \qquad \text{(since } d_{\gamma\gamma} = 0) \tag{75}$$

Using the above assumptions, zero normal velocity ($v_\beta = 0$) and the fact that the normal has components (0,1,0) in the local coordinate system, equation (67) in the local coordinate system becomes

$$\dot{\tau}_\alpha = \overset{*}{\sigma}_{\alpha\beta} + \sigma_{\beta\beta}\omega_{\alpha\beta} + \sigma_{\alpha\beta}d_{\alpha\alpha} \tag{76}$$

The other traction component is assumed to vanish, i.e.

$$\dot{\tau}_\beta = 0 \tag{77}$$

The final assumptions relate to friction

$$\overset{*}{\sigma}_{\alpha\beta} = \frac{G_s}{h}\mu v_\alpha \tag{78}$$

(where G_s and h are the shear modulus and the height of the interface element respectively, and μ is the coefficient of friction), and

$$\omega_{\alpha\beta} = \kappa v_\alpha \tag{79}$$

where κ is the local curvature.

The rate of the traction component, $\dot{\tau}_\alpha$, has now been obtained in terms of the tangential velocity v_α, its gradient in the tangential direction $v_{\alpha,\alpha}$, and the nonzero stress components $\sigma_{\alpha\beta}$ and $\sigma_{\beta\beta}$.

Solution strategy. The solution strategy, in essence, consists of marching foward in real time with suitable updating of the configuration of the body. The presence of velocity gradients in the boundary traction rates and in the last domain integral in equation (66) requires iteration within each time step. The elasticity problem is solved first to obtain the initial distribution of the state variables.

At each time step, an initial guess for the velocity gradient is fed in and iterated until convergence is achieved. Typically, the velocity gradient obtained at time '$t-\Delta t$' is used as the initial guess at time 't'. The extrusion process is nearly steady state and fast convergence could be achieved for reasonable values of Δt. The flow chart shown in Fig. 7 shows the solution procedure in detail.

The BEM program uses straight boundary elements and polygonal internal cells. The velocity and traction rate are taken to be piecewise linear over the boundary elements while the nonelastic deformation rate and the velocity gradients are assumed to be piecewise constant over the internal cells. The values of the boundary variables are assigned at nodes which lie at the intersections of boundary segments. Possible discontinuities in tractions are taken care of by placing a "zero length" element between nodes and assigning different values of traction at each of those nodes.

All integrations of kernels are carried out analytically. The last two terms in equation (69) are evaluated by first performing the integration over an internal cell for an arbitrary source point p_m and then differentiating the integral at p_m. Further details of this procedure are available in Mukherjee and Chandra (37).

Numerical results for plane strain extrusion. The primary purpose of obtaining the numerical results presented here is to demonstrate the ability of the boundary element method (BEM) in solving metal forming problems involving both geometric and material nonlinearities. Handling of the boundary conditions at the die interface requires special care for the extrusion problem. In the present BEM analysis, this can be incorporated quite easily through "load correction" (equations (67) and (74-79)).

The geometry of planar extrusion considered here is shown in Fig. 8 and numerical results are presented in Figs. 9-10. Fig. 9 shows the steady state distributions of the longitudinal stress σ_{11} for three different velocities in the absence of friction. In particular, the centre line of the workpiece is chosen to be the x_1 axis and the stress distributions are shown for material points in the deformed configuration which initially had the same relative ordinate (x_2/a = 0.11 and 0.89) in the billet. Maximum residual tensile stress in an extruded workpiece is of crucial importance in design, since this is the primary potential source for crack initiation and growth. It is seen from Fig. 3 that the rate dependence of this quantity is quite significant. As the piston velocity is tripled from 0.01"/sec (0.254 mm/sec) to 0.03"/sec (0.762 mm/sec), the BEM analysis predicts a change of 15.5 percent in the maximum longitudinal tensile stress in the workpiece. This compares well with the 17 percent change predicted by the FEM analysis Chandra and Mukherjee (11). The faster the billet is forced through the die, the less time there is for stresses to relax at material points in the workpiece, as they move through the die. Consequently, the maximum longitudinal tensile stress upon exit from the die increases substantially with the speed of extrusion.

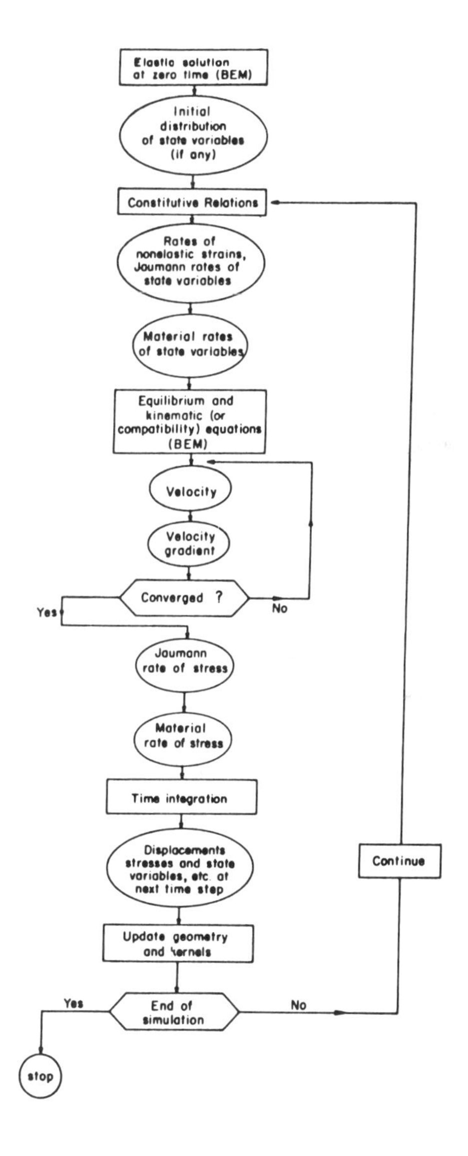

Figure 7. Solution strategy for large strain problems (from Mukherjee and Chandra 1984)

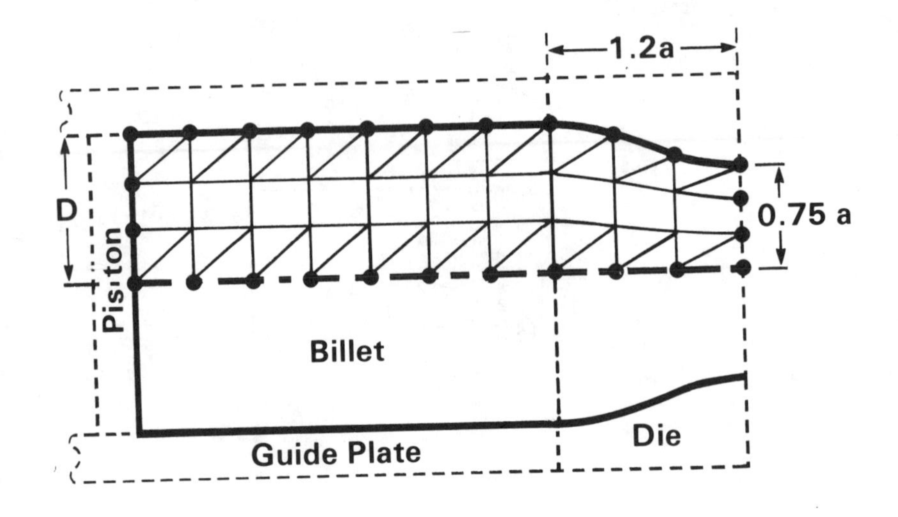

Figure 8. Extrusion geometry

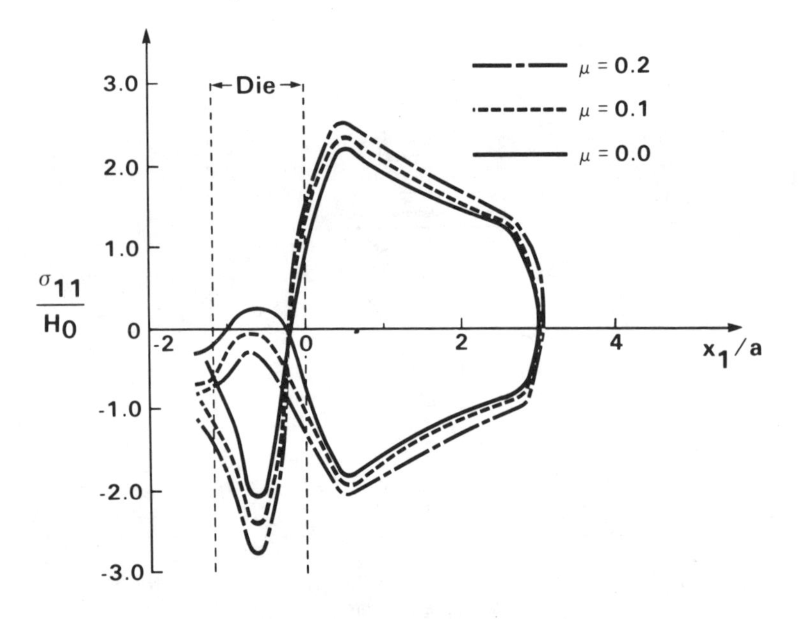

Figure 9. Steady state distributions of σ_{11} for the extrusion problem, for three different piston velocities (from Mukherjee and Chandra 1985b)

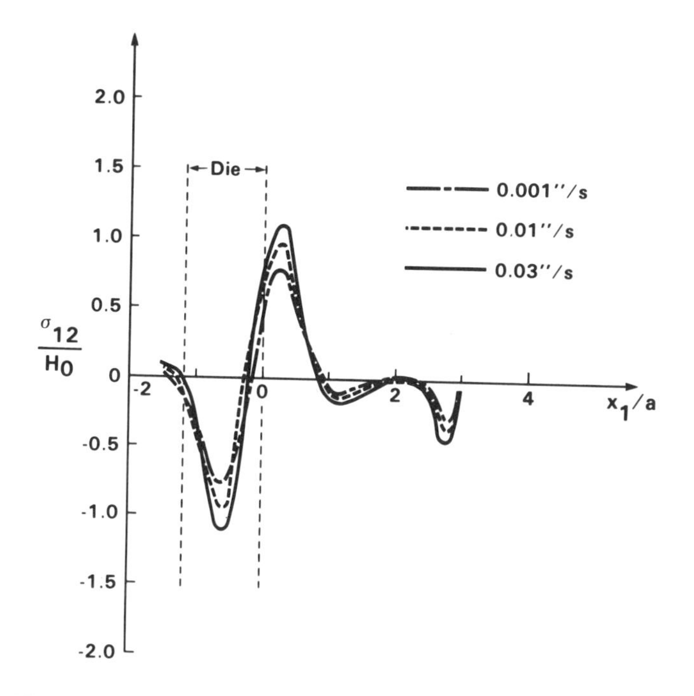

Figure 10. Steady state distributions of σ_{12} for the extrusion problem, for three different piston velocities (from Mukherjee and Chandra 1985b)

Predictions of crucial effects like rate dependence of residual stresses is only possible through a detailed analysis using a realistic elastic-viscoplastic model for material behaviour. It should be emphasized again that the material parameters for aluminium used in these calculations are those measured carefully by experiments.

Another important feature of elastic-viscoplastic analysis is that, following a peak value, the magnitude of σ_{11} decreases as a function of x_1 in most of the billet that has passed through the die. This is a result of stress relaxation in the workpiece after it is deformed, and the BEM predictions for such relaxations compare very well with those predicted by FEM analysis (Chandra and Mukherjee (11)).

The results for the steady state distributions of shearing stress σ_{12} are shown in Fig. 10. It is seen that there is a marked variation of shearing stress inside and in the neighborhood of the die.

The BEM has been demonstrated above to be able to solve the challenging problem of stress analysis in metal extrusion. Traditionally, the FEM has been used in such analyses and one of the main difficulties in solving any real life problem is the amount of CPU time necessary. This is where the application of the BEM becomes advantageous. Because of the reduced matrix size and the capability of obtaining secondary variables accurately (Mukherjee (35)) the BEM analysis took only about 55 percent of the time required by a similar FEM analysis performed by the authors. Of course, much improvement is still needed in both the FEM and the BEM to make them a viable day to day design tool for metal forming operations. However, the present study shows the strength and the potential of BEM and warrants that active research be pursued in this area in order to take advantage of the potential benefits of the boundary element method.

Acknowledgements

Mukherjee and Poddar's contribution to this research has been supported by grants from the U.S. National Science Foundation to Cornell University.

References

1. Banerjee, P.K. and Mustoe, G.C. W., "The boundary element method for two-dimensional problems of elasto-plasticity," Recent Advances in Boundary Element Methods, C.A. Brebbia (ed.), Pentech Press, Plymouth, Devon, UK, 1978, pp. 283-300.

2. Banerjee, P.K. and Cathie, D.N., "A direct formulation and numerical implementation of the boundary element method for two-dimensional problems of elasto-plasticity," International Journal for Mechanical Sciences, Vol. 22, 1980, pp. 233-245.

3. Banerjee, P.K. and Butterfield, R., "Boundary Element Methods in Engineering Science," McGraw Hill, UK, 1981.

4. Benitez, F.G., Alarcón, E. , Brebbia, C.A. and Telles, J., "Tridimensional plasticity using BIEM," Applied Mathematical Modelling, Vol. 5, 1981, pp.442-447.

5. Betti, E., "Teoria dell elasticita," Il Nuovo Ciemento, 1872-1873, pp. 7-10 .

6. Brebbia, C.A., Telles, J.C.F. and Wrobel, L.C., "Boundary Element Techniques Theory and Applications in Engineering," Springer Verlag, New York 1984.

7. Brunet, M., "Numerical analysis of cyclic plasticity using the boundary integral equation method," Boundary Element Methods, C.A. Brebbia (ed.), Springer-Verlag, Berlin, 1981, pp. 337-349.

8. Bui, H.D., "Some remarks about the formulation of three-dimensional thermoelastoplastic problems by integral equations," International Journal of Solids and Structures, Vol. 14, 1978, pp. 935-939.

9. Chandra, A. and Mukherjee, S., "Applications of the boundary element method to large strain large deformation problems of viscoplasticity," Journal of Strain Analysis, Vol. 18, 1983, pp. 261-270.

10. Chandra, A. and Mukherjee, S.,"Boundary element formulations for large strain-large deformation problems of viscoplasticity," International Journal of Solids and Structures, Vol. 20, 1984, pp. 41-53.

11. Chandra, A. and Mukherjee, S., "A finite element analysis of metal-forming problems with an elastic-viscoplastic material model," International Journal for Numerical Methods in Engineering, Vol. 20, 1984, pp. 1613-1628.

12. Chandra, A. and Mukherjee, S., "A boundary element formulation for sheet metal forming," Applied Mathematical Modelling, Vol. 9, 1985, pp. 175-182.

13. Chandra, A. and Mukherjee, S., "A boundary element formulation for large strain problems of compressible plasticity," Engineering Analysis, expected 1986.

14. Chandra, A. and Mukherjee, S., "An analysis of large strain viscoplasticity problems including the effects of induced material anisotropy," ASME Journal of Applied Mechanics, Vol. 53, 1986, pp. 77-82.

15. Chaudonneret, M., "Structure computation in viscoplasticity. Application to two-dimensional calculation of stress concentration," Addendum to Proceedings for the Second International Symposium on Innovative Numerical Analysis in Applied Engineering Science, Montreal, Canada. R.P. Shaw _et al_. eds., University of Virginia Press, Charlottesville, Virginia 1980. See also, Methode des équations integrales appliquées á la resolution de probléme de viscoplasticite, Journal de Mécanique Appliqué, Vol. 1, 1977, pp. 113-132.

16. Cordts, D. and Kollmann, F.G., "An implicit time integration scheme for inelastic constitutive equations with internal state variables," International Journal for Numerical Methods in Engineering, Vol. 23, 1986, pp. 533-554.

17. Cruse, T.A., "Numerical solutions in three dimensional elastostatics," International Journal of Solids and Structures, Vol. 5, 1969, pp. 1259-1274.

18. Cruse, T.A., "An improved boundary-integral equation method for three-dimensional elastic stress analysis," Computers and Structures, Vol. 4, 1974, pp. 741-754.

19. Cruse, T.A., Snow, D.W. and Wilson, R.B., "Numerical solutions in axisymmetric elasticity," Computers and Structures, Vol. 7, 1977, pp. 445-451.

20. Doblaré, M., Telles, J.C.F., Brebbia, C.A. and Alarcon, E., "Boundary element formulation for elastoplastic analysis of axisymmetric bodies," Applied Mathematical Modelling, Vol. 6, 1982, pp. 130-135.

21. Hart, E.W., "Constitutive relations for the non-elastic deformation of metals," ASME Journal of Engineering Materials and Technology, Vol. 98, 1976, pp. 193-202.

22. Jaswon, M.A. and Ponter, A.R., "An integral equation solution of the torsion problem," Proceedings of the Royal Society, London, Series A, 1963, pp. **273**, 237-246.

23. Jaswon, M.A. and Maiti, M., "An integral equation formulation of plate bending problems," Journal of Engineering Mathematics, Vol. **2**, 1968, pp. 83-93.

24. Jaumann, G., "Geschlossenes system physikalischer und chemischer differentialgesetze sitzber," Akad. Wiss. Wien (IIa), Vol. **120**, 1911, pp. 385-530.

25. Kumar, V. and Mukherjee, S., "A boundary-integral equation formulation for time-dependent inelastic deformation in metals," International Journal of Mechanical Sciences, Vol. **19**, 1977, pp. 713-724.

26. Kobayashi, S. and Nishimura, N., "Elastoplastic Analysis by the integral equation method," Memoirs of the faculty of engineering, Kyoto University, Vol. 42, 1980, pp. 324-334.

27. Lachat, J.C., "A Further Development of the Boundary-Integral Technique for Elastostatics," Dissertation, University of Southampton, England 1975.

28. Maier, G. and Novati, G., "Elastic-plastic boundary element analysis as a linear complementary problem," Applied Mathematical Modelling, Vol. **7**, 1983, pp. 74-82.

29. Mendelson, A., "Boundary Integral Methods in Elasticity and Plasticity," NASA TND-7418, 1973.

30. Morjaria, M. and Mukherjee, S., "Inelastic analysis of transverse deflection of plates by the boundary element method," ASME Journal of Applied Mechanics, Vol. **47**, 1980, pp. 291-296.

31. Mukherjee, S., "Corrected boundary integral equations in planar thermoelastoplasticity," International Journal of Solids and Structures, Vol **13**, 1977, pp. 331-335.

32. Mukherjee, S. and Kumar, V., "Numerical analysis of time-dependent inelastic deformation in metallic media using the boundary-integral equation method," ASME Journal of Applied Mechanics, Vol. **45**, 1978, pp. 785-790.

33. Mukherjee, S. and Morjaria, M., "Comparison of boundary element and finite element methods in the inelastic torsion of prismatic shafts," International Journal for Numerical Methods in Engineering, Vol. **17**, 1981, pp. 1576-1588.

34. Mukherjee, S., "Time-dependent inelastic deformation of metals by boundary element methods," Developments in Boundary Element Methods -2, P.K. Banerjee and R.P. Shaw, Eds., Elsevier Applied Science Publishers, Barking, Essex, England, 1982, pp 111-142.

35. Mukherjee, S., "Boundary Element Methods in Creep and Fracture," Elsevier Applied Science Publishers, Barking, Essex, England 1982.

36. Mukherjee, S. and Chandra, A.,"Boundary element formulations for large strain-large deformation problems of plasticity and viscoplasticity," Developments in Boundary Element Methods - III., P.K. Banerjee and S. Mukherjee (eds.), Elsevier Applied Science Publishers, Barking, Essex U.K., 1984, pp. 27-58.

37. Mukherjee, S. and Chandra, A., "Nonlinear Solid Mechanics.", Boundary Element Methods in Mechanics, D.E. Beskos (ed.), North Holland, Amsterdam, 1987.

38. Mukherjee, S. and Chandra, A., "A boundary element analysis of metal extrusion processes," Advanced Topics in Boundary Element Analysis", T.A. Cruse et al (eds.), ASME, N.Y., 1985, pp. 21-34.

39. Nishimura, N. and Kobayashi, S., "Elastoplastic analysis by indirect methods," Developments in Boundary Element Methods - 3, Eds: P.K. Banerjee and S. Mukherjee, Elsevier Appl. Sci. Publishers, England, 1984, pp. 59-81.

40. Oliveira F., Mota Soares, C.A., Seabra Pereira, M.F. and Brebbia, C.A., "Boundary elements in 2D plasticity using quadratic shape functions," Applied Mathematical Modelling, Vol. **5**, 1981, pp. 371-375.

41. Poddar, B. and Mukherjee, S.,"An integral approach to shell analysis," Submitted for publication.

42. Ponter, A.R.S., "On plastic torsion," International Journal of Mechanical Sciences, Vol. **8**, 1966, pp. 227-235.

43. Riccardella, P., "An Implementation of the Boundary Integral Technique for Plane Problems of Elasticity and Elasto-Plasticity," PhD Thesis, Carnegie Mellon University, Pittsburg, PA , 1973.

44. Rizzo, F.J., "An integral equation approach to boundary value problems of classical elastostatics," Quarterly of Applied Mathematics, Vol. **25**, 1967, pp. 83-95.

45. Sarihan, V. and Mukherjee, S., "Axisymmetric viscoplastic deformation by the boundary element method," International Journal of Solids and Structures, Vol. 12, 1982, pp 1113-1128.

46. Somigliana, C., "Sopra l'equilibrio di un corpo elastico isotropo," Il Nuovo Ciemento, 1885-1886, pp. 17-20.

47. Swedlow, J.L. and Cruse, T.A., "Formulation of boundary integral equations for three-dimensional elasto-plastic flow," International Journal of Solids and Structures, Vol. **7**, 1971, pp. 1673-1681.

48. Tanaka, M. and Tanaka, K., "On a new boundary element solution scheme for elastoplasticity," Ingenieur Archiv, Vol. **50**, 1981, pp. 289-295.

49. Telles, J.C.F. and Brebbia, C.A., "Viscoplasticity and creep using boundary elements," Progress in Boundary Element Methods - 2, Ed. C.A. Brebbia, Pentech Press, London, 1983, pp. 200-215.

50. Thomson, W. (Lord Kelvin), "Note on the integration of equations of equilibrium of an elastic solid," Cambridge and Dublin Mathematical Journal, III, 1848, pp. 87-89.

51. Tottemham, H., "The boundary element method for plates and shells," Developments in Boundary Element Methods-2, P.K. Banerjee and R. Butterfield (eds.) Elsevier Applied Science Publishers, Barking, Essex, England, 1979, pp. 173-205.

52. Willam K.J. (ed.), "Constitutive Equations - Macro and Computational Aspects," Publication #G00274. The Amercian Society of Mechanical Engineers, New York 1984.

53. Zienkiewicz, O.C., "The Finite Element Method. (3rd ed.)", McGraw Hill U.K., 1979, pp. 372.

Fracture Analysis with Interactive Computer Graphics

Anthony R. Ingraffea,[1] M. ASCE, Walter H. Gerstle,[2] A.M. ASCE, and Renato Perucchio,[3] A.M. ASCE

A system is described in which boundary elements, fracture mechanics, and interactive computer graphics are combined to address the problem of crack propagation in three dimensions. This system is composed of an interactive preprocessor, a multi-domain boundary element analysis code, an interactive post-processor, and a Fracture Editor. The Fracture Editor is the master control program which links all the system elements and controls automatic, local remeshing to accommodate each increment of crack advance. This system is illustrated with a number of practical example problems.

Introduction

A difficult class of problems in computational mechanics involves crack propagation in three dimensions. Even under the assumptions of linear elastic fracture mechanics (LEFM), difficulty arises because:

1. Preparation of the initial model for a three-dimensional structure containing one or more cracks usually requires substantial person-effort.

2. At each step of crack propagation, portions of the model must be modified to reflect changing boundary conditions or topology.

3. When symmetry conditions are not available or when crack propagation is non-planar, complex geometries can be created near the crack front.

4. Accurate computation of stress intensity factors at each point along the crack front is necessary.

5. It is difficult to visualize, and therefore to check, the model and the results of the analysis.

To circumvent these difficulties requires an integrated system involving selection of the proper numerical analysis method, an

---

[1]Associate Professor and Director, Computer Aided Design Instructional Facility, Cornell University, Ithaca, NY 14853
[2]Technical Staff, Sandia National Laboratories, Albuquerque, NM 87185
[3]Assistant Professor and Associate Director, Production Automation Project, University of Rochester, Rochester, NY 14627

appropriate computing environment, and effective means for problem description and result interpretation. The purpose of this chapter is to describe such a system. The important ingredients of the system are the boundary integral equation method, interactive mesh generation techniques, and a high level of interactive-adaptive computer graphics.

We will first describe an interactive graphic preprocessor which enables an analyst to create the initial boundary element model quickly and without bugs. The preprocessor requires minimal geometrical input from the analyst to create automatically boundary element meshes on each surface of the model. Moreover, the input is by way of a digitizing pen rather than a keyboard, and the result of each operation is immediately displayed on a vector graphics device. The facile mesh generation and modification afforded by this preprocessor effectively eliminates the first and second difficulties described above.

The boundary element method was selected to minimize difficulties three and four. When a crack, as a generally curved surface, moves through a finite element mesh, elements are sliced and distorted. The remeshing in the vicinity of this disturbance is certainly more challenging than that required for a boundary element mesh. In this case one is simply extending the crack surface into a void.

Although finite elements which reproduce the proper LEFM traction and displacement distributions near the crack front are available, placing the appropriate number of these with necessary size and shape all along the crack front is again a challenge. Traction singular, quarter-point boundary elements of simple quadrilateral and triangular shape can be more readily arrayed along an arbitrarily shaped crack front. We will show by an example problem how local remeshing and accurate stress intensity factor computation is performed in our system.

Finally, the problem of visualization for checking and for behavioral appreciation is best accomplished through interactive, three-dimensional computer graphics. The use of this tool is pervasive in our system. We have attempted through interactive graphics to maintain control of the solution of a complex problem in the hands, literally, of the user, rather than the CPU.

Preprocessing System

The development of an interactive computer graphic system for generating and editing three-dimensional meshes for boundary element analysis is described in this section. The various difficulties inherent in the generation of solid three-dimensional geometries for both finite and boundary element analysis are examined to explain the advantages of the present approach. The surface generation algorithm used for geometric modeling is based on the assumption that the three-dimensional region can be subdivided in a series of contiguous subdomains, which in turn can be separately described by a set of planar cross sections. Discrete transfinite mapping and cubic spline blending algorithms are used, respectively, to create plane cross-sectional meshes, and to generate the three-dimensional geometry by interpolating between the cross sections.

Background

Experience with the application of the finite element method has shown that the high cost of performing numerical analysis on three-dimensional domains is due largely to the effort required in defining and checking the geometrical data, element topology, boundary conditions, material properties, and loadings. In fact, the complexity of error detection and geometry modification when three-dimensional meshes are involved can substantially reduce the cost effectiveness and applicability of any finite element analysis program.

Fewer difficulties are encountered in creating a geometrical model and related database for boundary element analysis of a three-dimensional domain. In fact, the boundary element method as applied here requires only a discretization of a set of boundary surfaces which enclose the solid domain, thus eliminating the necessity of discretizing the inner field. The absence of internal nodes represents a significant relaxation of constraints on the mapping algorithms which are usually applied in the automatic generation of discrete meshes. Now nodes and topologies can be defined independently on each boundary surface and consistency requirements are limited to the nodes on the contours of adjacent surfaces.

Experience has also taught that an analysis preprocessor must be regarded as a combination of code and input techniques which best relieve the analyst from the tedious side of database generation and checking while enhancing the creative aspects of designing and modeling. From this point of view, interactive computer graphic techniques offer a highly efficient way of generating three-dimensional surfaces with the minimum amount of direct digital input, Romans, et al. (28); Grieger and Kamel (16); Kamel and Navabi (22). The interactive graphic preprocessor described here runs on a super-mini-computer and a high speed refresh vector display. Data are fed to the machine by a pen and digitizing tablet and are immediately displayed on the vector scope. The surface generation algorithm used for geometric modeling is based on the assumption that the three-dimensional region can be subdivided into a series of contiguous subdomains, which in turn can be separately described by a set of planar cross sections. The boundary surfaces of each subdomain are then approximated by lofting between the cross sections with a cubic spline interpolation scheme.

Interactive Mesh Generation

Of the two tasks that a three-dimensional interactive preprocessor is required to fulfill, i.e. generate a geometrical model and assign boundary values and material properties, the former is certainly the more complicated. Since both tablet and vector scope are basically two-dimensional devices and therefore ill-suited for handling three-dimensional geometries some appropriate strategy must be devised to permit input of geometrical and topological data without extensive use of a keyboard to type in nodal related information. Keeping in mind also that, contrary to the finite element method, the boundary element method as considered here requires only a discretization of the surfaces enclosing the three-dimensional domain, the geometrical problem can be restated simply: to create interactively a family of discretized, generally curved, surfaces.

A number of approaches toward this end have been explored. For a limited number of geometries, the discretization of an entire domain surface could be done at a global level by the interactive distortion of regular meshes previously stored, Haugerud (18); Hagen et al. (17). This fitting procedure may become cumbersome in a three-dimensional environment since the analyst will have to relocate a large set of key nodes and eventually may have to enter manually each spatial coordinate to achieve an adequate matching.

In another, more general approach mapping algorithms allow the discretization of a three-dimensional surface based on its boundary and some cross-sectional curves, Cook (7); Stanton et al. (32). However, the analyst is still left with the nontrivial task of entering a number of key curves in a three-dimensional space. This could be achieved by combining the information from a set of planar engineering views which are input through a digitizing tablet, Sagawa (29); Aarnaes (1). The spatial coordinates of key nodes can be computed automatically and a suitable mapping algorithm is then applied to generate nodes and elements on the domain surfaces. However, being strictly dependent on a set of accurate technical drawings makes this approach too rigid and time consuming for practical interactive preprocessing.

In the present approach the geometrical requirement that the mapping of each three-dimensional surface be based on a discretization of the complete family of boundary defining curves has been relaxed by assuming that a set of planar curves can describe adequately each surface. This is equivalent to the procedure used in Perucchio et al. (26) to generate solid, three-dimensional meshes for finite element analysis by discretizing cross sections into planar meshes and then using blending functions to interpolate between the sections. An important difference in the present case is that only the first and the last cross section need to be meshed. The interior cross sections are simply discretized, planar contours. The problem of generating a family of three-dimensional general surfaces which bound a solid domain is reduced to the simpler one of creating planar contours and two-dimensional meshes for which computer graphic techniques have proved extremely effective.

The analyst controls the surface discretization process by selecting and positioning the cross sections, and by grading the relative size of the lofted surface elements. After completing the lofting procedure, he can examine separately one surface at a time using the dynamic viewing capabilities and depth cueing built in the high-speed refresh vector display.

Full details of the mesh generation algorithms employed here are given in Perucchio (24); Perucchio and Ingraffea (25). The following section illustrates these mapping algorithms as implemented in a computer graphic preprocessor.

Interactive System for Preprocessing

The structure of the interactive program for creating and editing three-dimensional surface meshes for boundary element analysis is illustrated in Fig. 1. The program organization is modular in the sense that independent modules perform the three basic tasks of inputting cross-sectional data, creating a three-dimensional mesh, and assigning attributes. After selecting a task from the main control page, the analyst performs interactively all the steps required for completing the task. For instance, as illustrated in Fig. 1, to create cross-sectional data, first a grid must be defined on the x-y plane, then the outline is traced, and eventually, an element partition is generated. Since some steps may require several functions, the program structure is actually a level deeper than shown in the figure.

The modular organization has been adopted to enhance the flexibility of the preprocessor. Any module can be substituted independently from the other ones and new modules can be added to perform specific tasks. This allows the program to incorporate future developments in mesh generating techniques as well as to handle problems in applied physics which require an attribute editor different from the one currently implemented for elastostatics.

Examples

Two examples of the preprocessor modeling capability are presented in the following sections to illustrate additional features of the program, namely: the use of non-parallel cross sections, and the joining together of several models lofted independently to create a single domain.

Fontana Dam

The plan and elevation view of Tennessee Valley Authority Fontana concrete dam is shown in Fig. 2. A satisfactory modeling of the curvature between sections 5 and 6 is obtained after some experimenting with the selection of the cross sections in the proximity of the curved portion and with the definition of the related cross-sectional parametric values for the spline interpolation. A typical cross section is shown in Fig. 3 and the complete set of cross sections located in a three-dimensional reference system is illustrated in Fig. 4. Cross sections 1 through 5 are normal to the z axis and therefore are positioned by specifying their distance from the xy plane. For cross sections 6 through 8 the analyst must select three points on the two-dimensional cross section, then provide the three-dimensional coordinates of the selected points to generate the transformation matrix for the entire section.

The contours of the lofted meshes are shown in Fig. 5. Figures 6 and 7 illustrate the downstream side and the complete geometrical model, respectively. All the surface elements shown are 8-node quadratic elements.

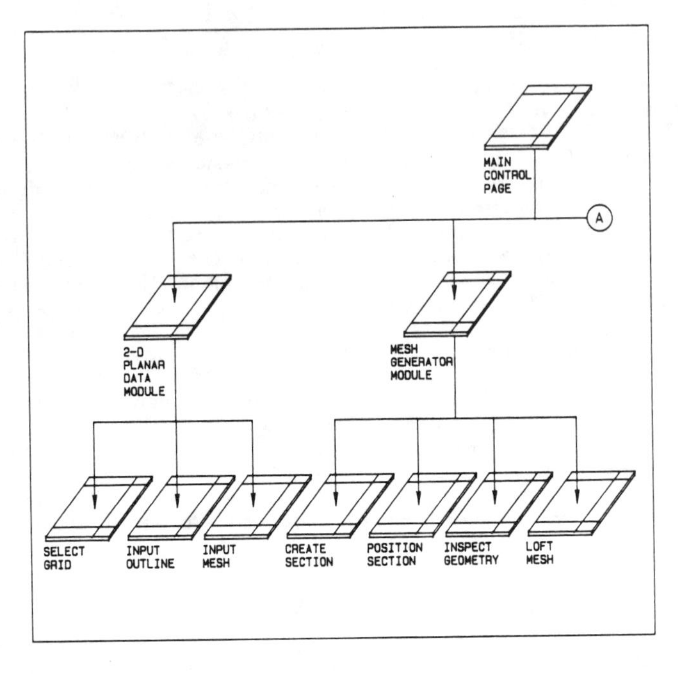

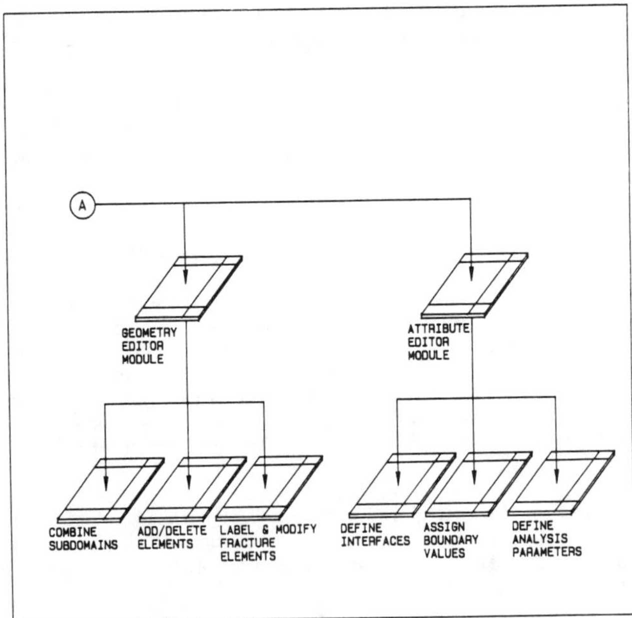

Figure 1. Preprocessor Program Structure

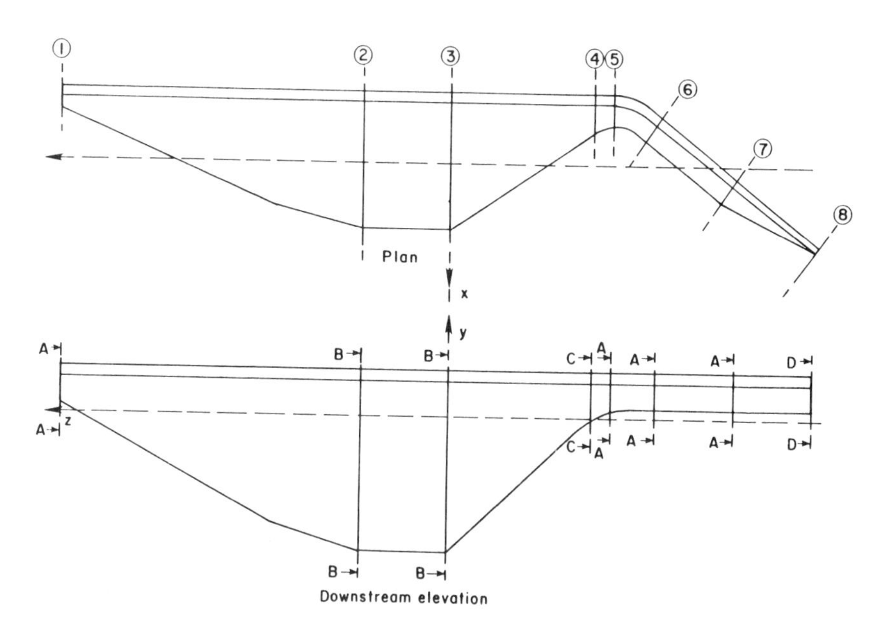

Figure 2 Plan View and Elevation of TVA Fontana Concrete Dam

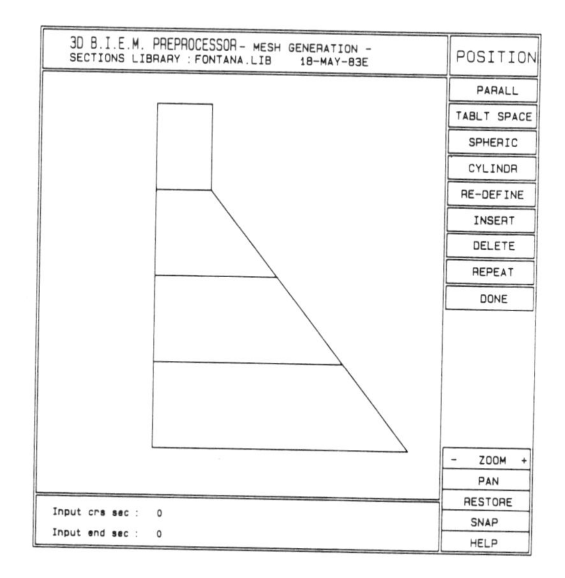

Figure 3. Typical Cross Section Used for Lofting Fontana Dam

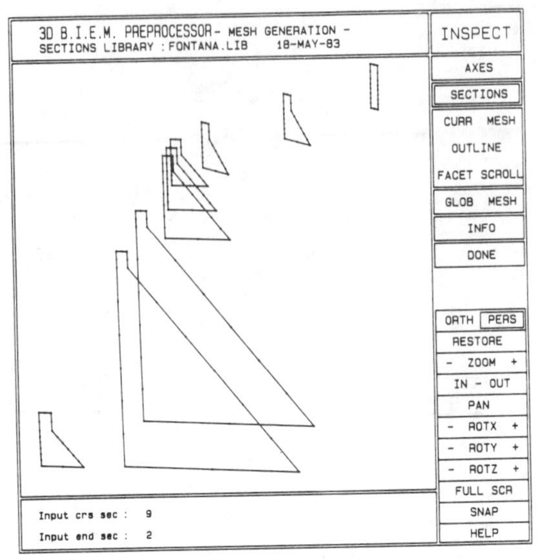

Figure 4. Spatial Arrangement of Dam Cross Sections

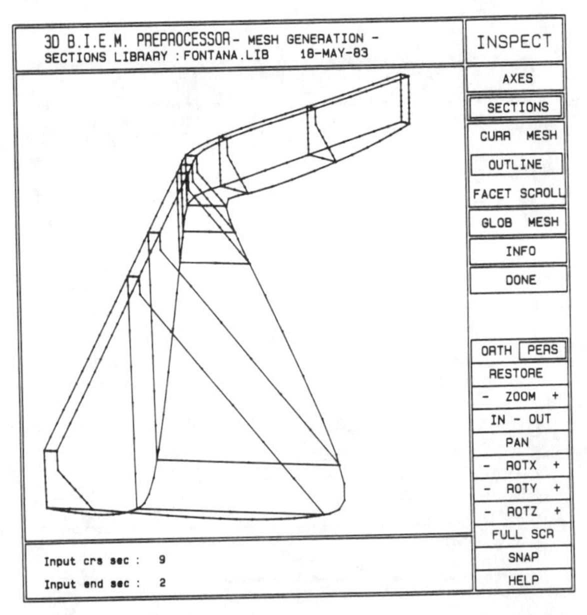

Figure 5. Dam Cross Sections and Boundary Curves of Lofted Surfaces

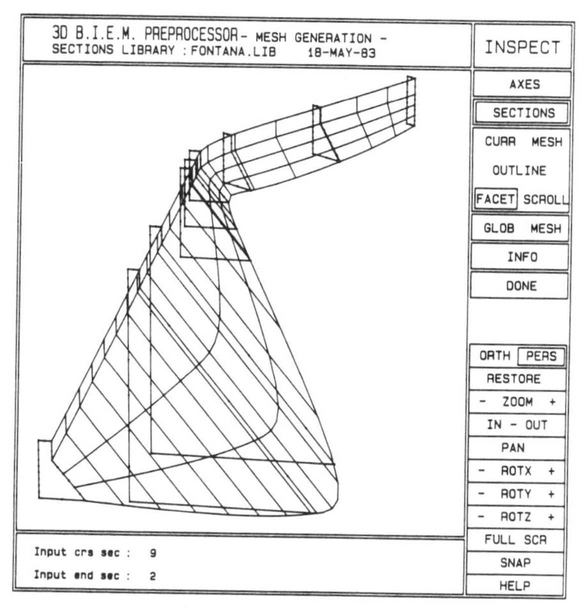

Figure 6. Dam Cross Sections and Lofted Surface Modeling the Downstream Side

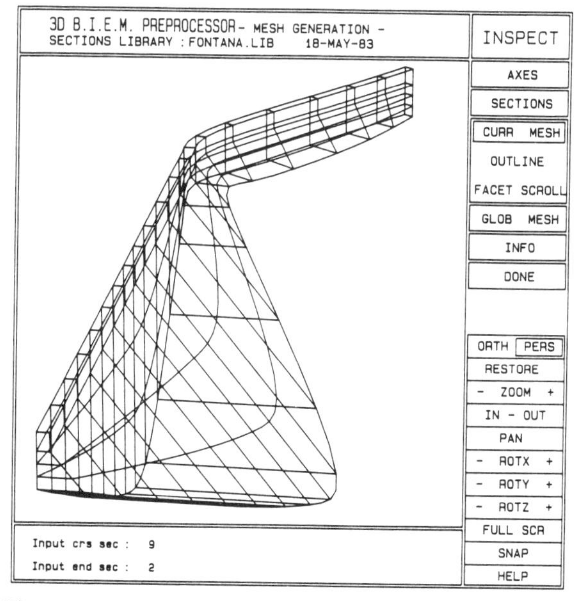

Figure 7. Global Geometrical Model of Fontana Dam

Peachtree Center Station Excavation

A perspective view of the excavation for the Peachtree Center Station of the Metropolitan Atlanta Rapid Transport Authority is shown in Fig. 8. The geometry of the tunnel system generated with the interactive preprocessor is a replica of the finite element model used in Azzouz et al. (3) for a three-dimensional analysis of the excavation.

To accommodate the geometrical discontinuities due to the intersection of different tunnels, the boundary element mesh is lofted in five pieces which are then joined together to provide a single mesh. The lofting of each piece requires the positioning of two vertical cross sections. Figure 9 shows one of the end cross sections, while a lofted mesh is illustrated in Fig. 10. The boundaries of the five partial meshes are visible within the contour of the global model, Fig. 11. Finally, the entire boundary element mesh is presented in Fig. 12.

The planning stage and the actual preprocessing required little more than six man hours. For the equivalent finite element model used in Azzouz et al. (3), the input preparation -- mesh generation and debugging runs -- was performed in two man months, as reported in Swartz et al. (33).

Stress-Intensity Factor Analysis

The next step after generation of the initial mesh and crack surfaces is to perform a stress intensity factor analysis. The stress intensity factors at each point along the crack front are required to predict the rate and duration of extension at that point. If the crack plane belongs to a plane of geometric symmetry, and the boundary conditions on the domain do not violate the symmetry conditions, then only a portion of the domain -- and only one of the crack surfaces -- needs to be modeled. Appropriate boundary conditions are attached to the symmetry planes to simulate the presence of the cut out parts of the domains which are not discretized. This approach has been followed extensively in the application of the BEM to three-dimensional problems in fracture mechanics, as by Cruse and Vanburen (9), Cruse and Meyers (8), Cruse and Wilson (10), Lange (23), Tan and Fenner (35), and Brebbia (6), among others. The drawback is that only symmetric -- mode I and mode III -- crack problems can be analyzed.

In the present work, domain substructuring has been used to separate numerically the two crack surfaces by including the crack plane within the interface of two adjoining subdomains. In this way, each crack surface belongs to a different subdomain and is treated numerically as part of a subdomain boundary. The resulting coefficient matrices at both subdomain and global level are non-singular. The present approach, while not precluding the use of symmetry conditions where appropriate, allows the BEM modeling of general -- mode I, mode II, and mode III -- problems in linear-elastic fracture mechanics. The use of domain substructuring to avoid the flat crack modeling problem was introduced by Blandford (5) for two-dimensional problem and is here extended to three-dimensional domains.

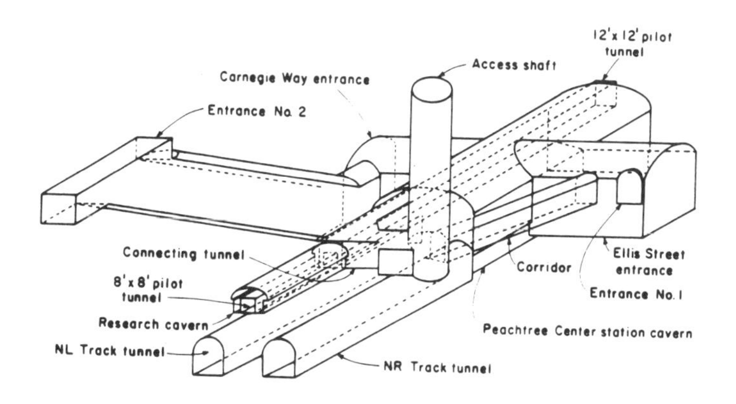

Figure 8. View of Excavation for Peachtree Center Station of Metropolitan Atlanta Rapid Transport Authority

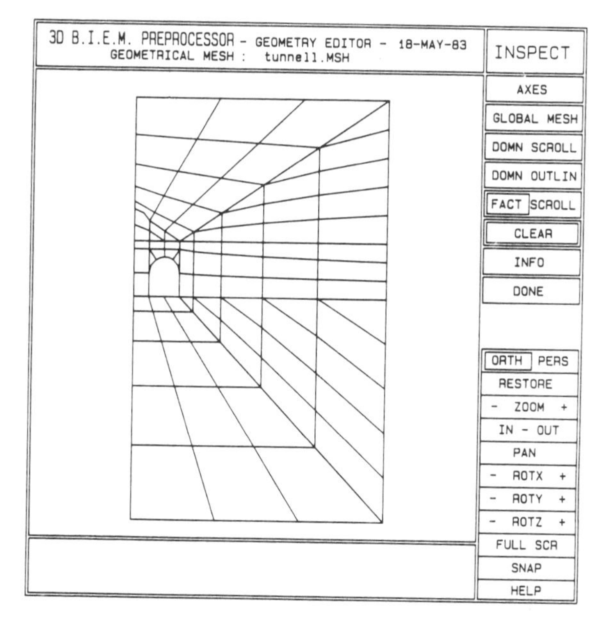

Figure 9. Planar Mesh Providing End Facet and Contours for Interior Cross Sections

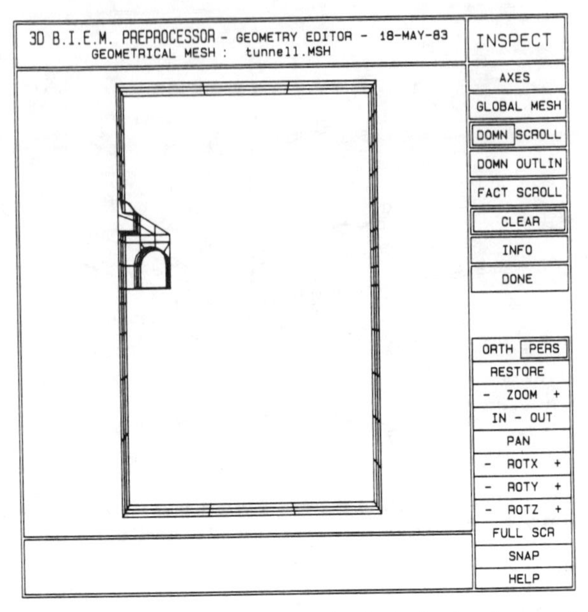

Figure 10. Perspective View of Lofted Mesh for Inner Subdomain

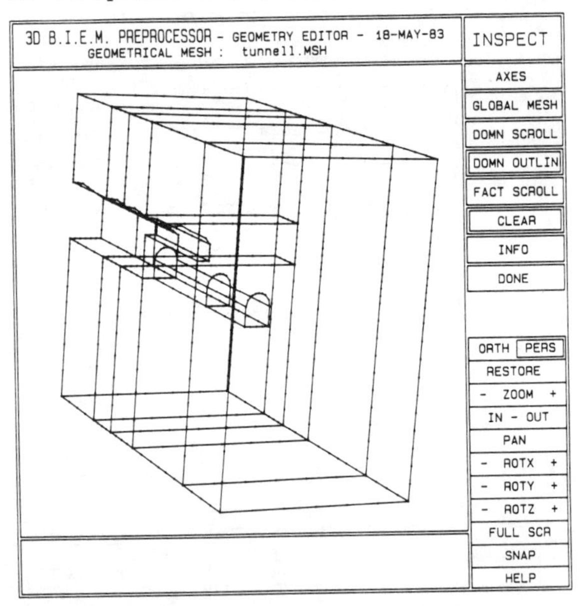

Figure 11. Boundary Curves of Global Model Showing Station Cavern, Research Chamber, Running Tunnel, and Cross-Tunnel. Outlines of Subdomains are also Displayed.

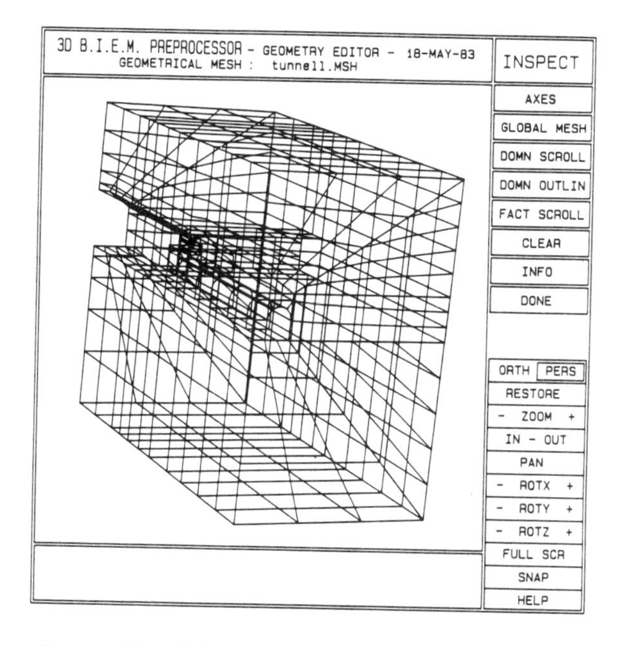

Figure 12. Global Geometrical Model of Excavation

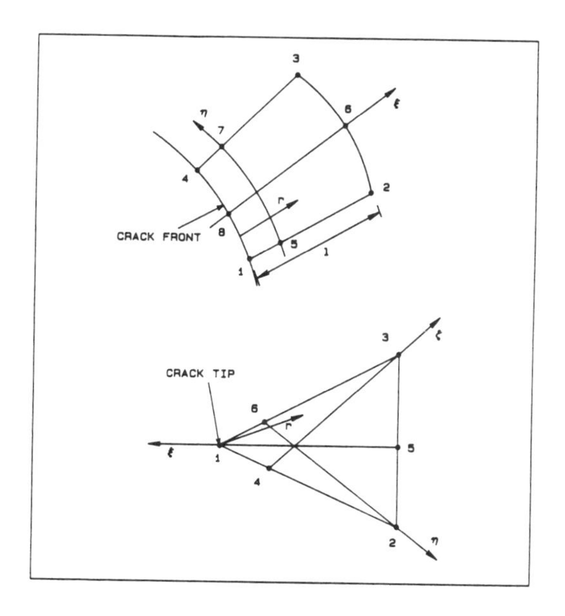

Figure 13. Quarter-Point Surface Elements

BEM Crack Elements

The modeling of the plane containing the crack surface requires the approximation of near crack tip tractions and displacements in terms of the interpolation functions used in the formulation of the boundary elements. If quadratic elements are used around the crack front, the traction and displacement fields will have at most a quadratic variation in r, r being the normal distance as defined in Fig. 13. Since the linear-elastic solution in the vicinity of the crack front requires a variation of the crack opening displacements as the square root of r and a variation of the tractions ahead of the crack as the inverse square root of r, considerable mesh refinement is necessary to reach acceptable modeling accuracy. However, since the interpolation functions do not contain the above r terms, the correct elastic solution cannot be achieved even with extreme mesh refinement.

Barsoum (4) has shown that prismatic, quadratic, isoparametric, finite elements can be made to embody the correct crack tip traction and displacement variation by repositioning certain midside nodes at the quarter-point. A similar approach can be used for transforming 6- and 8-node quadratic, isoparametric, boundary elements into fracture elements. As indicated in Fig. 13, the midpoint nodes of the element sides, normal to the crack front for the 8-node quadrilateral, or emanating from the crack tip for the 6-node triangle, are repositioned at the quarter-points on the crack side. If the sides of the 8-node element carrying the quarter-point nodes are straight segments normal to the crack front and of equal length l, the distance r from the crack front is related to the natural coordinate ξ as

$$\frac{r}{l} = \frac{(\xi + 1)^2}{4} \tag{1}$$

For the 6-node triangle, if all the sides are straight segments, as suggested by Freese and Tracey (12) for the equivalent finite element, the relation between the radial distance r and the natural coordinate ξ is given by

$$\frac{r}{c} = (1 - \xi)^2 \tag{2}$$

where c indicates the ratio η/ξ which is constant for a given direction r. Substituting Eqs. 1 and 2 into the quadratic shape functions, the relation between field variables and distance r becomes

$$\begin{Bmatrix} u(r) \\ t(r) \end{Bmatrix} = A_1 + A_2\sqrt{r} + A_3 r \tag{3}$$

The displacement variation given in Eq. 3 contains the correct square root of r term for modeling the near-front displacement field. This results in the correct inverse square root of r singularity for the traction field in the finite element applications, in which stresses are derived by differentiating the displacement interpolation functions. However, since tractions and displacements are independently modeled in the BEM, the quarter-point distortion does not produce the correct traction singularity ahead of the crack. The inclusion of the correct singular term for the 8-node element is accomplished, Cruse and Wilson (10), by multiplying the shape functions for the element tractions by

$$\phi(r) = \sqrt{\left(\frac{1}{r}\right)}\ \frac{2}{\xi + 1} \tag{4}$$

Similarly, for the 6-node triangle, the traction shape functions are multiplied by

$$\psi(r) = \sqrt{\left(\frac{c}{r}\right)} = \frac{1}{1 - \xi} \tag{5}$$

The functions $\phi(r)$ and $\psi(r)$ are $O(1/\sqrt{r})$ for r sufficiently small and assume the value of one at the element side opposite to the crack front or to the crack tip, for the 8- or 6-node element, respectively. This ensures that the correct singularity is achieved at the crack and traction continuity is preserved along the element boundaries.

The combination of quarter-point distortion and shape function modification results in a quarter-point, traction-singular boundary element characterized by the following displacement and traction variations

$$\begin{aligned} u(r) &= A_1 + A_2\sqrt{r} + A_3 r \\ t(r) &= B_1/\sqrt{r} + B_2 + B_3\sqrt{r} \end{aligned} \tag{6}$$

Special care must be taken in integrating these modified elements, because the Jacobian determinant varies rapidly in the ξ-direction, and also because the traction shape functions are singular. In an effort to produce more accurate stress-intensity factor solutions, midside fracture elements were developed, in which the geometrical, traction, and displacement shape functions in the crack-front elements are specified separately. These elements are presented in the next section.

The crack-front element integrations which have the largest effect on the stress-intensity solution are the singular integrations in which the source point and the field point lie on the same element. Error in the regular integrations, while they may be large, have a much smaller effect on the stress-intensity factor solution. It seems to be worthwhile to use 8x8 or even 12x12 regular integration of crack-front elements, because these elements contain singular traction shape-functions. Also, this will in most cases not increase the solution time unreasonably.

The singular integration of crack-front elements was found to be all important. These integrations have a very large effect on the stress-intensity solution, and also are the most difficult integrals to compute. This is because, in addition to a singularity in the traction shape functions along the crack front, the singularity in the kernel functions exists. The kernel function singularity in some cases rests on the crack front. It has been found that 12x12 or 24x24, or even adaptive integration in this case is advisable, Gerstle (13); Gerstle and Ingraffea (15).

Computation of the Stress-Intensity Factors

The stress-intensity factors are evaluated along the crack front using the displacement correlation technique developed for two-dimensional problems with quarter-point finite elements by Shih, deLorenzi, and German (30) and extended to three-dimensions by Ingraffea and Manu (21). The 8-node quarter-point boundary element, located on the crack surface and with the side AG modeling the crack front, Fig. 14, is used to illustrate how the numerical results for the displacements are correlated to the distribution of the elastic displacement fields around the crack to yield K_I, K_{II}, and K_{III}. The crack opening displacement v at a point within the element is related to the nodal values of v through quadratic interpolation functions of the parametric coordinates ξ and η. Following the notation of Fig. 14, this relationship takes the form

$$v = \frac{1}{4}(1 - \xi^2)(1 - \eta)(2v_B - v_C) + \frac{1}{4}(1 - \xi^2)(1 + \eta)(2v_E - v_F) + \frac{1}{4}(1 + \xi)\eta[(1 + \eta)\, v_F - (1 - \eta)\, v_C] + \frac{1}{2}(1 + \xi)(1 - \eta^2)\, v_D \tag{7}$$

Since the midside nodes B and E have been relocated to the quarter-point position and the geometry of the element is such that a line η = constant projects into a normal to the crack front, Eq. 1 can be used to replace ξ in terms of the distance from the crack front r and the element length L_1 to yield

$$v = [2v_B - v_C + 2v_E - v_F + v_D + \frac{1}{2}\eta(-4v_B + v_C + 4v_E - v_F) + \frac{1}{2}\eta^2\,)v_F + v_C - 2v_D)]\sqrt{\left(\frac{r}{L_1}\right)} + [(\eta - 1)(2v_B - v_c) - (1 + \eta)(2v_E - v_F)]\frac{r}{L_1} \tag{8}$$

To derive K_I, consider the near-crack-front displacement fields associated with the three modes of deformation. At θ = 180 deg. the contribution of mode II to the v displacement vanishes. Since no contribution to v comes from mode III, the crack opening displacement reduces to

$$v = \frac{1 - \nu}{G}\sqrt{\left(\frac{2r}{\pi}\right)}\, K_I \tag{9}$$

Equating like powers of r in Eqs. 8 and 9 the following expression is obtained for K_I

$$K_I = \frac{G}{1 - \nu} \sqrt{\left(\frac{\pi}{2L_I}\right)} [2v_B - v_C + 2v_E - v_F + v_D$$

$$+ \frac{1}{2} \eta(-4v_B + v_C + 4v_E - v_F)$$

$$+ \frac{1}{2} \eta^2(v_F + v_C - 2v_D)]. \tag{10}$$

Following the same procedure for the crack-sliding, u, and the crack-tearing, w, displacements, similar expressions are derived for K_{II} and K_{III}. For the general problem in which symmetry cannot be invoked and both crack surfaces have to be modeled, the stress-intensity factors are given by

$$K_I = \frac{E}{4(1 - \nu^2)} \sqrt{\left(\frac{\pi}{2L_I}\right)}$$

$$[2v_B - v_C + 2v_E - v_F + v_D - 2v_{B'} + v_C + v_{F'} - v_{D'}$$

$$+ \frac{1}{2} \eta(-4v_B + v_C + 4v_E - v_F + 4v_{B'} - v_{C'} - 4v_{E'} + v_{F'})$$

$$+ \frac{1}{2} \eta^2(v_F + v_C - 2v_D - v_{F'} - v_{C'} + 2v_{D'})] \tag{11}$$

$$K_{II} = \frac{E}{4(1 - \nu^2)} \sqrt{\left(\frac{\pi}{2L_I}\right)}$$

$$[2u_B - u_C + 2u_E - u_F + u_D - 2u_{B'} + u_C - 2u_{E'} + u_{F'} - u_{D'}$$

$$+ \frac{1}{2} \eta(-4u_B + u_C + 4u_E - u_F + 4u_{B'} - u_{C'} - 4u_{E'} + u_{F'})$$

$$+ \frac{1}{2} \eta^2(u_F + u_C - 2u_D - u_{F'} - u_{C'} + 2u_{D'})] \tag{12}$$

$$K_{III} = \frac{E}{4(1 + \nu)} \sqrt{\left(\frac{\pi}{2L_I}\right)}$$

$$[2w_B - w_C + 2w_E - w_F + w_D - 2w_{B'} + w_{C'} + 2w_{E'} + w_{F'} - w_{D'}$$

$$+ \frac{1}{2} \eta(-4w_B + w_C + 4w_E - w_F + 4w_{B'} - w_{C'} - 4w_{E'} + w_{F'})$$

$$+ \frac{1}{2} \eta^2(w_F + w_C - 2w_D - w_{F'} - w_{C'} + 2w_{D'})] \tag{13}$$

where the prime refers to the nodes on the crack surface element opposite to the one shown in Fig. 14.

As demonstrated in Ingraffea and Manu (21) for the finite element application, this technique allows the functional computation of stress-intensity factors along curved as well as straight crack fronts modeled with isoparametric quarter-point elements. Since quadratic shape functions are used to interpolate the displacements in the η direction, a quadratic variation of the stress-intensity factors can also be accommodated within the element.

Example Problem: Buried Circular Crack at 60 Degree Angle to Applied Load

A buried circular crack angled with respect to the direction of the load applied at infinity experiences all three modes of deformation. The exact solution for the intensity factors, Tada (34), is given by

$$K_I = \frac{2}{\pi} (\sigma \sin^2 \gamma)\sqrt{(na)} = A$$

$$K_{II} = \frac{4}{\pi(2 - \nu)} (\sigma \sin \gamma \cos \gamma) \cos \omega\sqrt{(\pi a)} = B \cos \omega \qquad (14)$$

$$K_{III} = \frac{4(1 - \nu)}{\pi(2 - \nu)} (\sigma \sin \gamma \cos \gamma) \sin \omega\sqrt{(\pi a)} = C \sin \omega$$

where a is the crack radius, γ the angle between crack plane and load direction, ω the polar angle as shown in Fig. 15, and σ the nominal load. The numerical modeling of the problem with either finite or boundary elements constitutes a formidable task. In fact, since no symmetry is present, the entire domain has to be discretized into a finite mesh. To represent accurately the condition on infinity, the overall size of the model has to be large with respect to the crack. At the same time, a refined partitioning may be necessary in the crack region to accommodate the sinusoidal variation of K_{II} and K_{III}. Clearly, finite element modeling of this problem is prohibitive unless some form of graphic preprocessing is available. Moreover, experience with quarter-point solid elements for analysis involving mode I, mode II, and mode III combined is almost non-existent and, therefore, there is no guarantee that a very refined mesh will not be needed to assure acceptable results.

The complete geometrical model, given in Fig. 16, is a parallelepiped in which the ratio between minor side and crack radius is set to 20 to reproduce with good approximation the conditions of infinity. The controlling geometrical feature is provided by the planar cross section, containing the crack, which separates the model into two subdomains. The mesh on this cross section is created according to the experience developed with previous problems, i.e. each singular element bordering the crack front covers a 45. degree arc and the ratio L/a is equal to 0.3. Moreover, the elements on the section are organized into two different topologies: the crack surface and the interface between subdomains. This cross section, positioned into space at an angle γ = 60. degree with respect to the longitudinal axis of the model, becomes

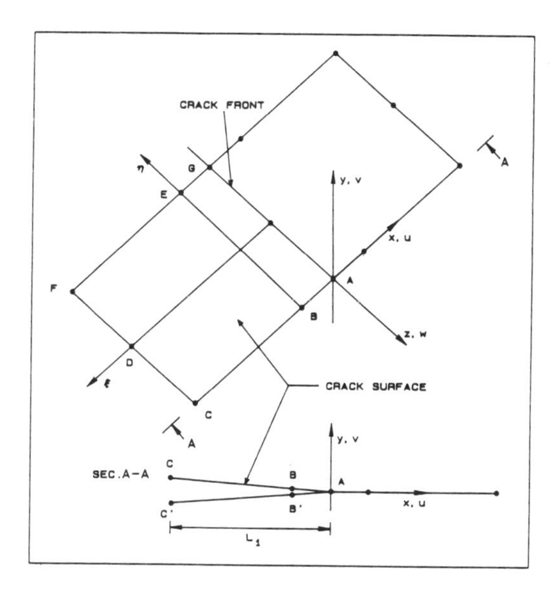

Figure 14. Nodal Lettering for Quarter-Point Elements Along Crack Front

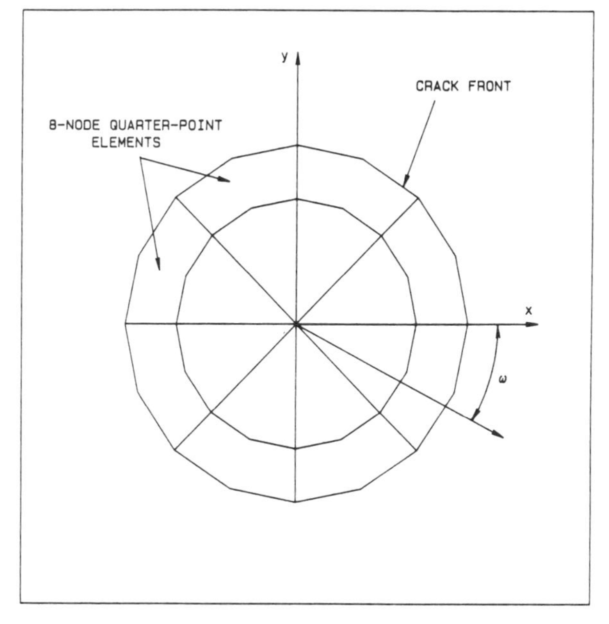

Figure 15. Angled Circular Crack: Element Distribution on the Crack Surface

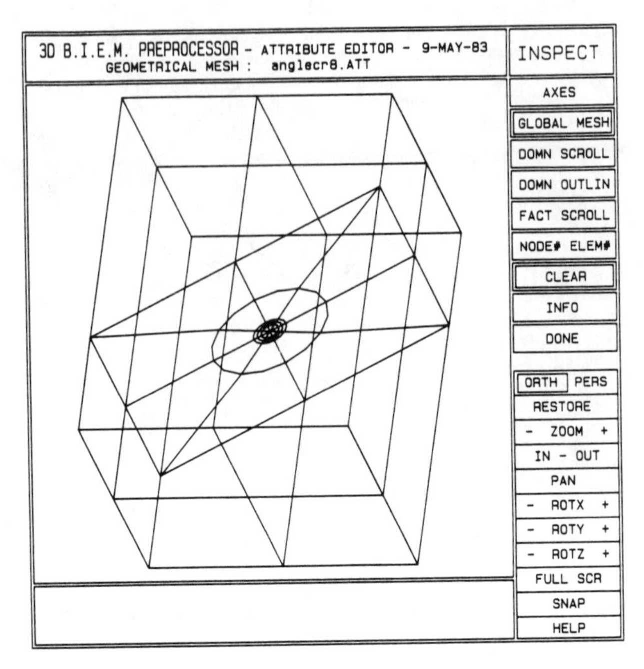

Figure 16. Global Element Mesh for Buried Circular Crack Angled with Loading Direction

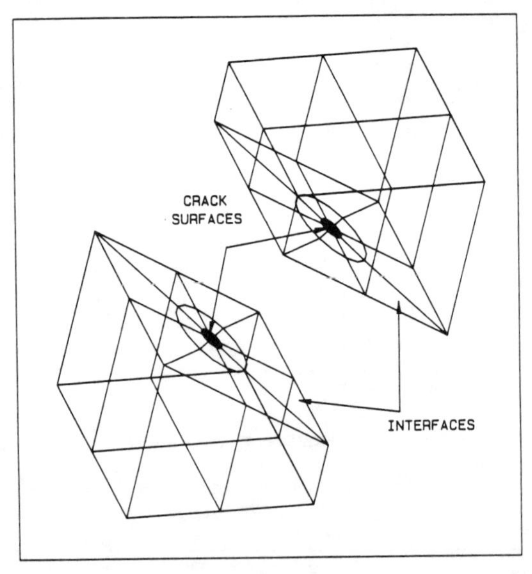

Figure 17. Exploded View of Subdomain Meshes for Angled Circular Crack Problem

part of both subdomains such that exactly the same nodes and elements are produced on the slanted side of each subdomain mesh, Fig. 17. Interface nodes on one subdomain are numerically bound to the matching nodes on the interface of the other subdomain. Instead, corresponding nodes on the two crack surfaces are left free to displace one with respect to the other in order to model the discontinuity due to the crack. The preprocessing time for mesh generation and attribute editing is about 35 minutes. Each subdomain is made up of 150 nodes and 52 quadratic elements, with 88 nodes belonging to the interface. The computer stress-intensity factors are normalized as K_I/A, K_{II}/B, and K_{III}/C and plotted as functions of the polar angle ω in Fig. 18. The scaling factors A, B, and C are evaluated as in Eq. 14. The results reproduce accurately the distribution of the stress-intensity factors along the crack front. The percent error for K_I ranges between 1.0% and 1.6%, while the maximum computed values for K_{II} and K_{III} differ from the equivalent exact values of 1.2% and 7.1%, respectively.

Fracture Propagation System

We have to this point shown how quick mesh generation and accurate calculation of stress intensity factors are accomplished The combination of these two capabilities is the kernel of a crack propagation processing system: model the initially cracked structure, compute its stress intensity factors, from those predict the new crack shape, re-model the cracked structure, and continue. Control of this sequential process is embedded in a program called the Fracture Editor, Gerstle (13).

Fracture Propagation Processor

The Fracture Editor is the tool by which interactive mesh modification is accomplished. It has two primary functions: mesh modification, and post-processing. Thus, after each crack propagation increment, the results of the previous analysis may be viewed by the user, and then, based upon the input of the user, the mesh may be interactively modified to allow the crack to propagate or to refine the mesh. Once the problem definition has been updated the problem is re-analyzed using BEM3D.

The concepts of pre-processing and post-processing, Abel (2), become somewhat blurred in the Fracture Editor. The Fracture Editor can be viewed as simply a boundary element processor, with both pre- and post-processing capabilities. Each of its important features are discussed in the next two sections, making use of Figs. 19 to 26, which show most of its menu pages.

The first menu page (Fig. 19) is the main Fracture Editor page. In this page the user scrolls to the domain on which he wants to work. Also, on this page, he can manipulate the object using the transformation menu options. These options include viewing the mesh from any desired position by using the rotate, pan, and zoom features. Also, these include choice of orthogonal or perspective views of the mesh, resetting the viewing transformations, and taking a hardcopy

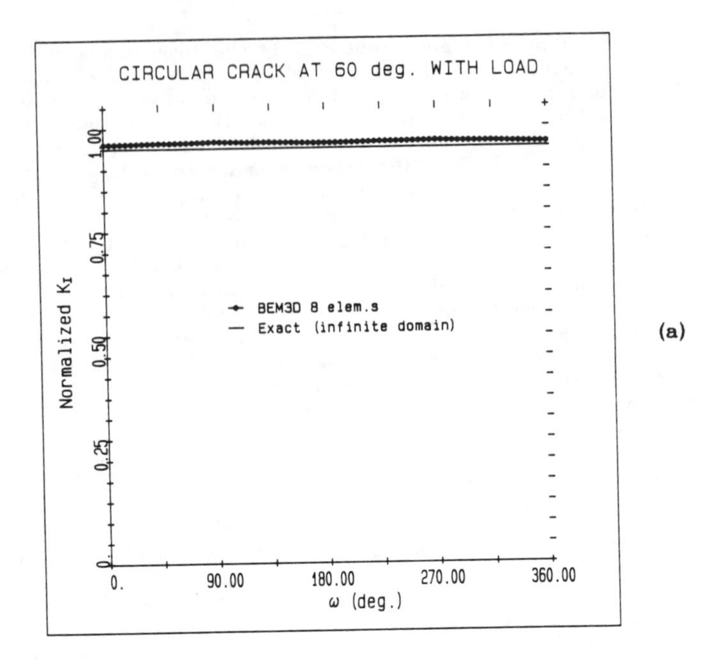

(a)

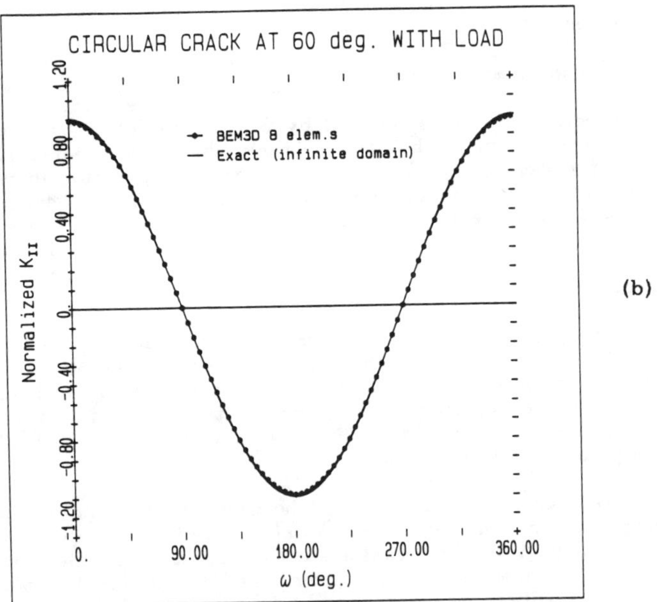

(b)

Figure 18. Angled Circular Crack: a) Distribution of Normalized K_I; b) Distribution of Normalized K_{II}

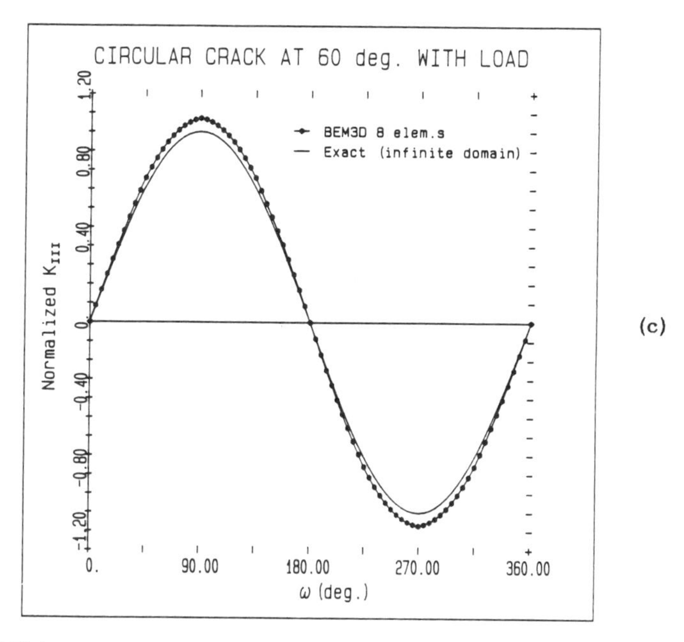

(c)

Figure 18(c): Angled Circular Crack: Distribution of Normalized K_{III}

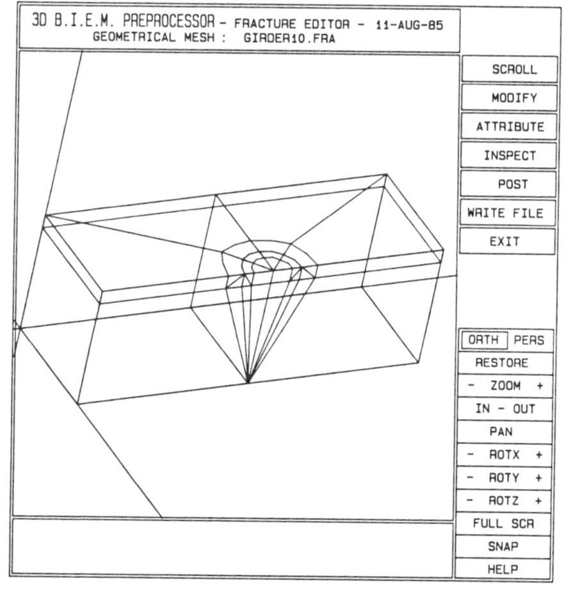

Figure 19. Main Menu Page of the Fracture Editor

picture of the picture on the vector scope. These features are common to all menu pages called from this menu page.

From the main Fracture Editor page, the user can go to any of four menu pages, entitled "MODIFY," "ATTRIBUTE," "INSPECT," and "POST." These menu pages are described in turn next.

The "MODIFY" page (shown in Fig. 20) has the purpose of allowing the user to interactively modify the mesh. This page allows the user to scroll through the facets of the current domain, and calls subsequent menu pages which are described next.

The "CRACK" menu page, shown in Fig. 21, allows the user to point at the curve in space which is the crack front. This page has the function of performing node-dragging operations necessary to satisfy the requirements of the crack-front elements. To extract stress-intensity factors using the displacement correlation method outlined in the previous section, these elements must have sides radiating from the crack front normal to the crack front. Also, the midside nodes must be shifted to the quarter points, and a constraint is placed upon the position of the midside node situated farthest from the crack front. These operations are all performed automatically.

The "EDIT FACET" page, shown in Fig. 22, allows the user to modify the mesh on any facet chosen. The facet to be edited is first transformed into the viewing plane. Then, the user can drag nodes (using either the pen or the keyboard to input the new coordinates of the node), add nodes, delete nodes, add elements, delete elements, and update the nodal connectivity of the curves contained by this facet. After the facet is edited, it is transformed back into three-dimensional space, and necessary modifications are made to the data base to connect the facet to the domain mesh.

The "MIDSIDE" option on the "MODIFY" page automatically moves all side nodes to their midside locations in the entire mesh (except for the midside nodes on the crack front, if it has already been defined). This feature may occasionally be useful in remeshing.

The "PROPAGATE" page is shown in Fig. 23. This page allows the user to choose a set of fracture propagation criteria. Based on these criteria, the crack front is automatically advanced. Presently, the growth models available are the Paris fatigue crack growth model, Rolfe and Barsom (27), and a simple quasi-static crack front incrementing model. The angle models determine which mixed-mode crack propagation theory is to be used: planar propagation, maximum tensile circumferential stress theory, Erdogan and Sih (11); minimum strain energy density theory, Sih (31); or maximum energy release rate theory, Hussain et al. (19). The cyclic model, used only in the case of fatigue crack growth, allows either constant or varying amplitude load application input.

The "PROPAGATE" page has elements of both pre- and post-processing. This is because the crack front position is changed (pre-processing) based upon the stress-intensity factors (post-processing) from the previous crack increment analysis results.

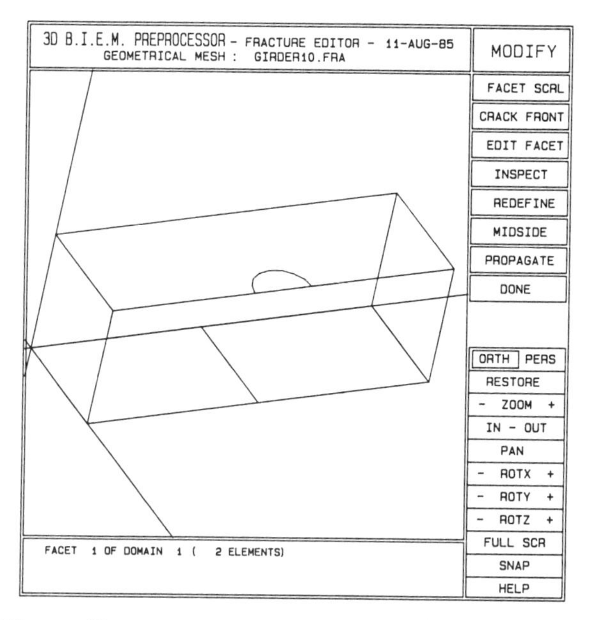

Figure 20. MODIFY Menu Page of the Fracture Editor

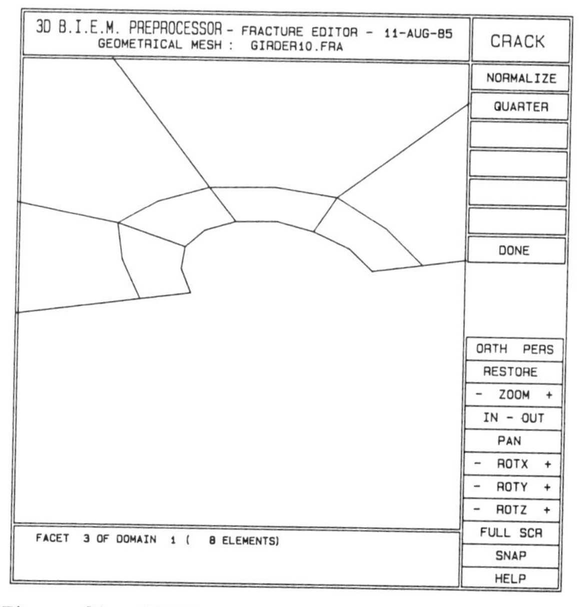

Figure 21. CRACK Menu Page of the Fracture Editor

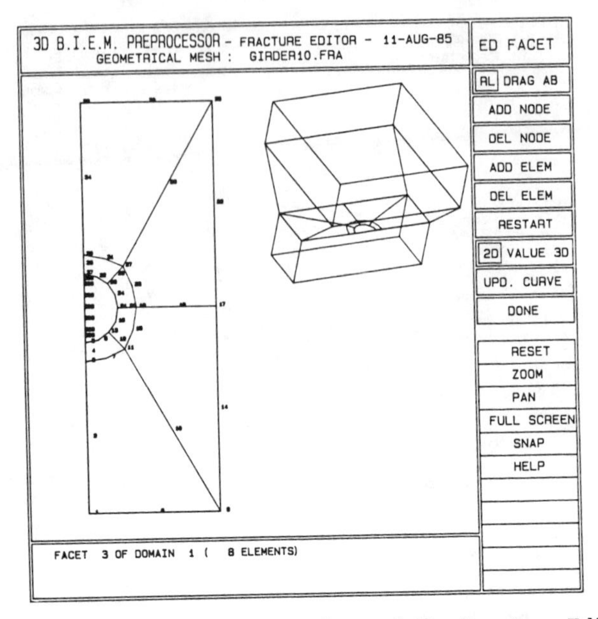

Figure 22. EDIT FACET Menu Page of the Fracture Editor

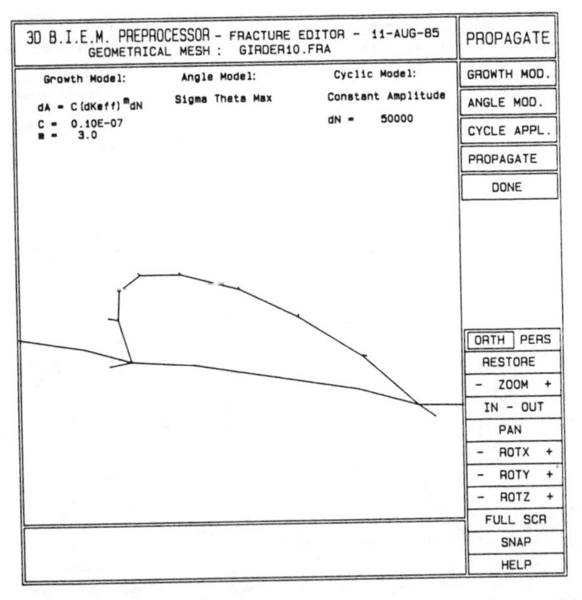

Figure 23. PROPAGATE Menu Page of the Fracture Editor

The "ATTRIBUTE" option on the "MAIN MENU" page calls the attribute editor, written by Perucchio in Perucchio and Ingraffea (25). This option is necessary because after the mesh has been modified new elements and nodes must have their attributes (specified boundary conditions) defined or redefined.

The "POST" option, shown in Fig. 24 is for post-processing the results of an analysis, and is discussed in the next section.

Post-processing Features

The post-processing menu page of the Fracture Editor is used to view the results of an analysis.

The deformed mesh may be displayed to any desired amplification. The deformed mesh is very useful for rapidly viewing the results of an analysis to determine that the results are plausible, and that no mistake has been made in applying boundary conditions.

The stress-intensity factors, K_I, K_{II}, and K_{III} may be displayed along the crack front, as shown on Fig. 24. These quantities are displayed at each node along the crack front as both numbers and as vectors whose lengths are proportional to the stress-intensity factor magnitude being displayed at that point.

The "Tx Ty Tz" option flashes the tractions as vectors at each node in the domain mesh.

The "SCRL POST2D" option allows the user to scroll through the facets of the current domain, select one of them, and then enter the "POST 2D" page. This page, shown in Fig. 25, transforms the selected facet into the viewing plane. Then the user can specify a straight line between two points on the facet, and a two-dimensional plot of a specified quantity (a traction or a displacement component) will be automatically created, as shown in Fig. 25.

Application of these Fracture Editor capabilities to a specific problem is shown in Figures 20 through 25. These are images created during the solution of the example problem presented next.

Example Problem: Crack Propagation in a Crane Runway Girder

Approach to Analysis

The idealized problem specification is shown in Fig. 26. A crane wheel is assumed to roll along a 175 lb. rail, over a certain location along a crane runway girder. An initial semicircular flat crack is assumed to exist at this location, at the bottom toe of the full penetration weld between the top flange and the web. The resultant of the load applied by the wheel has an eccentricity of one inch with respect to the centroidal axis of the girder. The girder is stiffened vertically at 72.0 in. spacing.

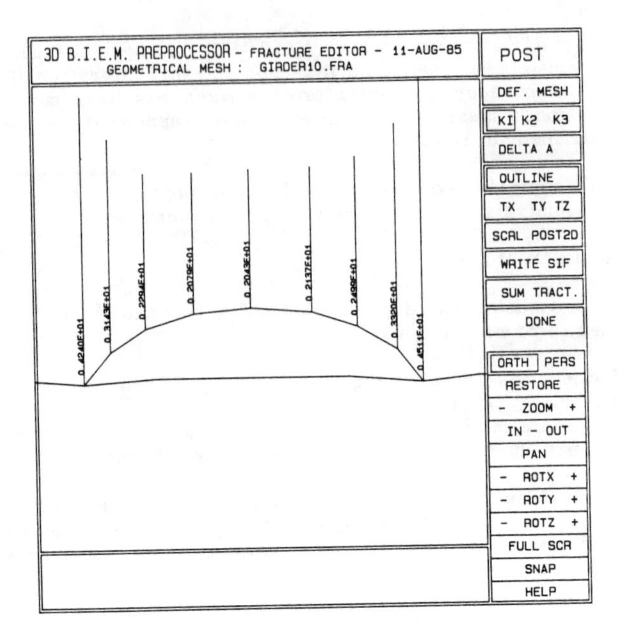

Figure 24. POST Menu Page of the Fracture Editor

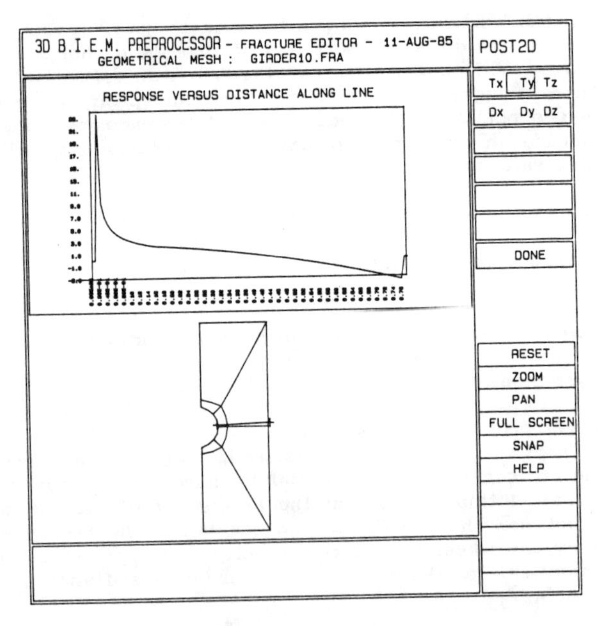

Figure 25. POST2D Menu Page of the Fracture Editor

The semi-circular edge crack at the weld toe, of assumed initial radius 0.25 inches, is assumed to grow according to the Paris model, even though the stresses resulting from applied loading in this region are compressive. Because very high residual tensile stresses are known to exist in the weld material of this region, the total vertical normal stresses are probably tensile through a large portion of this connection even as the wheel rolls directly over the crack.

A report and a paper for the Association of Iron and Steel Engineers, Gerstle and Ingraffea (14); Ingraffea et al. (20), document analysis and experiments to determine the vertical normal stresses in the web-to-flange junction of crane runway girders. In Gerstle and Ingraffea (14), several design curves were presented. These design curves give the vertical stress- and moment-distributions between the web and flange due to applied vertical forces and moments to the top of the crane rail. For the specific loading of 100 kips and applied moment of 100 kip-inches, these design aids give the stress distribution through the thickness of the web indicated in Fig. 26. These stresses were applied as traction boundary conditions to the bottom of the web in the small subobject, as shown in Fig. 26. The top of the subobject, which is cut out of the top flange, was fixed at all locations where the cut was made. The ends of the object were given traction-free boundary conditions.

Crack propagation was modelled, as explained in the next section, using the boundary element fracture processor. Complete details of the analysis process can be found in Gerstle (13).

Crack Propagation Analysis

Figures 27 to 29 show the boundary element mesh at crack propagation increment numbers 1, 3, 5, 7, 9, 10, and 11. One-half million cycles of load were applied during each increment, except in increments 10 and 11, where one million cycles were applied. The crack is seen to dip slightly into the web as it advances. The crack propagation direction was determined using the maximum circumferential tensile stress theory.

Figure 30 shows a projection onto the horizontal plane of the crack front shape at each increment. Also shown in this figure are the calculated mode I stress-intensity factor ranges at each increment of crack growth. The Paris model is valid for this material for stress-intensity fluctuations up to 90 ksi-$\sqrt{in}$ and the lower threshold for fatigue propagation is about 5 ksi-$\sqrt{in}$. Therefore, for the specified applied loadings, the use of the Paris model was justified.

Figure 31 shows a plot of normalized mode I stress-intensity at the center of the crack front versus normalized crack length. This plot can be used to determine for a web-flange connection of any absolute size, but geometrically similar to the one under consideration, what the through-web crack growth rate would be.

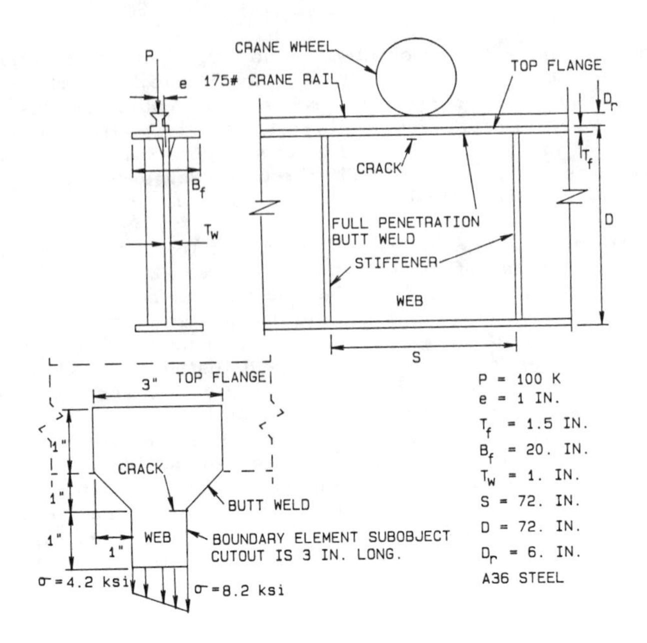

Figure 26. Crane Runway Girder Problem Specification

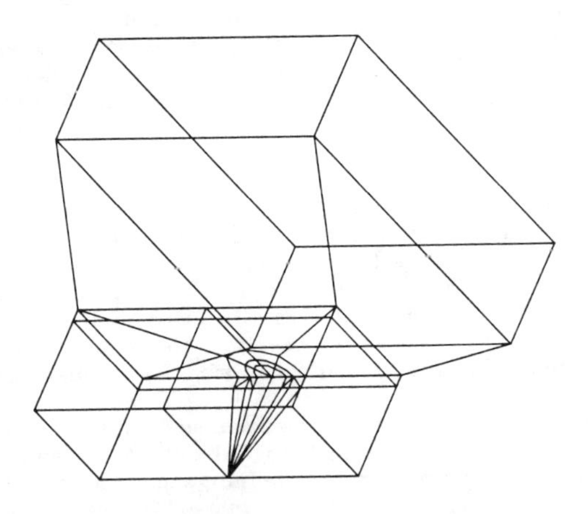

Figure 27. Boundary Element Mesh of Cracked Crane Girder. Initial Crack Radius is 0.25 Inches. Only Outline of Top Domain is Shown for Clarity.

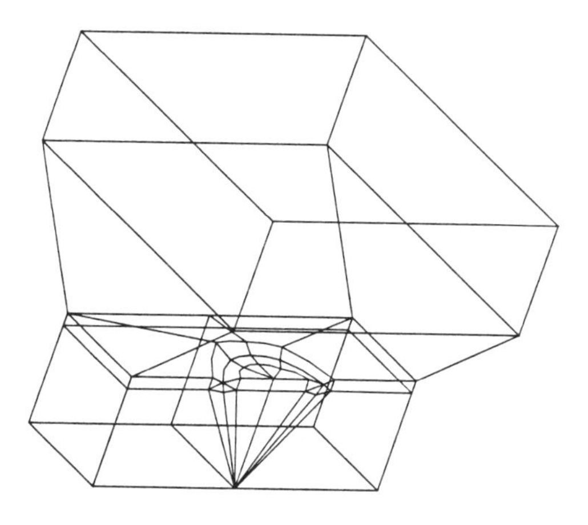

Figure 28. Boundary Element Mesh After Three Million Cycles of Applied Load. Seventh Crack Propagation Increment.

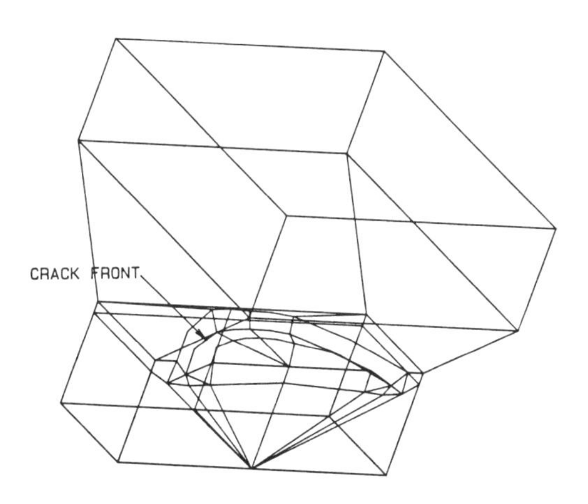

Figure 29. Boundary Element Mesh After 6.5 Million Cycles of Applied Load. Eleventh Crack Propagation Increment.

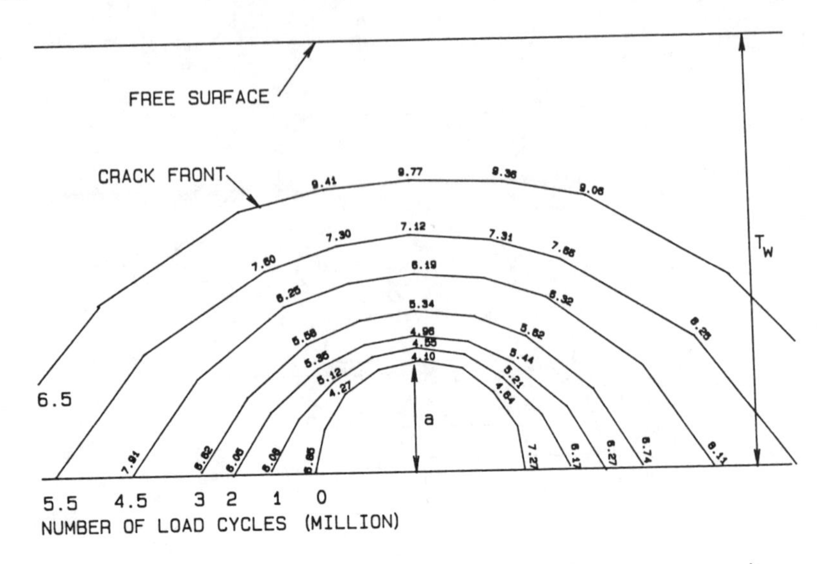

Figure 30. Projected Crack Shapes and Mode I Stress-Intensity Factors for Crack Increments 1, 3, 5, 7, 9, 10, and 11. Number of Cycles to Each Crack Increment Also Shown.

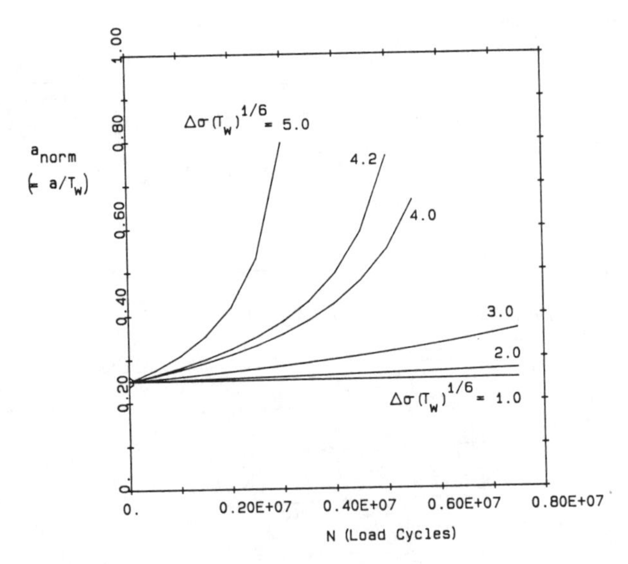

Figure 31. Normalized Crack Size Versus Number of Load Cycles N for Crack Propagation in Crane Runway Girder.

DISCUSSION

In this chapter we hope to have shown how an integrated, software/hardware system approach can be effective in solving problems of crack propagation in three dimensions. Analysis code for computing mode I stress intensity factors is only one of the major components of such a system. Equally important for attacking practical problems efficiently are:

- An interactive-graphical preprocessor which is used both to generate the initial model and to assist in mesh modification which accompanies crack growth.
- An accurate and quick means for computing *mixed-mode* stress intensity factors functionally along the crack front.
- An interactive-graphical post-processor which is necessary for checking the accuracy of the model and for effective communication of intermediate and final results.
- A Fracture Editor, or master control program, which links all the system elements together, and which controls automatic, local remeshing to accommodate each increment of crack advance.
- A graphics hardware environment which provides visualization of the complex, changing model and allows a real-time appreciation of predicted behavior.

This system is not a set of unconnected programs. Rather, its single data base and computational environment provide for a continuum of analyst interaction with the model.

Of course what we have described can benefit greatly from both technical and computational environment advances. On the technical side we would like to be able to include additional realistic effects such as residual stress fields in the Example Problem presented. Currently used models for both rate and direction of crack advance stem from two-dimensional idealizations: more rigorous models accounting for truly three-dimensional behavior are desirable.

The system described here operates in a *super-minicomputer* environment. For simple single-domain problems, preprocessing, post-processing, and fracture editing would typically require just a few minutes each for each crack step. However, actual analysis time for even a simple problem could well take an hour per step. In the multi-domain Example Problem presented, each analysis step required about five hours of wall-clock time. Such a delay effectively negates all the control and behavior appreciation advantages otherwise afforded by an interactive-graphical analysis environment. A solution to this problem, one being currently implemented at Cornell, is to add a *mini-supercomputer* to the modeling system. All of the interactive-graphical tasks are still performed in a graphics workstation environment; but, when number crunching time comes, the boundary element analysis is performed on an attached super-processor. In tests to date, such a processor has achieved solution time reduction factors of 100 to 200: the larger the problem the higher the reduction factor.

As a consequence of both hardware and software advances such as those described here, we predict that fully three-dimensional crack propagation modeling in complex structures will be available in near real-time within the next five years. We will be able to add more physical realism, more accuracy, and more excitement to the simulation process.

Acknowledgements

The development of the fracture analysis system reported here was made possible by Grants CME 79-16818, CEE 83-16730, and NSF 83-51914 by the National Science Foundation. The research was performed in the laboratories of the Cornell University Program of Computer Graphics.

References

1. Aarnaes, E., "Intersection Between Two Free Form Surfaces. Digitalization of 3-D Curves," Proceedings of the 4th International Conference and Exhibition on Computers in Design Engineering, CAD80, Brighton, UK, 1980, pp. 243-250.

2. Abel, J. F., "Interactive Computer Graphics for Applied Mechanics," Proceedings of the Ninth U.S. National Congress of Applied Mechanics, Cornell University, Ithaca, New York, June 21-25, 1982.

3. Azzouz, A. S., Schwartz, C. W., and Einstein, H. H., "Finite Element Analysis of the Peachtree Center Station in Atlanta," Report No. UMTA-MZA-06-0100-80-6, Department of Civil Engineering, Massachusetts Institute of Technology, Cambridge, Massachusetts, 1980.

4. Barsoum, R. S. "On the Use of Isoparametric Finite Elements in Linear Fracture Mechanics," International Journal for Numerical Methods in Engineering, Vol. 10, 1976, pp. 25-37.

5. Blandford, G. E., "Automatic Two-Dimensional Quasi-Static and Fatigue Crack Propagation Using the Boundary Element Method," Ph.D. Thesis and Research Report 81-3, Department of Structural Engineering, Cornell University, Ithaca, New York, 1981, 276 pp.

6. Brebbia, C. A. and Walker, S., Boundary Element Techniques in Engineering, Newnes-Butterworths, London, 1980.

7. Cook, W. A., "Body Oriented (Natural) Co-ordinates for Generating Three-Dimensional Meshes," International Journal for Numerical Methods in Engineering, Vol 8, 1974, pp. 27-43.

8. Cruse, T. A. and Meyers, G. J., "Three-Dimensional Fracture Mechanics Analysis," Journal of the Structural Division, ASCE, Vol. 103, 1977, pp. 309-320.

9. Cruse, T. A. and Van Buren, W., "Three-Dimensional Elastic Stress Analysis of a Fracture Specimen with an Edge Crack," International Journal of Fracture Mechanics in Engineering, Vol. 7, 1971, pp. 1-15.

10. Cruse, T. A. and Wilson, R. B., Boundary-Integral Equation Method for Elastic Fracture Mechanics Analysis, AFOSR-TR-78-0355, 1977.

11. Erdogan, F. and Sih, G. C., "On the Crack Extension in Plates Under Plane Loading and Transverse Shear," ASME J. Basic Engineering, Vol. 85, 1963, pp. 519-527.

12. Freese, C. E. and Tracey, D. M., "The Natural Isoparametric Triangle Versus Collapsed Quadrilateral for Elastic Crack Analysis," International Journal of Fracture, Vol. 12, 1976, pp. 767-770.

13. Gerstle, W. H., "Finite and Boundary Element Modelling of Crack Propagation in Two- and Three- Dimensions Using Interactive Computer Graphics," Ph.D. Thesis and Research Report 85-8, Department of Structural Engineering, Cornell University, Ithaca, New York, 1985, 220 pp.

14. Gerstle, W. H., and Ingraffea, A. R., "Numerical Modelling of Forces Transmitted to the Web to Flange Junction of Crane Runway Girders Due to Wheel Loads," AISE/Cornell University Crane Rail Girder Project, Document 84-3, Task II, Report No. 1, May 12, 1984.

15. Gerstle, W. H. and Ingraffea, A. R., "Error Control in Three-Dimensional Crack Modelling Using the Boundary Element Method," to appear in Proceedings of the Symposium on Advanced Topics in Boundary Element Analysis, ASME, Orlando, Florida, November 1985.

16. Grieger, I. and Kamel, H. A., "Application of Interactive Graphics and Other Programming Aids," in Special AMD Publication on State-of-the-Art Surveys in F.E.M., A. H. Noor ed., 1981.

17. Hagen, K., Henriksen, T., Pihlfeldt, A.-K., and Pahle, E., "3D Geometry Generator. A Study of Various Alternatives of Interactive Element Mesh Generation Systems for Solids and Shell Structures," Computas, Det Norske Veritas, Technical Report CP 77-902, Hovik, Norway, 1977.

18. Haugerud, M. H., "Interactive 3D Mesh Generation by an Idealization and Mapping Technique," Proceedings of the 3rd International Conference on Computers in Engineering and Building Design, CAD78, Brighton, UK, 1978, pp. 753-766.

19. Hussain, M. A., Pu, S. L., and Underwood, J. H., "Strain Energy Release Rate for a Crack Under Combined Mode I and Mode II," Fracture Analysis, ASTM, STJP 560, 1974, pp. 2-28.

20. Ingraffea, A. R., Gerstle, W. H., Mettam, K. I., Wawrzynek, P., and Hellier, A. K., "Cracking of welded crane runway girders: Physical testing and computer simulation," Iron and Steel Engineer, Vol. 62, No. 12, December 1985, pp. 46-52.

21. Ingraffea, A. R. and Manu, C., "Stress-Intensity Factor Computation in Three Dimensions with Quarter-Point Elements," International Journal for Numerical Methods in Engineering, Vol. 15, 1980, pp. 1427-1445.

22. Kamel, H. A. and Navabi, Z., "Digitizing for Computer-Aided Finite Element Generation," Journal of Mechanical Design, ASME, Special Edition, 1980.

23. Lange, D., "3-D Fracture Analysis Using the Boundary Integral Equation Method," Proceedings of the International Conference on Numerical Methods in Fracture Mechanics, A. R. Luxmore and D.R.J. Owen eds., Swansea, Wales, January 1978, pp. 115-127.

24. Perucchio, R., An Integrated Boundary Element Analysis System with Interactive Computer Graphics for Three-Dimensional Linear-Elastic Fracture Mechanics, Ph.D. Thesis and Research Report 84-2, Department of Structural Engineering, Cornell University, Ithaca, NY, 1984, 226 pp.

25. Perucchio, R. and Ingraffea, A. R., "Interactive Computer Graphics Preprocessing for Three-Dimensional Boundary Integral Element Analysis," Journal of Computers and Structures, Vol. 16, No. 1-4, 1983, pp. 153-166.

26. Perucchio, R., Ingraffea, A. R., and Abel, J. F., "Interactive Computer Graphic Preprocessing for Three-Dimensional Finite Element Analysis," International Journal for Numerical Methods in Engineering, Vol. 18, 1982, pp. 909-926.

27. Rolfe, S. T. and Barsom, J. M., "Fracture and Fatigue Control in Structures; Applicability of Fracture Mechanics," Prentice-Hall, 1977.

28. Romans, G., Iveson, F., and de Groot, H., "Graphic Systems and Their Effect on the Finite Element Method," in Finite Element Methods in the Commercial Environment, J. Robinson ed., Proceedings of the 2nd World Congress on F.E.M., Bournemouth, England, October 1978.

29. Sagawa, K., "Automatic Mesh Generation for Three Dimensional Structures Based on Their Three Views," in Theory and Practice in Finite Element Structural Analysis, Y. Yamada and R. H. Gallagher eds., University of Tokyo Press, 1973, pp. 687-702.

30. Shih, C. F., DeLorenzi, H. G., and German, M. D., "Crack Extension Modeling with Singular Quadratic Isoparametric Elements," International Journal of Fracture, Vol. 12, 1976, pp. 647-651.

31. Sih, G. C., "Strain Energy Density Factor Applied to Mixed Mixed-Mode Crack Problems," International Journal of Fracture, Vol. 10, 1974, pp. 305-321.

32. Stanton, E. L., Crain, L. M., and Neu, T. F., "A Parametric Cubic Modelling System for General Solids of Composite Material," International Journal for Numerical Methods in Engineering, Vol. 11, 1977, pp. 653-670.

33. Swartz, C. W., Azzouz, A. S., and Einstein, H. H., "Example Cost of 3-D FEM for Underground Openings," Journal of the Geotechnical Engineering Division, ASCE, Vol. 108, No. GT9, September 1982, pp. 1186-1191.

34. Tada, H., Paris, P., and Irwin, G., The Stress Analysis of Crack Handbook, Del Research Corporation, Hellerton, Pennsylvania, 1973.

35. Tan, C. L. and Fenner, R. T., "Elastic Fracture Mechanics Analysis by the Boundary Integral Equation Method," Proceedings of the Royal Society of London, Ser. A 369, 1979, pp. 243-260.

Dynamic Soil-Structure Interaction

John L. Tassoulas,* A.M. ASCE

Introduction

The study of dynamic soil-structure interaction addresses the effects of the flexibility of the soil on the response of structures to dynamic loads. For a long time, these effects were disregarded. In the analysis of structures subjected to earthquakes, for example, the motion at any point on the soil-structure interface was assumed to be equal to an expected or recorded ground motion at the site before construction. Of course, this assumption is strictly correct, if the soil is very stiff, almost rigid. However, if the soil is deformable, there may be significant spatial variation of the motion even before the structure is built. The behavior of the structure may be altered substantially. Furthermore, the structure may modulate the motion of the ground. If the excitation of a soil-structure system originates somewhere in the soil, as in the case of an earthquake, the deviation of the interface motion from what would have occured had the structure not been there can be defined as a measure of the interaction. On the other hand, if the structure is the driving component of the system, e.g., because of a vibrating machine located in the structure, the interface motion can be thought of as an indicator of the magnitude of the interaction. In the last fifteen years, the behavior of soil-structure systems has been a subject of intensive research. The effects of the coupling between the structure and the flexible soil have been found to be particularly significant when considering very stiff and massive structures supported on relatively soft soil.

The analysis of dynamic soil-structure interaction is not an easy matter. Perhaps the most significant complication arises from the difficulty in characterizing the behavior of the soil. The experimental data is often inadequate. Also, it is hard to devise constitutive equations that capture all the material nonlinearities. Another complication appears inevitable because of the possible occurence of separation between the soil and the structure. Of course, separation may be followed by reattachment. The modelling of such events is by no means straightforward. Complications due to nonlinear structural behavior are not to be left out

*Associate Professor, Department of Civil Engineering, The University of Texas, Austin, Texas 78712.

of the list. Moreover, if the response of a soil-structure system to an earthquake is to be analyzed, the definition of the seismic input is far from being clear.

Unfortunately, even if the analyst is willing to get started accepting an idealization of the soil as a linearly elastic, or linearly viscoelastic solid, assuming no separation between the soil and the structure, neglecting the nonlinearities due to the structure and adopting a seismic input composed of some incident waves that match a ground acceleration time history, the task is still not an easy one. At this level of simplification of the analysis, the major complication is due to the extent of the soil. Perhaps the most reasonable assumption is that the soil region is unbounded. Thus, widely accepted idealizations of the soil domain include the half space and the stratum, the latter being appropriate, if the soil is underlain by very stiff rock at some depth. For realistic modelling, horizontal layering is often considered. Of course, the solution of a problem posed on an infinite domain is not a trivial matter. Most of the techniques employed in the study of dynamic soil-structure interaction are based on the finite element method. Certainly, finite elements are versatile, especially when dealing with complex geometries or material inhomogeneities. However, the treatment of an infinite domain using finite elements is not straightforward. In most of the applications of the finite element method to problems of dynamics of soil-structure systems, a neighborhood of the structure, the near field, is defined by introducing an artificial boundary. The complementary soil region, the far field, remains unbounded. Although the near field is readily dealt with using finite elements, the success of the analysis relies heavily on the conditions imposed on the artificial boundary. The ground motion due to soil-structure interaction must satisfy the radiation condition. Thus, waves impinging on the artificial boundary must be transmitted into the far field. This transmission is at best only simulated by applying the so-called transmitting, absorbing, nonreflecting or silent conditions on the artificial boundary. By their derivation, these conditions are aimed at capturing the damping characteristics of the far field. More often than not, this is done well. However, the stiffness and mass of the far field are left out of the picture and the results of the analysis may be in substantial error.

In recent years, the boundary element method has emerged as a powerful tool in the analysis of dynamic soil-structure interaction. It is founded on the theorem of reciprocity between elastodynamic states. Fundamental solutions, i.e., Green functions, are used as virtual states. The method is particularly appealing in problems posed on an infinite domain. Unlike the finite element method, it requires discretization of the boundary of the domain, not of the interior. Moreover, the solution obtained by applying the reciprocity theorem satisfies the radiation condition. If use is made of the fundamental solutions corresponding to the particular idealization of the soil region, e.g., a layered half space or stratum, only discretization of the soil-structure interface is necessary. Thus, the computational effort may be reduced significantly.

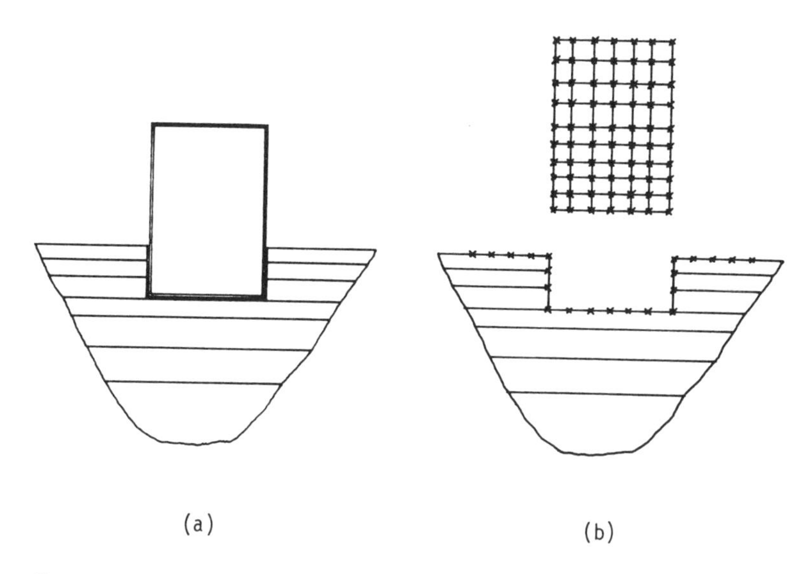

Figure 1: (a) a soil-structure system, (b) discretization of the system.

This chapter deals with the use of the boundary element method in the study of dynamic soil-structure interaction. For the purposes of the discussion, a typical soil-structure system is sketched in Figure 1a. The soil and the structure are usually treated separately, perhaps by different methods, and then coupled at the interface. It is not within the scope of this chapter to present arguments against or for any method of treatment of the structure. The finite element method has been used extensively in structural analysis and the discretization shown in Figure 1b stems from this fact. Of course, the choice depends largely on the structure under consideration. It should be noted that it may be convenient to treat some of the soil around the structure by the same method, especially if severe soil inhomogeneities are present in a neighborhood of the structure. In regard to the complementary infinite soil region, the preference towards the boundary element method cannot be concealed in the present discussion. Attention will be focused on this region. Boundary elements will be used on the soil-structure interface and, if necessary, on the ground surface as shown in Figure 1b. The soil will be assumed linearly elastic or linearly viscoelastic. In most of the applications to date, the boundary element method has been used in time-transformed problems, i.e., in the frequency domain, followed by inversion back to the time domain. Recently, the method was applied directly in the time domain. Both time-harmonic and transient vibrations will be considered in this chapter. Examples of applications to problems of dynamics of foundations and underground structures will be discussed.

Also, the coupling of boundary elements and finite elements will be examined.

There are several summaries of the progress that has been made in the study of dynamic soil-structure interaction, e.g., References 8, 15, 27, 28, 36, 37, 38, and 40. The reader is referred to these sources for enlightenment on the various facets of the subject.

Notation

The infinite soil region will be denoted by Ω and its boundary by S. In most of the discussion, rectangular Cartesian coordinate systems (x_1,x_2) and (x_1,x_2,x_3) will be employed for two and three dimensions respectively. Accordingly, indicial notation will be used. Repetition of an index will imply summation over the range of the index. The displacement and traction components or their amplitudes (Fourier or Laplace transforms) will be denoted by u_i and T_i respectively. Furthermore, B_i will denote the body force density components or their amplitudes. The components of a unit vector in the direction of the outward normal to the boundary S will be denoted by n_i. A point with coordinates x_i will be identified by $\underset{\sim}{x}$. Finally, $\hat{u}_i$ and $\hat{T}_i$ will denote the displacement and traction components or their amplitudes in the free field, e.g., due to an earthquake. The assumption is that this free field motion would be observed at the site, if the structure were not built there. Of course, the traction components are equal to zero on the ground surface but not necessarily on the soil-structure interface, if the structure is embedded or underground.

Frequency Domain

The reciprocity relation between the transforms of any two elastodynamic states in Ω is given by:

$$\int_S T_i^{(1)} u_i^{(2)}\, dS + \int_\Omega B_i^{(1)} u_i^{(2)}\, d\Omega = \int_S T_i^{(2)} u_i^{(1)}\, dS + \int_\Omega B_i^{(2)} u_i^{(1)}\, d\Omega \qquad (1)$$

Consider the state corresponding to a force with components f_j concentrated at a point $\underset{\sim}{X}$. This state is usually referred to as a fundamental solution. Let the resulting displacements and tractions be given by $U_{ji}f_j$ and $T_{ji}f_j$ respectively, where U_{ji} and T_{ji} denote the x_i-components of the displacement and traction vectors for a unit force in the x_j-direction. Also, consider the state which is due to soil-structure interaction, i.e., the difference between the actual state and the state in the free field. The associated displacements and tractions are given by $u_i - \hat{u}_i$ and $T_i - \hat{T}_i$ respectively. According to equation (1), the reciprocity relation between the two states can be written as:

$$C_{ji}(\underset{\sim}{X}) f_j [u_i(\underset{\sim}{X}) - \hat{u}_i(\underset{\sim}{X})] + \int_S T_{ji} f_j (u_i - \hat{u}_i)\, dS = \int_S U_{ji} f_j (T_i - \hat{T}_i)\, dS \quad (2)$$

assuming that the body force density is equal to zero in Ω. If $\underset{\sim}{X}$ is in the interior of Ω, then $C_{ji} = \delta_{ji}$, where $\delta_{ji} = 1$, if $i = j$, and $\delta_{ji} = 0$, if $i \neq j$. On the other hand, if $\underset{\sim}{X}$ is on S, then C_{ji} depends on the geometry of S at $\underset{\sim}{X}$ and on the conditions that the fundamental solution may be satisfying on S. This matter will be taken up later in the discussion.

Equation (2) forms the basis for applications of the boundary element method. The boundary S is covered with elements and some interpolation of u_i and T_i is assumed in each element in terms of the values at the nodes of the element. As it can be seen in equation (2), continuity of the approximations of u_i and T_i on S in not required. The equation is applied as many times as the number of nodal values of the displacement or traction components by considering the fundamental solutions corresponding to forces of unit amplitude in each one of the coordinate directions at each node. Thus, a system of algebraic equations relating the nodal values of the traction components to those of the displacement components is obtained:

$$\underset{\sim}{H}\,(\underset{\sim}{U} - \underset{\sim}{\hat{U}}) = \underset{\sim}{G}\,(\underset{\sim}{T} - \underset{\sim}{\hat{T}}) \tag{3}$$

where: $\underset{\sim}{H}$ and $\underset{\sim}{G}$ are square matrices of order 2N or 3N for two or three dimensions respectively, N being the number of nodes; $\underset{\sim}{U}$ and $\underset{\sim}{T}$ are the vectors of nodal values of the displacement and traction components; $\underset{\sim}{\hat{U}}$ and $\underset{\sim}{\hat{T}}$ are the vectors of nodal values of the displacement and traction components in the free field. It should be noted that the interpolation of $\hat{u}_i$ and $\hat{T}_i$ is usually convenient but not necessary since they are assumed to be known everywhere. In fact, the contribution of the free field motion can be evaluated directly as implied by equation (2). The resulting system of algebraic equations is:

$$\underset{\sim}{H}\,\underset{\sim}{U} - \underset{\sim}{G}\,\underset{\sim}{T} = \underset{\sim}{F} \tag{4}$$

where $\underset{\sim}{F}$ is a vector corresponding to the free field motion. In equation (3), $\underset{\sim}{F}$ is approximated by $\underset{\sim}{H}\,\underset{\sim}{\hat{U}} - \underset{\sim}{G}\,\underset{\sim}{\hat{T}}$. For the sake of completeness, the calculation of $\underset{\sim}{H}$ and $\underset{\sim}{G}$ will be discussed briefly. Let the matrices be partitioned into 2 by 2 or 3 by 3 submatrices for two or three dimensions respectively. As usual, it is assumed that the numbering of the rows in $\underset{\sim}{H}$ and $\underset{\sim}{G}$ is identical to the numbering of the displacement and traction components and that rows and columns corresponding to a node are contiguous. The submatrices $\underset{\sim}{H}^{mn}$ and $\underset{\sim}{G}^{mn}$ (row m, column n) can be calculated by applying equation (2) with the fundamental solution corresponding to node m and evaluating the contribution associated with node n. Let N_n be the interpolation function corresponding to node n. It is understood that this function is identically equal to zero on boundary elements which are not connected to node n, it is equal to 1 at node n, it vanishes at other nodes of the elements connected to node n but may be nonzero at other points of these elements. Let node m be at $\underset{\sim}{X}_m$. Equation (2) yields:

$$H_{ji}^{mn} = \int_{S_n} T_{ji}(\underset{\sim}{x};\underset{\sim}{X}_m) N_n(\underset{\sim}{x})\, dS_n \quad (m \neq n) \tag{5}$$

$$H_{ji}^{mn} = \int_{S_n} T_{ji}(\underset{\sim}{x};\underset{\sim}{X}_m)N_n(\underset{\sim}{x})\, dS_n + C_{ji}(\underset{\sim}{X}_m) \quad (m = n) \tag{6}$$

$$G_{ji}^{mn} = \int_{S_n} U_{ji}(\underset{\sim}{x};\underset{\sim}{X}_m)N_n(\underset{\sim}{x})\, dS_n \tag{7}$$

where S_n denotes the union of the boundary elements connected to node n while $U_{ji}(\underset{\sim}{x};\underset{\sim}{X}_m)$ and $T_{ji}(\underset{\sim}{x};\underset{\sim}{X}_m)$ are the values of $\underset{\sim}{U}_{ji}$ and $\underset{\sim}{T}_{ji}$ at $\underset{\sim}{x}$ (on S_n) for unit forces applied at X_m. The expressions in equations (5-7) appear deceptively simple. Although straightforward Gaussian integration may be sufficient for the calculation of submatrices which are off the diagonal, i.e., $m \neq n$, care must be exercised when dealing with the submatrices located on the diagonal, i.e., $m = n$. The difficulty arises because of the strong singularity of T_{ji} and the weak one of U_{ji} at $\underset{\sim}{X}_m$. Ways to treat these singularities will be discussed below.

After the displacement and traction components are determined on S, the displacements at any point $\underset{\sim}{X}$ in the interior of Ω can be calculated from equation (2):

$$u_j(\underset{\sim}{X}) = \int_S U_{ji}(\underset{\sim}{x};\underset{\sim}{X})(T_i - \hat{T}_i)\, dS - \int_S T_{ji}(\underset{\sim}{x};\underset{\sim}{X})(u_i - \hat{u}_i)\, dS + \hat{u}_j(\underset{\sim}{X}) \tag{8}$$

Of course, the calculations in this equation are not hampered by any singularities.

To complete the description of the procedure for applying the boundary element method in the frequency domain, the fundamental solutions which have been employed in various applications to problems of dynamic soil-structure interaction will be examined.

Fundamental Solution for the Full Space

Most of the applications to date have made use of the fundamental solution for the full space. This is known as the Stokes fundamental solution and applies to a homogeneous, isotropic, linearly viscoelastic medium. The convenience in using this solution is due to the fact that it is available explicitly in terms of elementary (three dimensions) or special (two dimensions) mathematical functions which can be evaluated easily. Of course, use of the Stokes fundamental solution requires discretization of the entire boundary. This means that, if the soil region is idealized as a homogeneous half space, the soil-structure interface as well as the ground surface must be covered with boundary elements. Examples of this use of the full space fundamental solution can be found in the work by Dominguez (12,13) who was apparently first to apply the boundary element method as described above in problems of dynamic soil-structure interaction. It was found in that work that only part of the surface of the half space near the soil-structure interface must be discretized because the influence of boundary elements covering the free surface of the half space decays reasonably fast with distance from the interface, especially in three dimensions. This point will be illustrated in an example later in this chapter. If the soil region is idealized as a stratum, the base of the region, i.e., the interface between the soil and the underlying rock which is assumed to be rigid must be discretized as well. This is the case in the recent study by Von Estorff and Schmid (14). If the soil region is layered or, in general, contains subregions of different materials, the interfaces between the subregions must be discretized and the boundary element method must be applied to each subregion. An example of this situation can be found in the investigation by Abascal and Dominguez (1).

The displacements and tractions at $\underset{\sim}{x}$ due to unit forces at $\underset{\sim}{X}$ in the full space (see References 10 and 11) are given by:

$$U_{ji} = \frac{1}{\alpha\pi\rho C_T^2} [\ \psi\delta_{ji} - \chi \frac{\partial r}{\partial x_i} \frac{\partial r}{\partial x_j}\ ] \tag{9}$$

$$\begin{aligned} T_{ji} = \frac{1}{\alpha\pi} [\ & (\frac{d\psi}{dr} - \frac{1}{r}\chi)\ (\delta_{ji} \frac{\partial r}{\partial n} - \frac{\partial r}{\partial x_i} n_j) \\ & - \frac{2}{r}\chi\ (n_i \frac{\partial r}{\partial x_j} - 2 \frac{\partial r}{\partial x_i} \frac{\partial r}{\partial x_j} \frac{\partial r}{\partial n}) \\ & - 2 \frac{d\chi}{dr} \frac{\partial r}{\partial x_i} \frac{\partial r}{\partial x_j} \frac{\partial r}{\partial n} \\ & + (C_L^2/C_T^2 - 2)\ (\frac{d\psi}{dr} - \frac{d\chi}{dr} - \frac{\alpha}{2r}\chi) \frac{\partial r}{\partial x_j} n_i] \end{aligned} \tag{10}$$

where: α is equal to 2 for a full space in plane strain and 4 in three dimensions, r is the distance between x and X:

$$r = [(x_i - X_i)(x_i - X_i)]^{1/2}$$

and

$$\frac{\partial r}{\partial n} = \frac{\partial r}{\partial x_i} n_i$$

ρ is the mass density, C_L and C_T are the longitudinal and transverse wave speeds respectively:

$$C_T = (G/\rho)^{1/2}$$

$$C_L = [2(1-\nu)/(1-2\nu)]C_T$$

G and ν being the shear modulus and Poisson's ratio. In the case of plane strain, the functions χ and ψ can be expressed as:

$$\chi = K_2(\frac{i\omega r}{C_T}) - (C_T^2/C_L^2)\, K_2(\frac{i\omega r}{C_L})$$

$$\psi = K_0(\frac{i\omega r}{C_T}) + (C_T/i\omega r)\, [K_1(\frac{i\omega r}{C_T}) - (C_T/C_L)K_1(\frac{i\omega r}{C_L})]$$

while for three-dimensional problems:

$$\chi = (-\frac{3C_T^2}{\omega^2 r^2} + \frac{3C_T}{i\omega r} + 1)\, \frac{\exp(-i\omega r/C_T)}{r}$$

$$- (C_T^2/C_L^2)(-\frac{3C_L^2}{\omega^2 r^2} + \frac{3C_L}{i\omega r} + 1)\, \frac{\exp(-i\omega r/C_L)}{r}$$

$$\psi = (1 - \frac{C_T^2}{\omega^2 r^2} + \frac{C_T}{i\omega r}\,)\, \frac{\exp(-i\omega r/C_T)}{r}$$

$$-(C_T^2/C_L^2)(-\frac{C_L^2}{\omega^2 r^2} + \frac{C_L}{i\omega r}\,)\, \frac{\exp(-i\omega r/C_L)}{r}$$

ω being the frequency (K_0, K_1, K_2 are modified Bessel functions of order 0, 1, 2). To simulate damping in the soil, the shear modulus is usually assumed to be of the form:

$$G = G_0[1 + 2i\beta \mathrm{sgn}(\omega)]$$

G_0 being the value at $\omega = 0$, i.e., the static value (sgn denotes the sign function). The constant β is the damping ratio.

For a full space in antiplane shear, the (out-of-plane) displacement and the traction are given by:

$$U_{33} = \frac{1}{2\pi\rho C_T^2}\, K_0(\frac{i\omega r}{C_T}) \tag{11}$$

$$T_{33} = -\frac{i\omega}{2\pi C_T}\, K_1(\frac{i\omega r}{C_T})\, \frac{\partial r}{\partial n} \tag{12}$$

It can be shown that, as $r \rightarrow 0$, the displacement and traction components, in the case of plane strain, behave as:

$$U_{ji} \sim -\frac{1}{8\pi G(1-\nu)}\, [\ (3-4\nu)\ \ln(r)\ \delta_{ji} - \frac{\partial r}{\partial x_i}\, \frac{\partial r}{\partial x_j}\ ] \tag{13}$$

$$T_{ji} \sim -\frac{1}{4\pi(1-\nu)}\frac{1}{r}\left[(1-2\nu)\frac{\partial r}{\partial n}\delta_{ji} + 2\frac{\partial r}{\partial x_i}\frac{\partial r}{\partial x_j}\frac{\partial r}{\partial n} - (1-2\nu)\left(\frac{\partial r}{\partial x_j}n_i - \frac{\partial r}{\partial x_i}n_j\right)\right] \quad (r \rightarrow 0) \tag{14}$$

In three dimensions:

$$U_{ji} \sim \frac{1}{16\pi G(1-\nu)}\frac{1}{r}\left[(3-4\nu)\delta_{ji} + \frac{\partial r}{\partial x_i}\frac{\partial r}{\partial x_j}\right] \tag{15}$$

$$T_{ji} \sim -\frac{1}{8\pi(1-\nu)}\frac{1}{r^2}\left[(1-2\nu)\frac{\partial r}{\partial n}\delta_{ji} + 3\frac{\partial r}{\partial x_i}\frac{\partial r}{\partial x_j}\frac{\partial r}{\partial n} - (1-2\nu)\left(\frac{\partial r}{\partial x_j}n_i - \frac{\partial r}{\partial x_i}n_j\right)\right] \quad (r \rightarrow 0) \tag{16}$$

Finally, in antiplane shear:

$$U_{33} \sim -\frac{1}{2\pi G}\ln(r) \tag{17}$$

$$T_{33} \sim -\frac{1}{2\pi}\frac{1}{r}\frac{\partial r}{\partial n} \quad (r \rightarrow 0) \tag{18}$$

Thus, as is readily recognized, the singular part of the fundamental solution, for any frequency, is identical to the static version of the solution ($\omega = 0$), commonly known as the Kelvin fundamental solution. Let the displacements and tractions in this solution be denoted by U^0_{ji} and T^0_{ji}. Certainly, $U_{ji} - U^0_{ji}$ and $T_{ji} - T^0_{ji}$ are nonsingular. According to equations (6) and (7), the submatrices $\underset{\sim}{H}^{mm}$ and $\underset{\sim}{G}^{mm}$ on the diagonals of the partitioned H and G can be written as:

$$H^{mm}_{ji} = \int_{S_m}(T_{ji} - T^0_{ji})N_m\, dS_m + \int_{S_m}T^0_{ji}N_m\, dS_m + C_{ji}(\underset{\sim}{X}_m) \tag{19}$$

$$G^{mm}_{ji} = \int_{S_m}(U_{ji} - U^0_{ji})N_m\, dS_m + \int_{S_m}U^0_{ji}N_m\, dS_m \tag{20}$$

The static versions ${}^0\underset{\sim}{H}^{mm}$ and ${}^0\underset{\sim}{G}^{mm}$ of these submatrices are given by:

$${}^0H^{mm}_{ji} = \int_{S_m}T^0_{ji}N_m\, dS_m + C^0_{ji}(\underset{\sim}{X}_m) \tag{21}$$

$${}^0G^{mm}_{ji} = \int_{S_m}U^0_{ji}N_m\, dS_m \tag{22}$$

Consider the calculations in (20). The first integral can be evaluated by straightforward Gaussian integration because $U_{ji} - U^0_{ji}$ is nonsingular. On the other hand, the second integral in (20), the same as that in (22), is improper but exists in the ordinary sense because U^0_{ji} is only weakly singular. In fact, this integral can be obtained in closed form for various commonly used simple boundary elements (see Reference 7). A general procedure for the numerical evaluation of the integral is given by Lachat and Watson (26) and Rizzo and Shippy (33). The calculations in (19) are somewhat more demanding. Again, the first integral can be computed by Gauss quadrature since $T_{ji} - T^0_{ji}$ is nonsingular. However, the second integral, the same as that in (21), is improper and exists only in the sense of a Cauchy Principal Value because of the strong singularity of T^0_{ji}. An elegant procedure to deal with this difficulty has been devised by Rizzo and Shippy (34). It makes use of the fact that

$$C_{ji} = C^0_{ji} \tag{23}$$

and, moreover,

$${}^0\underset{\sim}{H}^{mm} = -\sum_{\substack{n=1 \\ n\neq m}}^{N} {}^0\underset{\sim}{H}^{mn} \tag{24}$$

since static rigid-body translations in any direction are admissible with tractions equal to zero. Of course, the calculation of ${}^0H^{mn}$, $n \neq m$, is straightforward since there is no singularity involved. Thus, equation (19) can be rewritten as:

$$H^{mm}_{ji} = \int_{S_m} (T_{ji} - T^0_{ji}) N_m \, dS_m - \sum_{\substack{n=1 \\ n\neq m}}^{N} \int_{S_n} T^0_{ji} N_n \, dS_n \tag{25}$$

so that no improper integrals are encountered. It should be noted that this procedure renders the computation of C_{ji} unnecessary. This is convenient especially in cases of nodes located at points where the boundary is not smooth, i.e., corners. It is well known that, for the full space fundamental solution, $C_{ji} = (1/2)\delta_{ji}$ at points where the boundary is smooth. At corners, there is another difficulty stemming from the fact that the traction vector is not unique there. A way to deal with this complication has been suggested by Chaudonneret (6). It exploits the symmetry of the stress tensor and the invariance of its trace. However, in most applications, the nonuniqueness of the traction vector is not treated, since its influence on the results sufficiently far from the corner has been found to be negligible (see Reference 7).

Constant boundary elements, i.e., elements on which the displacements and the tractions are assumed to be constant, have been employed in most of the applications. Rectangles in three dimensions (7, 12, 13) and straight-line segments in plane problems (1, 7, 14) with a single node, usually at the center of the element, have been the most common. Higher-order elements have been used by Manolis and Beskos (30) and by Rizzo et al. (35). The required order of the Gaussian integration on an element is governed by the oscillatory character of the fundamental solution (the smallest wavelength in the fundamental solution is the one corresponding to transverse waves: $2\pi C_T/\omega$) and by the order of the interpolation functions.

Fundamental Solution for the Layered Half Space

If the infinite soil region is idealized as a horizontally layered half space, it is desirable to apply the boundary element method using the fundamental solution which satisfies the condition that the surface of the half space be free of tractions as well as the necessary continuity conditions at layer interfaces. The method then becomes most appealing, since boundary elements are necessary only on the soil-structure interface. Intensive research efforts have been devoted to the task of devising methods to calculate the solution (see Reference 2).

Recent work by Luco and Apsel (29) has led to the development of an efficient procedure for the computation of the elastodynamic state in a layered half space due to any time-harmonic source. Integral representations of the displacement and stress components are employed, the integrands being calculated by means of a novel technique (2, 29). The integrals are of the form:

$$\int^{\infty} F(k)\; J_{\ell}(kr)\; dk \tag{26}$$

where J_{ℓ} denotes the Bessel function of order ℓ (ℓ = 0, 2, ...). For example, the vertical displacement of a point on the surface of a homogeneous half space at a distance r from the point of application of a unit vertical force, also on the surface, is given by:

$$\frac{1}{2\pi\rho C_T^2}\int_0^{\infty}\frac{(1 - A_L A_T)A_L(1 - A_T^2)}{A_L A_T(1 - A_T^2)^2 - (2A_L A_T - A_T^2 - 1)^2}\; J_0(kr)\; dk \tag{27}$$

with

$$A_L = [1-(\omega/kC_L)^2]^{1/2}$$

$$A_T = [1-(\omega/kC_T)^2]^{1/2}$$

An accurate procedure for the evaluation of such integrals is described by Apsel and Luco (3). The integration is conveniently carried out along the real axis, as damping in the half space shifts the singularities away from the axis.

Recently, the fundamental solution for the half space, as calculated in the work referenced above (2, 3, 29), was used by Rizzo et al. (35) in some applications of the boundary element method to problems of dynamic soil-structure interaction.

Fundamental Solution for the Layered Stratum

For a layered stratum, it is most convenient to use the fundamental solution which satisfies the condition that the surface of the stratum be traction-free, the condition that the displacements vanish on the base of the stratum (soil-rock interface) and, of course, the continuity conditions at layer interfaces. This solution can be obtained by the procedure developed in the work referred to earlier (2, 3, 29). However, in the case of the stratum, a semidiscrete approximation of the fundamental solution can be derived, essentially in closed form, in terms of the modes of wave propagation. The development is due to Kausel (23) (also see Reference 25). It is built on the elegant work by Waas (42) and Kausel (21) on consistent transmitting boundaries (a review can be found in Reference 41). The finite element method is employed in order to calculate the modes of vibration. These modes as well as the resulting fundamental solution are discrete with respect to depth but continuous in any horizontal direction. It is instructive to examine the derivation in some detail.

Consider the layered stratum shown in Figure 2. A cylindrical coordinate system is most suitable (see Figure 2). The surface and the base of the stratum are located at $z = z_1$ and $z = z_{L+1}$ respectively, L being the number of layers. Layer j is the region between the planes $z = z_j$ and $z = z_{j+1}$ ($z_1 < z_2 < \dots < z_{L+1}$). Let u, w and v denote the amplitudes of the radial, axial (vertical) and tangential displacements, respectively. Also, let the corresponding stress components acting on horizontal planes be identified by τ_{zr}, σ_z and $\tau_{\theta z}$. A Fourier series expansion and a transformation with respect to the radial coordinate lead to the following representation:

$$\begin{bmatrix} u & \tau_{zr} \\ w & \sigma_z \\ v & \tau_{\theta z} \end{bmatrix} (r,\theta,z) = \sum_{\ell=0}^{\infty} \underset{\sim}{\Theta}^s(\ell\theta) \int_0^{\infty} k\, \underset{\sim}{C}_\ell(kr) \begin{bmatrix} U & R \\ W & Z \\ V & T \end{bmatrix}_\ell^s (k,z)\, dk$$

$$+ \sum_{\ell=0}^{\infty} \underset{\sim}{\Theta}^a(\ell\theta) \int_0^{\infty} k\, \underset{\sim}{C}_\ell(kr) \begin{bmatrix} U & R \\ W & Z \\ V & T \end{bmatrix}_\ell^a (k,z)\, dk \qquad (28)$$

where the superscripts s and a identify, respectively, symmetric and antisymmetric components, while:

$$\underset{\sim}{C}_\ell(kr) = \begin{bmatrix} J_\ell'(kr) & 0 & \frac{\ell}{kr} J_\ell(kr) \\ 0 & -J_\ell(kr) & 0 \\ \frac{\ell}{kr} J_\ell(kr) & 0 & J_\ell'(kr) \end{bmatrix}$$

$$\underset{\sim}{\Theta}^s(\ell\theta) = \text{diag}\,[\cos(\ell\theta), \cos(\ell\theta), -\sin(\ell\theta)]$$

$$\underset{\sim}{\Theta}^a(\ell\theta) = \text{diag}\,[\sin(\ell\theta), \sin(\ell\theta), \cos(\ell\theta)]$$

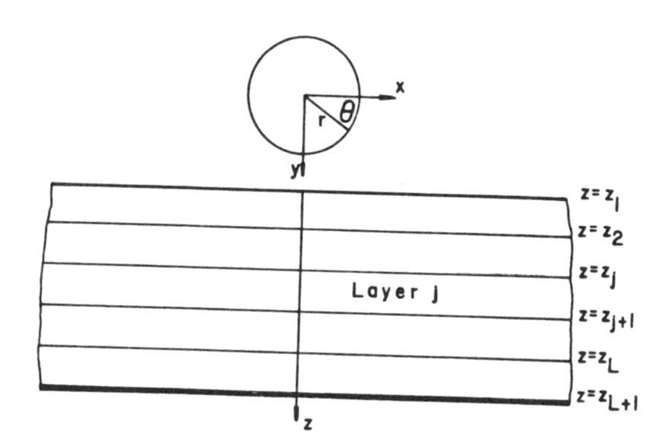

Figure 2: A layered stratum.

and

$$\begin{pmatrix} U & R \\ W & Z \\ V & T \end{pmatrix}_{\ell}^{s|a} (k,z) = a_{\ell} \int_0^{\infty} r\, \underset{\sim}{C}_{\ell}(kr)\, dr \int_0^{2\pi} \underset{\sim}{\Theta}^{s|a}(\ell\theta) \begin{pmatrix} u & \tau_{zr} \\ w & \sigma_z \\ v & \tau_{\theta z} \end{pmatrix} d\theta \qquad (29)$$

with

$$a_0 = \frac{1}{2\pi}$$

$$a_{\ell} = \frac{1}{\pi} \, , \quad \ell \neq 0$$

For a derivation of (28, 29), see References 21 and 22.

In layer j, U and W must satisfy the equations:

$$k^2(\lambda_j+2G_j)\, U + ik(\lambda_j+G_j)\, \frac{dW}{dz} - G_j\, \frac{d^2U}{dz^2} - \omega^2\rho_j\, U = 0 \qquad (30a)$$

$$k^2G_j\, W + ik(\lambda_j+G_j)\, \frac{dU}{dz} - (\lambda_j+2G_j)\, \frac{d^2W}{dz^2} - \omega^2\rho_j\, W = 0 \qquad (30b)$$

while the governing equation for V is:

$$k^2 G_j \, V - G_j \frac{d^2V}{dz^2} - \omega^2 \rho_j \, V = 0 \tag{31}$$

where λ_j, G_j and ρ_j denote the Lamé moduli and the mass density corresponding to layer j. Furthermore, it can be shown that:

$$R = G_j \left(\frac{dU}{dz} - kW\right) \tag{32a}$$

$$Z = (\lambda_j + 2G_j) \frac{dW}{dz} + k\lambda_j \, U \tag{32b}$$

$$T = G_j \frac{dV}{dz} \tag{32c}$$

At layer interfaces, continuity of u, w and v is required:

$$U(k, z_j^-) = U(k, z_j^+) \tag{33a}$$

$$W(k, z_j^-) = W(k, z_j^+) \tag{33b}$$

$$V(k, z_j^-) = V(k, z_j^+) \tag{33c}$$

$$2 \le j \le L$$

On the other hand, the tractions on the bottom of layer j-1 and those on the top of layer j must be in equilibrium with the loads acting on the interface. This implies that

$$R(k, z_j^-) - R(k, z_j^+) = F_r(k, z_j) \tag{34a}$$

$$Z(k, z_j^-) - Z(k, z_j^+) = F_z(k, z_j) \tag{34b}$$

$$Z(k, z_j^-) - Z(k, z_j^+) = F_\theta(k, z_j) \tag{34c}$$

$$2 \le j \le L$$

The relationship between F_r, F_z, F_θ and the loads f_r, f_z, f_θ is identical to the one between R, Z, T and τ_{zr}, σ_z, $\tau_{\theta z}$, given by equations (28, 29). For example, the loads corresponding to a concentrated unit vertical force at $r = 0$, $z = z_j$ are:

$$f_r = 0, \quad f_z = \frac{\delta(r)}{2\pi r}, \quad f_\theta = 0 \tag{35a}$$

and the transform of the only nonzero Fourier component is:

$$\begin{Bmatrix} F_r \\ F_z \\ F_\theta \end{Bmatrix}_0^s (k,z_j) = \begin{Bmatrix} 0 \\ -\frac{1}{2\pi} \\ 0 \end{Bmatrix} \tag{35b}$$

while for a horizontal force:

$$f_r = \frac{\delta(r)}{2\pi r}\cos(\theta),\ f_z = 0,\ f_\theta = -\frac{\delta(r)}{2\pi r}\sin(\theta) \tag{36a}$$

and the only nonvanishing transform is:

$$\begin{Bmatrix} F_r \\ F_z \\ F_\theta \end{Bmatrix}_1^s (k,z_j) = \begin{Bmatrix} \frac{1}{2\pi} \\ 0 \\ \frac{1}{2\pi} \end{Bmatrix} \tag{36b}$$

δ being the delta function. It should be pointed out that the materials on either side of an interface $z = z_j$, $2 \le j \le L$, may be identical. In the present derivation, such interfaces are necessary, if loads like (35) or (36) are applied at locations other than the interfaces between different materials.

On the surface of the stratum, the tractions must be equal to the applied loads:

$$-R(k,z_1^+) = F_r(k,z_1) \tag{37a}$$

$$-Z(k,z_1^+) = F_z(k,z_1) \tag{37b}$$

$$-T(k,z_1^+) = F_\theta(k,z_1) \tag{37c}$$

Finally, on the base, the displacements must vanish:

$$U(k,z_{L+1}^-) = 0 \tag{38a}$$

$$W(k,z_{L+1}^-) = 0 \tag{38b}$$

$$V(k,z_{L+1}^-) = 0 \tag{38c}$$

It is important to note that the differential equations (30), the relationships (32), the interface conditions (33, 34) and the boundary conditions (37, 38) are independent of the Fourier number ℓ. Moreover, they are the same for symmetric and antisymmetric Fourier components. The development by Kausel (23) proceeds by applying the finite element method to the problem defined by these equations and conditions. Each layer is divided into sublayers. Linear interpolation of U, W and V is assumed in each sublayer. Of course, any other continuous approximation can be used as well. For the purposes of the derivation, the distinction between layers and sublayers is not necessary. In the sequel, all sublayers will be referred to as layers and the notation will not be altered. For layer j, application of the finite element method (see References 21, 41 and 42) yields:

$$(k^2\underset{\sim}{A}^j + k\underset{\sim}{B}^j + \underset{\sim}{G}^j - \omega^2\underset{\sim}{M}^j)\begin{Bmatrix} U(k,z_j^+) \\ W(k,z_j^+) \\ V(k,z_j^+) \\ U(k,z_{j+1}^-) \\ W(k,z_{j+1}^-) \\ V(k,z_{j+1}^-) \end{Bmatrix} = \begin{Bmatrix} -R(k,z_j^+) \\ -Z(k,z_j^+) \\ -T(k,z_j^+) \\ R(k,z_{j+1}^-) \\ Z(k,z_{j+1}^-) \\ T(k,z_{j+1}^-) \end{Bmatrix} \tag{39}$$

$\underset{\sim}{A}^j$, $\underset{\sim}{B}^j$, $\underset{\sim}{G}^j$ and $\underset{\sim}{M}^j$ are symmetric block-tridiagonal matrices given by:

$$\underset{\sim}{A}^j = \frac{1}{6} h_j \begin{bmatrix} 2(\lambda_j + 2G_j) & 0 & 0 & \lambda_j + 2G_j & 0 & 0 \\ 0 & 2G_j & 0 & 0 & G_j & 0 \\ 0 & 0 & 2G_j & 0 & 0 & G_j \\ \lambda_j + 2G_j & 0 & 0 & 2(\lambda_j + 2G_j) & 0 & 0 \\ 0 & G_j & 0 & 0 & 2G_j & 0 \\ 0 & 0 & G_j & 0 & 0 & 2G_j \end{bmatrix}$$

$$\underset{\sim}{B}^j = \frac{1}{2} \begin{bmatrix} 0 & -(\lambda_j - G_j) & 0 & 0 & \lambda_j + G_j & 0 \\ -(\lambda_j - G_j) & 0 & 0 & -(\lambda_j + G_j) & 0 & 0 \\ 0 & 0 & 0 & 0 & 0 & 0 \\ 0 & -(\lambda_j + G_j) & 0 & 0 & \lambda_j - G_j & 0 \\ \lambda_j + G_j & 0 & 0 & \lambda_j - G_j & 0 & 0 \\ 0 & 0 & 0 & 0 & 0 & 0 \end{bmatrix}$$

$$\underset{\sim}{G}^j = \frac{1}{h_j}\begin{bmatrix} G_j & 0 & 0 & -G_j & 0 & 0 \\ 0 & \lambda_j + 2G_j & 0 & 0 & -(\lambda_j + 2G_j) & 0 \\ 0 & 0 & G_j & 0 & 0 & -G_j \\ -G_j & 0 & 0 & G_j & 0 & 0 \\ 0 & -(\lambda_j + 2G_j) & 0 & 0 & \lambda_j + 2G_j & 0 \\ 0 & 0 & -G_j & 0 & 0 & G_j \end{bmatrix}$$

$$\underset{\sim}{M}^j = \frac{1}{6}\rho_j h_j \begin{bmatrix} 2 & 0 & 0 & 1 & 0 & 0 \\ 0 & 2 & 0 & 0 & 1 & 0 \\ 0 & 0 & 2 & 0 & 0 & 1 \\ 1 & 0 & 0 & 2 & 0 & 0 \\ 0 & 1 & 0 & 0 & 2 & 0 \\ 0 & 0 & 1 & 0 & 0 & 2 \end{bmatrix}$$

where h_j denotes the thickness of layer j: $h_j = z_{j+1} - z_j$. Equations (39) together with the conditions (33, 34, 37, 38) yield:

$$(k^2\underset{\sim}{A} + k\underset{\sim}{B} + \underset{\sim}{G} - \omega^2\underset{\sim}{M})\ \underset{\sim}{X}(k) = \underset{\sim}{P}(k) \tag{40}$$

in which:

$$X_{3j-2} = U(k,z_j)$$
$$X_{3j-1} = W(k,z_j)$$
$$X_{3j} = V(k,z_j)$$

$$P_{3j-2} = F_r(k,z_j)$$
$$P_{3j-1} = F_z(k,z_j)$$
$$P_{3j} = F_\theta(k,z_j)$$

$$1 \le j \le L$$

$\underset{\sim}{A}$, $\underset{\sim}{B}$, $\underset{\sim}{G}$ and $\underset{\sim}{M}$ are obtained from the assemblages of $\underset{\sim}{A}^j$, $\underset{\sim}{B}^j$, $\underset{\sim}{G}^j$ and $\underset{\sim}{M}^j$, $1 \le j \le L$, by deleting the last three rows and columns, thus imposing (38). Equation (40) must be solved for $\underset{\sim}{X}(k)$. As shown by Kausel (23), the inversion can be carried out in closed form in terms of the solutions of the quadratic eigenvalue problem:

$$(\xi^2 \underset{\sim}{A} + \xi \underset{\sim}{B} + \underset{\sim}{G} - \omega^2 \underset{\sim}{M})\, \underset{\sim}{\phi} = \underset{\sim}{0} \tag{41}$$

where $\underset{\sim}{\phi}$ and ξ denote an eigenvector and the corresponding eigenvalue respectively. This problem can be solved easily (see Reference 42). Clearly, U and W are not coupled with V. Thus, for any eigenvalue-eigenvector pair $(\xi, \underset{\sim}{\phi})$, either $\phi_{3j} = 0,\ 1 \le j \le L$, or $\phi_{3j-2} = \phi_{3j-1} = 0,\ 1 \le j \le L$. Furthermore, if $(\xi, \underset{\sim}{\phi})$ is a pair, $(-\xi, \underset{\sim}{\psi})$ is another one with $\psi_{3j-2} = \phi_{3j-2}$, $\psi_{3j-1} = -\phi_{3j-1}$, $\psi_{3j} = \phi_{3j}$. It is possible to choose 3L pairs $(\xi_m, \underset{\sim}{\phi}_m)$, $1 \le m \le 3L$, such that $\mathrm{Im}(\xi_m) < 0$, or $\mathrm{Re}(\xi_m) > 0$ and $\mathrm{Im}(\xi_m) = 0$. The eigenvectors can be normalized so that

$$\underset{\sim}{\Xi}\, \underset{\sim}{\Psi}^T \underset{\sim}{A}\, \underset{\sim}{\Phi}\, \underset{\sim}{\Xi} - \underset{\sim}{\Psi}^T (\underset{\sim}{G} - \omega^2 \underset{\sim}{M})\, \underset{\sim}{\Phi} = 2\, \underset{\sim}{\Xi}\, \underset{\sim}{\Xi} \tag{42}$$

where:

$$\underset{\sim}{\Xi} = \mathrm{diag}[\xi_m,\ 1 \le m \le 3L]$$

$$\underset{\sim}{\Phi} = [\underset{\sim}{\phi}_m,\ 1 \le m \le 3L]$$

$$\underset{\sim}{\Psi} = [\underset{\sim}{\psi}_m,\ 1 \le m \le 3L]$$

It is convenient to group rows 1, 4, 7, ..., 3L-2 of $\underset{\sim}{X}$, $\underset{\sim}{P}$ and $\underset{\sim}{\Phi}$ into $\underset{\sim}{U}$, $\underset{\sim}{P}^r$ and $\underset{\sim}{\Phi}^r$ respectively, rows 2, 5, 8, ..., 3L-1 into $\underset{\sim}{W}$, $\underset{\sim}{P}^z$ and $\underset{\sim}{\Phi}^z$ and rows 3, 6, 9, ..., 3L into $\underset{\sim}{V}$, $\underset{\sim}{P}^\theta$ and $\underset{\sim}{\Phi}^\theta$. Following Kausel (23), it can be shown that

$$\begin{Bmatrix} \underset{\sim}{U} \\ \underset{\sim}{W} \\ \underset{\sim}{V} \end{Bmatrix} = \begin{bmatrix} \underset{\sim}{\Phi}^r (k^2 \underset{\sim}{I} - \underset{\sim}{\Xi}\underset{\sim}{\Xi})^{-1} (\underset{\sim}{\Phi}^r)^T & k \underset{\sim}{\Phi}^r \underset{\sim}{\Xi}^{-1} (k^2 \underset{\sim}{I} - \underset{\sim}{\Xi}\underset{\sim}{\Xi})^{-1} (\underset{\sim}{\Phi}^z)^T & 0 \\ k \underset{\sim}{\Phi}^z (k^2 \underset{\sim}{I} - \underset{\sim}{\Xi}\underset{\sim}{\Xi})^{-1} \underset{\sim}{\Xi}^{-1} (\underset{\sim}{\Phi}^r)^T & \underset{\sim}{\Phi}^z (k^2 \underset{\sim}{I} - \underset{\sim}{\Xi}\underset{\sim}{\Xi})^{-1} (\underset{\sim}{\Phi}^z)^T & 0 \\ 0 & 0 & \underset{\sim}{\Phi}^\theta (k^2 \underset{\sim}{I} - \underset{\sim}{\Xi}\underset{\sim}{\Xi})^{-1} (\underset{\sim}{\Phi}^\theta)^T \end{bmatrix} \begin{Bmatrix} \underset{\sim}{P}^r \\ \underset{\sim}{P}^z \\ \underset{\sim}{P}^\theta \end{Bmatrix} \tag{43}$$

$\underset{\sim}{I}$ being the identity matrix of order 3L. It is now straightforward to calculate the displacements due to any loads such as those in (35,36) using equation (28). For a unit horizontal force (in the plane $\theta = 0$) at $(r = 0)$ $z = z_j$, the displacements on the plane $z = z_q$ are:

$$u(r,\theta,z_q) = \frac{1}{4i}\left\{\sum_{m=1}^{3L} \frac{1}{\xi_m} \Phi^r_{qm}\Phi^r_{jm} \frac{d}{dr}[H_1^{(2)}(\xi_m r)] + \frac{1}{r}\sum_{m=1}^{3L} \frac{1}{\xi_m} \Phi^\theta_{qm}\Phi^\theta_{jm} H_1^{(2)}(\xi_m r)\right\} \cos(\theta) \tag{44a}$$

$$w(r,\theta,z_q) = -\frac{1}{4i}\left[\sum_{m=1}^{3L} \Phi^z_{qm}\Phi^r_{jm} H_1^{(2)}(\xi_m r)\right] \cos(\theta) \tag{44b}$$

$$v(r,\theta,z_q) = \frac{1}{4i}\left\{\frac{1}{r}\sum_{m=1}^{3L} \frac{1}{\xi_m} \Phi^r_{qm}\Phi^r_{jm} H_1^{(2)}(\xi_m r) + \sum_{m=1}^{3L} \frac{1}{\xi_m} \Phi^\theta_{qm}\Phi^\theta_{jm} \frac{d}{dr}[H_1^{(2)}(\xi_m r)]\right\} [-\sin(\theta)] \tag{44c}$$

$$1 \le q \le L$$

while the displacements due to a unit vertical force are given by:

$$u(r,\theta,z_q) = \frac{1}{4i}\sum_{m=1}^{3L} \Phi^r_{qm}\Phi^z_{jm} H_1^{(2)}(\xi_m r) \tag{45a}$$

$$w(r,\theta,z_q) = \frac{1}{4i}\sum_{m=1}^{3L} \Phi^z_{qm}\Phi^z_{jm} H_0^{(2)}(\xi_m r) \tag{45b}$$

$$v(r,\theta,z_q) = 0 \tag{45c}$$

$$1 \le q \le L$$

$H_0^{(2)}$ and $H_1^{(2)}$ are the Hankel functions of the second kind of order zero and one, respectively. It should be noted that the summations in (44, 45) involve either 2L or L terms, since L columns in $\underset{\sim}{\Phi}^r$ and $\underset{\sim}{\Phi}^z$ have all entries equal to zero while the same holds for 2L columns in $\underset{\sim}{\Phi}^\theta$ (U and W are not coupled with V). Details of the derivation can be found in Reference 23 where, in addition, solutions for line, disk and ring loads are given.

Kausel and Peek (24) have examined the use of the semidiscrete fundamental solutions in applications of the boundary element method. They suggested a modification of the reciprocity relation (1), recognizing the fact that the tractions corresponding to the semidiscrete displacements are discontinuous at layer interfaces.

Hull and Kausel (17) have devised a condition that can be applied, instead of (38), on the base of the stratum in order to take into account, approximately, the flexibility of the underlying rock. The condition is obtained from the exact one, which involves transcendental functions of k (see References 17 and 22), by expanding in Taylor series and retaining only the terms of order 0, 1 and 2.

Waas et al. (43) have derived solutions for ring, point, line and disk loads following a different approach. The solutions are applicable to a layered stratum with material properties that may vary in the vertical direction in each layer. Also, isotropy is assumed only in horizontal planes.

Time Domain

Apparently, Cole et al. (9) and Niwa et al. (32) were the first to apply the boundary element method to problems of elastodynamics directly in the time domain. Manolis (31) examined the performance of a time domain formulation in comparison with frequency domain treatments combined with the Fourier or Laplace transform for a problem of transient elastodynamics. The first application of a time domain boundary element method to problems of dynamic soil-structure interaction is due to Beskos and Karabalis (4) and Beskos and Spyrakos (5). A brief description of the method as applied in their work will be given below (see References 4, 5, 18, 19, 20 and 39).

Let

$$U_{ji}\{\underset{\sim}{x},t;\ \underset{\sim}{X},[g(\tau),\ 0 \le \tau \le t]\}f_j$$

and

$$T_{ji}\{\underset{\sim}{x},t;\ \underset{\sim}{X},[g(\tau),\ 0 \le \tau \le t]\}f_j$$

be the displacements and tractions at $\underset{\sim}{x}$, at time $t > 0$, due to a concentrated force at $\underset{\sim}{X}$ with components $g(\tau)f_i$, $0 \le \tau \le t$. The past, $t < 0$, is assumed quiescent. This fundamental solution can be obtained in closed form for a homogeneous, isotropic full space. In three dimensions:

$$U_{ji}\{\underset{\sim}{x},t;\ \underset{\sim}{X},[g(\tau),\ 0 \le \tau \le t]\}$$

$$= \frac{1}{4\pi\rho}\ \{\ \frac{1}{r}\ (\ 3\ \frac{\partial r}{\partial x_i}\frac{\partial r}{\partial x_j} - \delta_{ji}\ )\ \int_{1/C_L}^{1/C_T} sg(t-rs)\ ds$$

$$+ \frac{1}{r} \frac{\partial r}{\partial x_i} \frac{\partial r}{\partial x_j} \left[\frac{1}{C_L^2} g(t-r/C_L) - \frac{1}{C_T^2} g(t-r/C_T) \right]$$

$$+ \frac{1}{r} \frac{1}{C_T^2} g(t-r/C_T) \} \qquad (46)$$

while for a full space in plane strain:

$$U_{ji}\{\underset{\sim}{x},t;\ \underset{\sim}{X},[g(\tau),\ 0 \leq \tau \leq t]\}$$

$$= \frac{1}{2\pi\rho} \{ \frac{\partial^2}{\partial x_i \partial x_j} \int_0^{t-r/C_L} \frac{dp}{[(t-p)^2 - r^2/C_L^2]^{1/2}} \int_0^p s g(p-s)\, ds$$

$$- \frac{\partial^2}{\partial x_i \partial x_j} \int_0^{t-r/C_T} \frac{dp}{[(t-p)^2 - r^2/C_T^2]^{1/2}} \int_0^p s g(p-s)\, ds$$

$$+ \delta_{ji} \frac{1}{C_T^2} \int_0^{t-r/C_T} \frac{g(p)dp}{[(t-p)^2 - r^2/C_T^2]^{1/2}} \} \qquad (47)$$

For a full space in antiplane shear, the (out-of-plane) displacement is given by:

$$U_{33}\{\underset{\sim}{x},t;\ \underset{\sim}{X},[g(\tau),\ 0 \leq \tau \leq t]\}$$

$$= \frac{1}{2\pi\rho C_T^2} \int_0^{t-r/C_T} \frac{g(p)dp}{[(t-p)^2 - r^2/C_T^2]^{1/2}} \qquad (48)$$

It is straightforward to obtain the tractions from the displacements (46-48) by calculating the strains and using the constitutive equations.

The reciprocity relation between any two elastodynamic states in Ω (see Reference 44) leads to the following integral representation (see Reference 16):

$$C_{ji}(\underset{\sim}{X}) f_j [u_i(\underset{\sim}{X},t) - \hat{u}_i(\underset{\sim}{X},t)]$$

$$= \int_S U_{ji}\{\underset{\sim}{x},t;\ \underset{\sim}{X},[T_i(\underset{\sim}{x},\tau) - \hat{T}_i(\underset{\sim}{x},\tau),\ 0 \leq \tau \leq t]\} f_j\, dS$$

$$- \int_S T_{ji}\{\underset{\sim}{x},t;\ \underset{\sim}{X},[u_i(\underset{\sim}{x},\tau) - \hat{u}_i(\underset{\sim}{x},\tau),\ 0 \leq \tau \leq t]\} f_j\, dS \qquad (49)$$

assuming that

$$u_i(\underset{\sim}{x},0) - \hat{u}_i(\underset{\sim}{x},0) = 0$$

$$\frac{\partial u_i}{\partial t}(\underset{\sim}{x},0^+) - \frac{\partial \hat{u}_i}{\partial t}(\underset{\sim}{x},0^+) = 0$$

and that the body force density vanishes.

In the work referred to above (4, 5, 19, 39), the integral representation (49) was employed by covering the boundary S with elements and assuming that the displacements and tractions are constant on each element. Furthermore, a piecewise constant approximation was adopted with respect to time. Let the time interval of interest be divided into Q subintervals, i.e., time steps. For the sake of simplicity, let us assume that all subintervals are of the same length, Δt. The displacement and traction vectors at element n in subinterval q, i.e., $(q-1)\Delta t \leq t \leq q\Delta t$ will be denoted by $\underset{\sim}{U}^{n,q}$ and $\underset{\sim}{T}^{n,q}$ respectively. Also, the corresponding free field vectors will be identified by $\hat{\underset{\sim}{U}}^{n,q}$ and $\hat{\underset{\sim}{T}}^{n,q}$. If the integral representation (49) is invoked at the end of each time subinterval and for each element, e.g., at the center of the element, a system of algebraic equations is obtained:

$$\frac{1}{2}\left(\underset{\sim}{U}^{n,q} - \hat{\underset{\sim}{U}}^{n,q}\right) = \sum_{m=1}^{N}\left\{\sum_{\ell=1}^{q}\left[\underset{\sim}{G}_n^{m,\ell}\left(\underset{\sim}{T}^{m,q-\ell+1} - \hat{\underset{\sim}{T}}^{m,q-\ell+1}\right) - \underset{\sim}{H}_n^{m,\ell}\left(\hat{\underset{\sim}{U}}^{m,q-\ell+1} - \hat{\underset{\sim}{U}}^{m,q-\ell+1}\right)\right]\right\} \quad (50)$$

$$1 \leq n \leq N,\ 1 \leq q \leq Q$$

(N is the number of boundary elements). It should be noted that the fact that all time subintervals were assumed to be of the same length has been exploited in writing (50). The matrices $\underset{\sim}{G}_n^{m,\ell}$ and $\underset{\sim}{H}_n^{m,\ell}$ are square of order 3 for three-dimensional problems, 2 in the case of plane strain and 1 for antiplane shear. They are given by:

$$(\underset{\sim}{G}_n^{m,\ell})_{ji} = \int_{S_m} U_{ji}\{\underset{\sim}{x},\ell\Delta t;\ \underset{\sim}{X}_n,[g(\tau),\ 0 \leq \tau \leq \ell\Delta t]\}\ dS_m \quad (51)$$

$$(\underset{\sim}{H}_n^{m,\ell})_{ji} = \int_{S_m} T_{ji}\{\underset{\sim}{x},\ell\Delta t;\ \underset{\sim}{X}_n,[g(\tau),\ 0 \leq \tau \leq \ell\Delta t]\}\ dS_m \quad (52)$$

where S_m denotes the boundary region occupied by element m, $\underset{\sim}{X}_n$ is the center of element n, while:

$$g(\tau) = 1,\ 0 \leq \tau \leq \Delta t$$

$$g(\tau) = 0,\ \Delta t < \tau \leq \ell\Delta t$$

As in the frequency domain, there are singularities that must be treated in the evaluation of the integrals (51) and (52). This matter is discussed in References 4, 5, 19 and 39.

Equation (50) couples the tractions and displacements at time $q\Delta t$ with their values at all previous times, i.e., Δt, $2\Delta t$, ..., $(q-1)\Delta t$. Of course, this implies a step-by-step procedure in applying the equation. Moreover, it means that the entire past history of boundary displacements and tractions must be available at all times.

Coupling of Boundary Elements and Finite Elements

As mentioned earlier, it may be advantageous to discretize the structure using finite elements while treating the infinite soil domain by the boundary element method. The question then arises as to how the finite elements will be coupled with the boundary elements at the soil-structure interface. Let us discuss this aspect of the analysis in a frequency domain formulation. Similar ideas are applicable to time domain treatments. For the sake of clarity, it will be assumed that structural degrees of freedom other than those on the interface have been eliminated. Thus, for the finite elements, the relationship between the vector of nodal displacement amplitudes $\underset{\sim}{\mathcal{U}}$ and consistent nodal forces $\underset{\sim}{\mathcal{F}}$ is given by:

$$\underset{\sim}{\mathcal{P}} + \underset{\sim}{\mathcal{F}} = \underset{\sim}{\mathcal{K}}\, \underset{\sim}{\mathcal{U}} \tag{53}$$

where $\underset{\sim}{\mathcal{K}}$ and $\underset{\sim}{\mathcal{P}}$ denote the dynamic stiffness matrix and the vector of consistent nodal forces corresponding to loads that are applied on the structure directly, e.g., due to vibrations of machines operating within the structure. On the other hand, for the boundary elements, the relationship between the vector of nodal tractions $\underset{\sim}{T}$ and the vector of nodal displacements $\underset{\sim}{U}$ is expressed by equation (4):

$$\underset{\sim}{H}\, \underset{\sim}{U} - \underset{\sim}{G}\, \underset{\sim}{T} = \underset{\sim}{F} \tag{54}$$

The conditions to be imposed are that the displacements must be continuous and the tractions acting on the interface must be in equilibrium. First, let us express $\underset{\sim}{\mathcal{F}}$ in terms of $\underset{\sim}{T}$. At any point $\underset{\sim}{x}$ on the inteface:

$$\underset{\sim}{p}(\underset{\sim}{x}) = \underset{\sim}{N}(\underset{\sim}{x})\, \underset{\sim}{T} \tag{55}$$

in which $\underset{\sim}{N}$ is the matrix of interpolation functions corresponding to the boundary elements and $\underset{\sim}{p}$ is the traction vector at $\underset{\sim}{x}$. Furthermore, $\underset{\sim}{\mathcal{F}}$ can be written as:

$$\underset{\sim}{\mathcal{F}} = - \int_S \underset{\sim}{\mathcal{N}}^T(\underset{\sim}{x})\, \underset{\sim}{p}(\underset{\sim}{x})\, dS \tag{56}$$

where $\underset{\sim}{\mathcal{N}}$ denotes the matrix of interpolation functions corresponding to the finite elements. Equations (55) and (56) imply that

$$\underset{\sim}{\mathcal{F}} = \underset{\sim}{\mathcal{R}}\, \underset{\sim}{T} \tag{57}$$

with

$$\underset{\sim}{\mathcal{R}} = - \int_S \underset{\sim}{\mathcal{N}}^T(\underset{\sim}{x})\, \underset{\sim}{N}(\underset{\sim}{x})\, dS$$

The next step is to express $\underset{\sim}{U}$ in terms of $\underset{\sim}{\mathcal{U}}$. Of course, if the boundary element nodes coincide with the finite element nodes, $\underset{\sim}{U} = \underset{\sim}{\mathcal{U}}$. In general, it is possible to write:

$$\underset{\sim}{U} = \underset{\sim}{R}\, \underset{\sim}{\mathcal{U}} \tag{58}$$

$\underset{\sim}{R}$ being composed of the values of $\underset{\sim}{N}$ at the boundary element nodes. Equation (54) can be solved for $\underset{\sim}{T}$ in terms of $\underset{\sim}{U}$:

$$\underset{\sim}{T} = \underset{\sim}{G}^{-1}\, \underset{\sim}{H}\, \underset{\sim}{U} - \underset{\sim}{G}^{-1} \underset{\sim}{F} \tag{59}$$

Finally, equations (59), (58), (57) and (53) lead to:

$$\underset{\sim}{P} - \underset{\sim}{\mathcal{R}}\, \underset{\sim}{G}^{-1}\, \underset{\sim}{F} = (\underset{\sim}{K} - \underset{\sim}{\mathcal{R}}\, \underset{\sim}{G}^{-1}\, \underset{\sim}{H}\, \underset{\sim}{R})\, \underset{\sim}{\mathcal{U}} \tag{60}$$

This equation can be solved for $\underset{\sim}{\mathcal{U}}$. Subsequently, equation (58) can be used to estimate $\underset{\sim}{U}$. Substitution in equation (59) yields $\underset{\sim}{T}$.

Further elaboration can be found in Reference 45. An application in which boundary elements are coupled with finite elements as outlined above is the analysis of flexible foundations on the surface of a half space reported in References 4, 5, 19 and 39.

Examples

To illustrate the use of the boundary element method in problems of dynamic soil-structure interaction, let us calculate the dynamic stiffness of a rigid foundation embedded in a homogeneous half space. This frequency domain application is encountered frequently, as it is often reasonable to assume that the soil-structure interface, i.e., the foundation, is rigid. In such a case, for the purposes of structural analysis, it suffices to determine the dynamic stiffness matrix and the response of the foundation to the prescribed seismic input, in the absence of the structure. The dynamic stiffness matrix relates the amplitudes of the forces F_x, F_y and F_z, and moments M_x, M_y and M_z acting on the foundation to the amplitudes of the rigid-body translations Δ_x, Δ_y and Δ_z, and rotations θ_x, θ_y and θ_z, (x,y,z) being a rectangular Cartesian system of coordinates. A rigid rectangular foundation of length 2L, width 2B and embedment E is shown in Figure 3. For this foundation, it is possible to write:

$$\begin{Bmatrix} F_x \\ F_y \\ F_z \\ M_x \\ M_y \\ M_z \end{Bmatrix} = \begin{bmatrix} K_{xx} & 0 & 0 & 0 & K_{xr_y} & 0 \\ 0 & K_{yy} & 0 & K_{yr_x} & 0 & 0 \\ 0 & 0 & K_{zz} & 0 & 0 & 0 \\ 0 & 0 & K_{r_x y} & K_{r_x r_x} & 0 & 0 \\ K_{r_y x} & 0 & 0 & 0 & K_{r_y r_y} & 0 \\ 0 & 0 & 0 & 0 & 0 & K_{r_z r_z} \end{bmatrix} \begin{Bmatrix} \Delta_x \\ \Delta_y \\ \Delta_z \\ \theta_x \\ \theta_y \\ \theta_z \end{Bmatrix}$$

(with respect to the coordinate system shown in Figure 3). Of course, the dynamic stiffness matrix is symmetric. Let us use the boundary element method to calculate the dimensionless stiffnesses:

$$\frac{K_{xx}}{GB}, \frac{K_{yy}}{GB}, \frac{K_{zz}}{GB}, \frac{K_{r_x r_x}}{GB^3}, \frac{K_{r_y r_y}}{GB^3}$$

in the case of a rectangular foundation with $L/B = 2$ and $E/B = 1$ in a homogeneous half space with Poisson's ratio $\nu = 1/3$ and damping ratio $\beta = 0$, for a range of values of the dimensionless frequency: $0 \le \omega B/C_T \le 2$. To this end, the fundamental solution for the full space is employed requiring that the foundation as well as the surface of the half space be covered with boundary elements. For the purposes of this example, square elements with constant interpolation of tractions and displacements are used. Numerical integrations are carried out using a 4 by 4 array of Gauss integration points in each element. Various discretizations are considered. In each mesh, all elements are of the same size. The meshes are identified by the element size and the extent of the half space surface that is covered with boundary elements. For example, (B/2; B) indicates a mesh of (B/2)-square elements extending to a distance B from each of the foundation edges on the half space surface. Figure 4 depicts the discretization (B/2; 2B). The real and imaginary parts of the stiffnesses are shown in Figures 5-9 for a variety of meshes. Clearly, the imaginary parts of the stiffnesses are not sensitive to the extent of the mesh on the free surface of the half space. This is hardly surprising: the imaginary part is related to the loss of energy due to radiation in the half space and is influenced by boundary elements on the free surface indirectly (tractions are equal to zero on such elements). The situation is markedly different in the case of the real parts of the stiffnesses. It can be seen that the results obtained with the discretization (B/2; 0), in which only the foundation is covered with boundary elements, differ substantially from the rest. Also, it is clear that the difference between the results calculated with the meshes (B/2; 2B) and (B/2; B) is small and even smaller between (B/2; 4B) and (B/2; 2B), indicating that there is reasonably fast convergence as the mesh extends further away from the edges of the foundation. Finally, the meshes (B/2; 2B) and (B/4; 2B) yield almost the same results, suggesting that

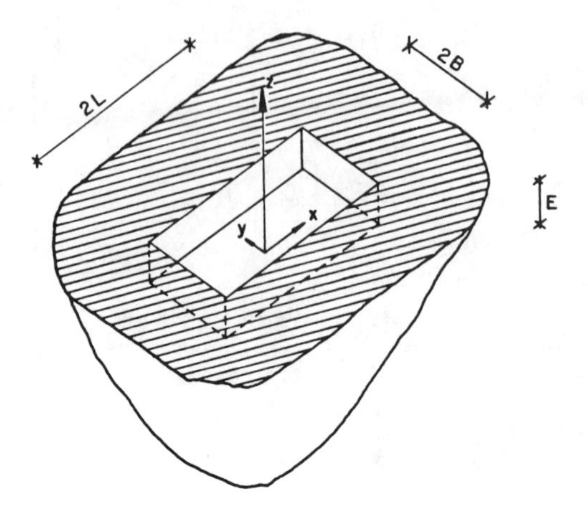

Figure 3: A rectangular foundation embedded in a half space.

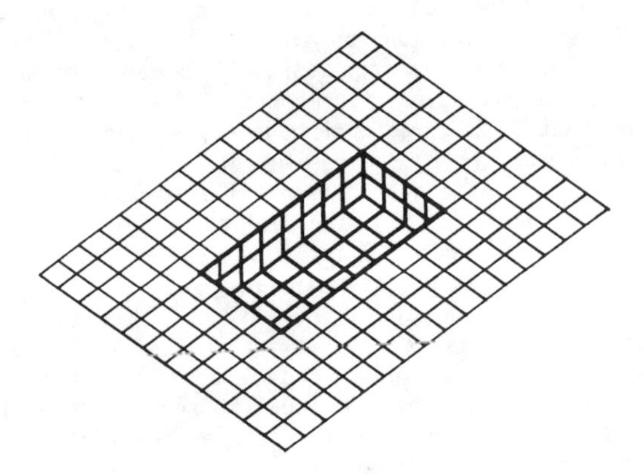

Figure 4: Discretization (B/2; 2B): all elements are B/2 by B/2 squares and the mesh extends to a distance 2B from each edge of the foundation on the half space surface.

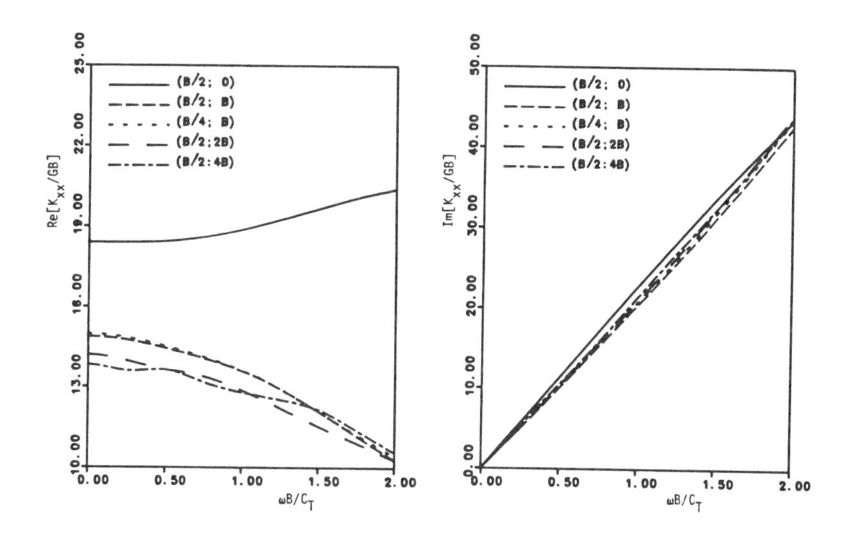

Figure 5: Dynamic horizontal stiffness (x-direction) of a rectangular foundation ($L/B = 2$, $E/B = 1$) embedded in a homogeneous half space ($\nu = 1/3$, $\beta = 0$).

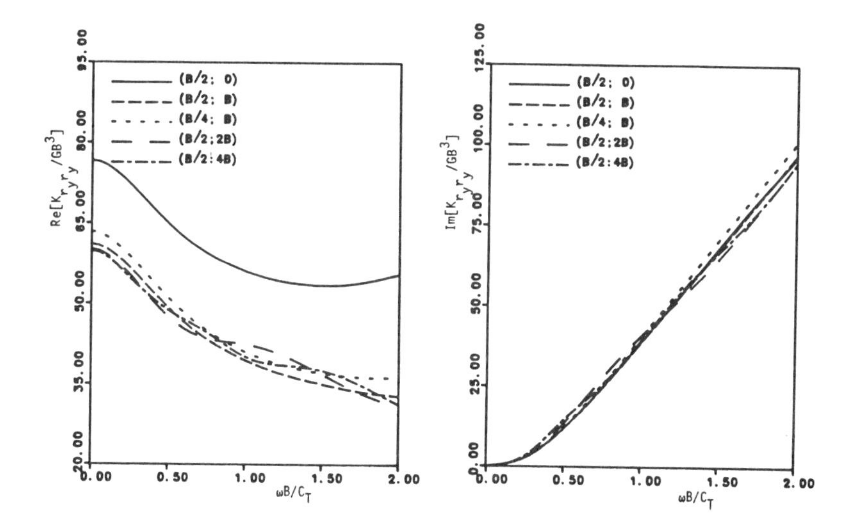

Figure 6: Dynamic rocking stiffness (y-axis) of a rectangular foundation ($L/B = 2$, $E/B = 1$) embedded in a homogeneous half space ($\nu = 1/3$, $\beta = 0$).

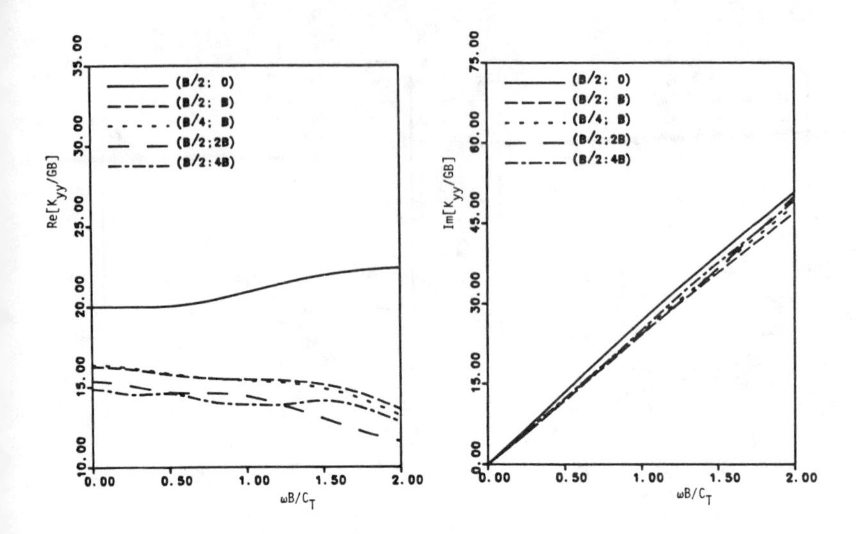

Figure 7: Dynamic horizontal stiffness (y-direction) of a rectangular foundation ($L/B = 2$, $E/B = 1$) embedded in a homogeneous half space ($\nu = 1/3$, $\beta = 0$).

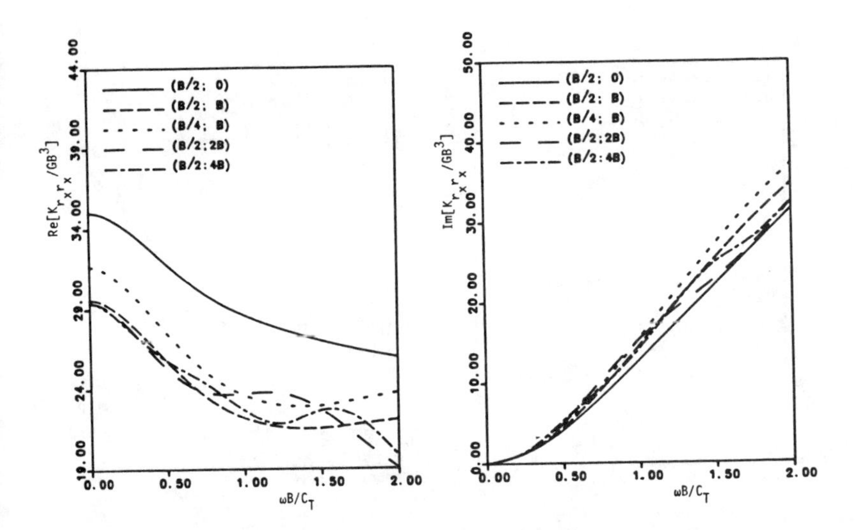

Figure 8: Dynamic rocking stiffness (x-axis) of a rectangular foundation ($L/B = 2$, $E/B = 1$) embedded in a homogeneous half space ($\nu = 1/3$, $\beta = 0$).

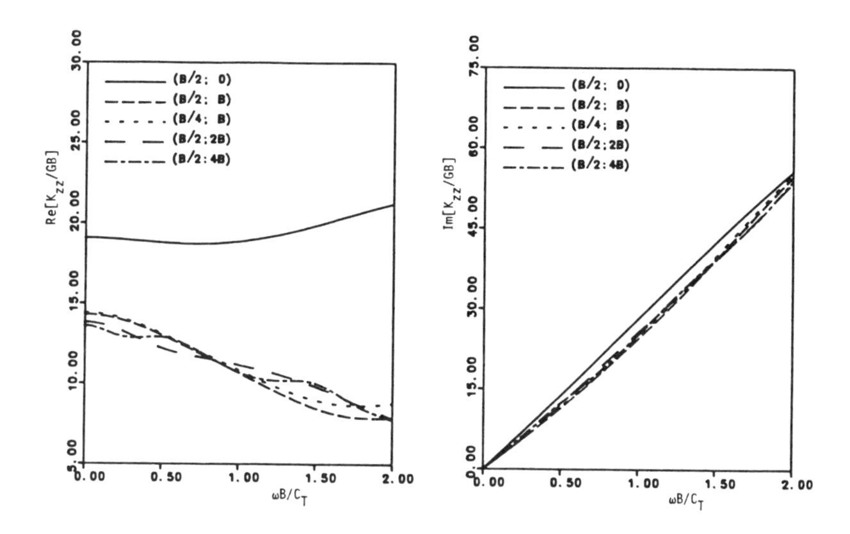

Figure 9: Dynamic vertical stiffness of a rectangular foundation ($L/B = 2$, $E/B = 1$) embedded in a homogeneous half space ($\nu = 1/3$, $\beta = 0$).

the (B/2)-square element is fine enough for frequencies in the range considered.

Another example will be drawn from the work by Manolis and Beskos (30) and deals with an underground structure. Figure 10 depicts a circular cylindrical lined tunnel in a homogeneous full space. The inner and outer diameters of the lining are denoted by d and D respectively. It is assumed that the liner is bonded perfectly to the surrounding medium. The tunnel is subjected to a longitudinal wave travelling in the negative x-direction (the coordinate systems are defined in Figure 10) with $\sigma_x = \sigma_0 H(x-D/2+C_L t)$, $\sigma_y = [\nu/(1-\nu)]\sigma_x$ ($\tau_{xz} = 0$), H being the Heaviside step function (C_L and ν are the longitudinal wave speed and Poisson's ratio of the full space medium). Figure 11 shows the dynamic stress concentration σ_x/σ_0 at $r = d/2$ (lining), $\theta = \pi/2$ and at $r = D/2^+$ (infinite medium), $\theta = \pi/2$ for a liner and an enclosing medium with mass densities 0.00073 lb sec^2/in^4 (7805 kg/m^3) and 0.00025 lb sec^2/in^4 (2673 kg/m^3), respectively, longitudinal wave speeds 221000 in/sec (5613 m/s) and 208000 in/sec (5283 m/s), Poisson's ratio equal to 0.25 for both materials, d = 408 in (10.4 m) and D = 424 in (10.8 m). The analysis was carried out (see Reference 30) using a frequency domain boundary element formulation combined with the Laplace transform. Both the inner surface of the tunnel and the interface between the liner and the infinite domain were treated using isoparametric three-node boundary elements (30). The results were compared with those of other solutions, numerical as well as analytical, and were found to be in good agreement.

An application of the time domain boundary element formulation referred to earlier (References 4, 5, 18, 19, 20 and 39) will be extracted from the work by Karabalis and Beskos (18). Along with other results, they have obtained the response W of a rigid square foundation of width 2B, on the surface of a homogeneous half space, to a rectangular impulse of duration Δt and magnitude Z in the vertical direction. The foundation is covered with 64 equal square elements and the analysis is simplified by imposing the so-called relaxed boundary conditions so that no elements are necessary on the free surface of the half space and only vertical displacements and tractions are considered in the analysis with practically negligible error (see References 4, 5, 12 and 13). Figure 12 shows the variation of GBW/Z with $t/\Delta t$ for $\Delta t = (1/4)B/(C_L \pi^{1/2})$ and $\nu = 1/3$, G, C_L and ν being the shear modulus, the longitudinal wave speed and Poisson's ratio of the half space. In fact, the duration of the impulse was set equal to the time step employed in the calculations. Such an impulse response can be subsequently utilized in computations for other vertical disturbances of the foundation. It should be noted that the time step was selected equal to the time required for longitudinal waves originating at the center of a boundary element to excite an area on the boundary equal to that of the element. Such a selection is convenient in evaluating some integrals (see Reference 18).

Concluding Remarks

Undoubtedly, the boundary element method shows promise for advancing the state of the art in the analysis of dynamic soil-structure interaction. The appeal of the method in problems of dynamics of soil-structure systems has been enhanced considerably by the development of procedures for the computation of fundamental solutions for layered media. Research efforts towards improving the efficiency of this computation are indispensable. The method has already provided the means of obtaining accurate solutions to a variety of problems that other methods cannot deal with satisfactorily. For problems posed on an infinite domain, like those arising in soil dynamics, the method seems natural. Of course, further work is desirable in several directions which are not particular of the applications considered herein. However, the method is under intensive development and it is reasonable to expect substantial improvements.

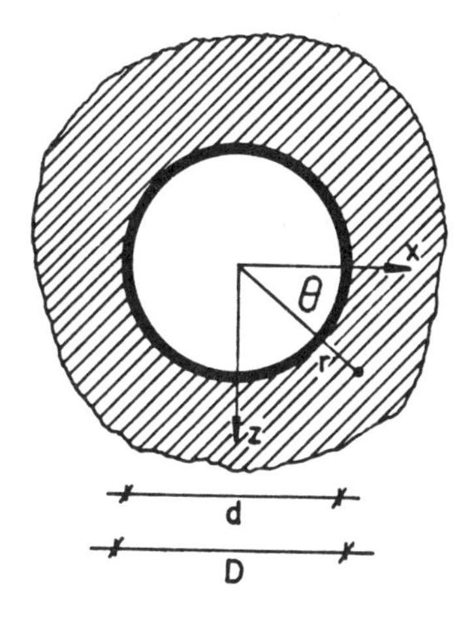

Figure 10: A circular cylindrical lined tunnel.

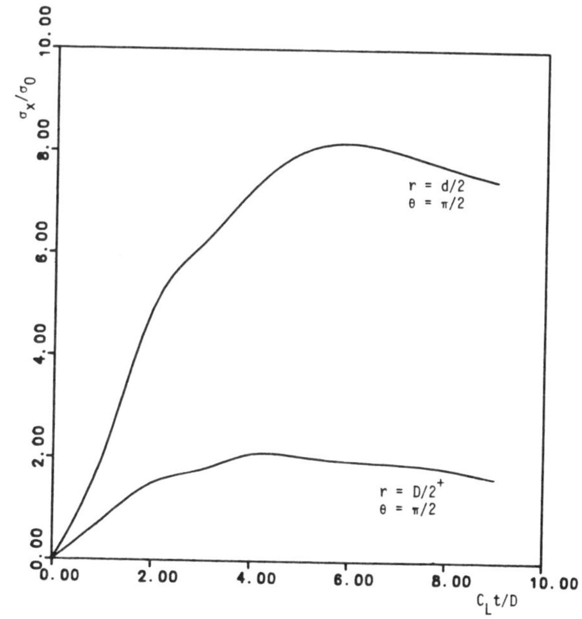

Figure 11: Dynamic stress concentration in the liner and the surrounding medium (Reference 30).

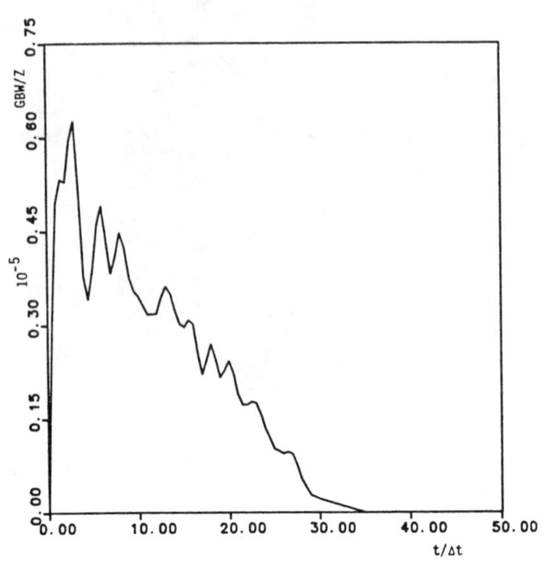

Figure 12: Response of a rigid square foundation on the surface of a homogeneous half space ($\nu = 1/3$) to a rectangular impulse ($\Delta t = (1/4)B(C_L\pi^{1/2})$, Reference 18).

References

1. Abascal, R. and Dominguez, J., "Dynamic Behavior of Strip Footings on Non-Homogeneous Viscoelastic Soils," Proceedings, International Symposium on Dynamic Soil-Structure Interaction, Minneapolis, Minnesota, 4-5 September 1984, pp. 25-35.

2. Apsel, R.J., "Dynamic Green's Functions for Layered Media and Applications to Boundary-Value Problems," Ph.D. Dissertation, University of California, San Diego, La Jolla, 1979.

3. Apsel, R.J. and Luco, J.E., "On the Green's Functions for a Layered Half-Space: Part II," Bulletin of the Seismological Society of America, Vol. 73, No. 4, August 1983, pp. 931-951.

4. Beskos, D.E. and Karabalis, D.L., "Dynamic Response of Three Dimensional Foundations," Final Report, Part A, National Science Foundation Research Grant CEE-8024725, January 1984.

5. Beskos, D.E. and Spyrakos, C.C., "Dynamic Response of Strip Foundations by the Time Domain BEM-FEM Method," Final Report, Part B, National Science Foundation Research Grant CEE-8024725, January 1984.

6. Chaudonneret, M., "On the Discontinuity of the Stress Vector in the Boundary Integral Equation Method for Elastic Analysis," Recent Advances in Boundary Element Methods, edited by C.A. Brebbia, Pentech Press, 1978, pp. 185-194.

7. Chen, H.-T., "Dynamic Stiffness of Nonuniformly Embedded Foundations," Ph.D. Dissertation, The University of Texas at Austin, 1984.

8. Christian, J.T., Hall, J.R. and Kausel, E., "Soil-Structure Interaction," Report UCRL-15230, Lawrence Livermore Laboratory, 1980.

9. Cole, D.M., Kosloff, D.D. and Minster, J.B., "A Numerical Boundary Integral Equation Method for Elastodynamics. I," Bulletin of the Seismological Society of America, Vol. 68, No. 5, October 1978, pp. 1331-1357.

10. Cruse, T.A. and Rizzo, F.J., "A Direct Formulation and Numerical Solution of the General Transient Elastodynamic Problem. I," Journal of Mathematical Analysis and Applications, Vol. 22, 1968, pp. 244-259.

11. Cruse, T.A., "A Direct Formulation and Numerical Solution of the General Transient Elastodynamic Problem. II," Journal of Mathematical Analysis and Applications, Vol. 22, 1968, pp. 341-355.

12. Dominguez, J., "Dynamic Stiffness of Rectangular Foundations," Research Report R78-20, Department of Civil Engineering, Massachusetts Institute of Technology, August 1978.

13. Dominguez, J., "Response of Embedded Foundations to Travelling Waves," Research Report R78-24, Department of Civil Engineering, Massachusetts Institute of Technology, August 1978.

14. Von Estorff, O. and Schmid, G., "Application of the Boundary Element Method to the Analysis of the Vibration Behavior of Strip Foundations on a Soil Layer," Proceedings, International Symposium on Dynamic Soil-Structure Interaction, Minneapolis, Minnesota, 4-5 September 1984, pp. 11-17.

15. Gazetas, G., "Analysis of Machine Foundation Vibrations: State of the Art," Soil Dynamics and Earthquake Engineering, Vol. 2, No. 1, 1983, pp. 2-42.

16. de Hoop, A.T., "Representation Theorems for the Displacement in an Elastic Solid and their Application to Elastodynamic Diffraction Theory," Doctoral Dissertation, Delft, 1958.

17. Hull, S.W. and Kausel, E., "Dynamic Loads in Layered Halfspaces," Engineering Mechanics in Civil Engineering, Proceedings, Fifth Engineering Mechanics Division Specialty Conference, ASCE , Vol. 1, August 1984, pp. 201-204.

18. Karabalis, D.L. and Beskos, D.E., "Dynamic Response of 3-D Rigid Surface Foundations by Time Domain Boundary Element Method," Earthquake Engineering and Structural Dynamics, Vol. 12, 1984, pp. 73-93.

19. Karabalis, D.L., "Dynamic Response of Three-Dimensional Foundations," Ph.D. Dissertation , University of Minnesota, 1984.

20. Karabalis, D.L., Spyrakos, C.C. and Beskos, D.E., "Dynamic Response of Surface Foundations by Time Domain Boundary Element Method," Proceedings, International Symposium on Dynamic Soil-Structure Interaction, Minneapolis, Minnesota, 4-5 September 1984, pp. 19-24.

21. Kausel, E., "Forced Vibrations of Circular Foundations on Layered Media," Research Report R74-11, Department of Civil Engineering, Massachusetts Institute of Technology, January 1974.

22. Kausel, E. and Roesset, J.M., "Stiffness Matrices for Layered Soils," Bulletin of the Seismological Society of America, Vol. 71, No. 6, December 1981, pp. 1743-1761.

23. Kausel, E., "An Explicit Solution for the Green Functions for Dynamic Loads in Layered Media," Research Report R81-13, Department of Civil Engineering, Massachusetts Institute of Technology, May 1981.

24. Kausel, E. and Peek, R., "Boundary Integral Method for Stratified Soils," Research Report R82-50, Department of Civil Engineering, Massachusetts Institute of Technology, June 1982.

25. Kausel, E. and Peek, R., "Dynamic Loads in the Interior of a Layered Stratum: An Explicit Solution," Bulletin of the Seismological Society of America, Vol. 72, No. 5, October 1982, pp. 1459-1481.

26. Lachat, J.C. and Watson, J.O., "Effective Numerical Treatment of Boundary Integral Equations: A Formulation for Three-Dimensional Elastostatics," International Journal for Numerical Methods in Engineering, Vol. 10, 1976, pp. 991-1005.

27. Luco, J.E., "Linear Soil-Structure Interaction," Report UCRL-15272, Lawrence Livermore Laboratory, 1980.

28. Luco, J.E., "Linear Soil-Structure Interaction: A Review," Earth-

quake Ground Motion and Its Effects on Structures, AMD-53, S.K. Datta, Ed., ASME, New York, 1982, pp. 41-57.

29. Luco, J.E. and Apsel, R.J., "On the Green's Functions for a Layered Half-Space. Part I," Bulletin of the Seismological Society of America, Vol. 73, No. 4, August 1983, pp. 909-929.

30. Manolis, G.D. and Beskos, D.E., "Dynamic Response of Lined Tunnels by an Isoparametric Boundary Element Method," Computer Methods in Applied Mechanics and Enginnering, Vol. 36, 1983, pp. 291-307.

31. Manolis, G.D., "A Comparative Study on Three Boundary Element Method Approaches to Problems in Elastodynamics," International Journal for Numerical Methods in Engineering, Vol. 19, 1983, pp. 73-91.

32. Niwa, Y., Fukui, T., Kato, S. and Fujiki, K., "An Application of the Integral Equation Method to Two-Dimensional Elastodynamics," Theoretical and Applied Mechanics, Vol. 28, University of Tokyo Press, 1980, pp. 281-290.

33. Rizzo, F.J. and Shippy, D.J., "An Advanced Boundary Integral Equation Method for Three-Dimensional Thermoelasticity," International Journal for Numerical Methods in Engineering, Vol. 11, 1977, pp. 1753-1768.

34. Rizzo, F.J., Shippy, D.J. and Rezayat, M., "A Boundary Integral Equation Method for Radiation and Scattering of Elastic Waves in Three Dimensions", International Journal for Numerical Methods in Engineering, Vol. 21, 1985, pp. 115-129.

35. Rizzo, F.J., Shippy, D.J. and Rezayat, M., "Boundary Integral Equation Analysis for a Class of Earth-Structure Interaction Problems," Final Report, National Science Foundation Research Grant CEE-8013461, February 1985.

36. Roesset, J.M., "A Review of Soil-Structure Interaction," Report UCRL-15262, Lawrence Livermore Laboratory, 1980.

37. Roesset, J.M. and Tassoulas, J.L., "Nonlinear Soil-Structure Interaction: An Overview," Earthquake Ground Motion and Its Effects on Structures, AMD-53, S.K. Datta, Ed., ASME, New York, 1982, pp. 59-76.

38. Seed, H.B. and Lysmer, J., "The Seismic Soil-Structure Interaction Problem for Nuclear Facilities," Report UCRL-15254, Lawrence Livermore Laboratory, 1980.

39. Spyrakos, C.C., "Dynamic Response of Strip Foundations by the Time Domain BEM-FEM Method," Ph.D. Dissertation, University of Min-

nesota, 1984.

40. Tassoulas, J.L., Roesset, J.M., and Kausel, E., "Consistent Boundaries for Semi-Infinite Problems," Computational Methods for Infinite Domain Media-Structure Interaction, The Applied Mechanics Division, ASME, Vol. 46, 1981, pp. 149-166.

41. Tassoulas, J.L., "Elements for the Numerical Analysis of Wave Motion in Layered Media," Research Report R81-2, Department of Civil Engineering, Massachusetts Institute of Technology, February 1981.

42. Waas, G., "Linear Two-Dimensional Analysis of Soil Dynamics Problems in Semi-Infinite Layered Media," Ph.D. Dissertation, University of California, Berkeley, 1972.

43. Waas, G., Riggs, H.R. and Werkle, H., "Displacement Solutions for Dynamic Loads in Transversely-Isotropic Stratified Media," Earthquake Engineering and Structural Dynamics, Vol. 13, 1985, pp. 173-193.

44. Wheeler, L.T. and Sternberg, E., "Some Theorems in Classical Elastodynamics," Archive for Rational Mechanics and Analysis, Vol. 31, 1968, pp. 51-90.

45. Zienkiewicz, O.C., Kelly, D.W. and Bettess, P., "The Coupling of the Finite Element Method and Boundary Solution Procedures," International Journal for Numerical Methods in Engineering, Vol. 11, 1977, pp. 355-375.

Computer Programs

George D. Manolis*, A.M. ASCE

The boundary element method (BEM) has recently emerged as a strong candidate for the solution of non-trivial problems in structural mechanics. Although integral equation formulations for problems in elasticity appeared in the literature over a century ago, construction of numerical algorithms based on these formulations took place during the last decade. BEM procedures are very promising because (a) in most cases it is possible to model the surface of the problem only, thus substantially reducing the size of the resulting system of algebraic equations and (b) since numerical quadrature techniques are directly applied to the boundary integral equations which are a closed-form solution of the problem at hand, high levels of accuracy can be expected. The BEM yields a system of algebraic equations in much the same way as domain-type techniques such as the finite element method (FEM) and the finite difference method (FDM). The difference is that the BEM system matrix is smaller, but fully populated and non-symmetric. In general, much of the numerical machinery developed for the FDM and especially the FEM can be used for the construction of BEM - based computer programs.

Introduction

Compared to the FEM and FDM, very few general purpose computer programs based on the BEM exist as of this date. The author is aware of the following codes only: BEASY (14), BEST 3D (3), and CASTOR (8). All three will be discussed in detail later. In addition, there is in all likelihood a large number of special purpose BEM programs in existence today. It is difficult, however, to monitor them because of lack of exposure. The special purpose programs that the author was able to locate in the literature will also be discussed subsequently.

An overview of the BEM was given in the previous chapters where its fields of application were identified, the theory behind the method explained, its numerical implementation discussed, and its advantages as well as disadvantages demonstrated through examples. Construction of computer codes based on the BEM is conceptually simple once the method becomes established. It is, however, far from easy: the work is tedious and requires cooperation among a number of in-

* Associate Professor, Department of Civil Engineering, State University of New York at Buffalo, Buffalo, N.Y. 14260.

dividuals, some of whom are knowlegable about the method and some who are computer programming experts. The whole process is time consumming and one can never be too careful: an endless verification process is required before the program can be released to the general public. Without general purpose programs, however, the method cannot reach the general engineering community and it remains only within a closed circle of researchers and specialists. There is also a marketing aspect to this business because a general purpose program must be advertised and supported on a continuous basis. Finally, selling or leasing of the program must generate enough revenue to cover the initial investment and cost of upkeeping the program.

Typical general and special purpose programs are modular in form to facilitate periodic updating as well as future expansion. The difference between these two types of programs is in scope. As their name implies, general purpose programs are able to solve a wide variety of problems within a particular discipline. Occasionally, they venture into more than one disciplines, especially if they have to solve coupled problems. Special purpose programs have a narrow scope because they are built to solve a particular problem. There is also a question of size, with special purpose programs being much smaller than general purpose ones. Most programs are written in Fortran, although some early small programs are in Basic and some newer ones are in Pascal. Most commercially - released programs are machine independent, i.e., they can be adapted with minor modifications to run on a number of medium to large size computing machines.

A complete general purpose BEM - based program for structural analysis must have the following features: (a) Preprocessing facilities for automatic mesh generation and ease of data input. (b) A library of one, two, and three dimensional elements of various types such as beams, plates, shells, membranes, solid elements and crack elements. (c) Efficient numerical quadrature schemes for forming the system matrices by integrating fundamental solution-shape function products over elements. (d) Efficient reduction schemes for solving the resulting system matrices in-core for small problems and out-of-core for large ones and (e) Post - processing facilities for an intelligent presentation of the results. In addition, a general purpose program must have static, steady-state dynamic and transient dynamic capabilities and to be able to handle both material and geometric nonlinearities. It is good if the program can solve governing equations other than the ones for structural mechanics such as the ones for scalar wave propagation case , because it then becomes possible to solve coupled problems such as fluid-structure interaction, consolidation, thermoelasticity, etc.

In what follows, the theoretical foundations of the BEM will be presented and numerical implementation aspects will be discussed. This background material is necessary in view of the fact that the structure of a typical BEM-based program will be subsequently discussed in detail. Next, three general-purpose programs are presented and their capabilities gaged through examples. Finaly, an extensive list of special purpose programs based on the BEM is given.

Basic Equations and Formulation

The basic equations of linear dynamic elasticity are the equations of motion

$$\sigma_{ij,i} + \rho f_j = \rho \ddot{u}_j \quad (1)$$

the kinematical relationships

$$\varepsilon_{ij} = \tfrac{1}{2}(u_{i,j} + u_{j,i}) \quad (2)$$

and the constitutive law

$$\sigma_{ij} = \lambda \varepsilon_{\kappa\kappa} \delta_{ij} + 2\mu \varepsilon_{ij} \quad (3)$$

In the above, $u_i(\underline{x},t)$ is the displacement vector at $\underline{x}$ and at time t, σ_{ij} and ε_{ij} are the stress and strain, respectively, and f_j the body force per unit mass. Furthermore, λ and μ are the Lamé constants and ρ the mass density. The Cartesian coordinate system is used and the summation convention is implied for repeated indices. Furthermore, commas indicate spatial differentiation, dots indicate time derivatives and Kronecher's delta δ_{ij} is equal to unity if $i = j$ and is zero otherwise. Equations 1-3 can be combined to give a system of governing differential equations (Navier- Cauchy) in terms of the displacement u_i as

$$(\lambda + \mu)\, u_{i,ij} + \mu\, u_{j,ii} + \rho f_j = \rho \ddot{u}_j \quad (4)$$

Equation 4 is for a three-dimensional body of surface S and volume V. Plane stress, plane strain, and antiplane strain are all special cases. By going back to the basic equations (1)-(3) and introducing various simplifying assumptions, the governing equations for beam, membrane, plate and shell vibrations can be obtained. If the time dependence of all variables is of the form $\exp(i\omega t)$, where $i = \sqrt{-1}$ and ω is the frequency, then the steady-state or harmonic case is recovered. The same situation results if the Fourier transform is applied to the governing differential equations. Under certain conditions (20), the Laplace transformed governing equations can be directly obtained from the Fourier transformed ones by replacing $i\omega$ with the Laplace transform parameter s. If the inertia effects manifested by the accelaration term in the right-hand side of Eqs. 1 or 4 are neglected, then the static case results. Material nonlinearities are captured by introducing new constitutive laws in place of Eq. 3, while geometric nonlinearities are reproduced by appropriate modifications of the kinematical relationships, Eq. 2.

Construction of boundary integral equations completely equivalent to the governing differential equations as a mathematical statement of the problem at hand is now standard process (2). Their general form is:

$$c_{ij}(\underline{\xi})u_i(\underline{\xi},t) = \int_S \{G_{ij}(\underline{x},\underline{\xi},t) * t_j(\underline{x},t) - F_{ij}(\underline{x},\underline{\xi},t) * u_i(\underline{x},t)\}\, dS(\underline{x})$$
$$+\int_V \rho\{G_{ij}(\underline{x},\underline{\xi},t) * f_i(\underline{x},t) + G_{ij}(\underline{x},\underline{\xi},t)\mathring{v}_i(\underline{x}) + \dot{G}_{ij}(\underline{x},\underline{\xi},t)\mathring{u}_i(x)\}\, dV(\underline{x})$$
$$+\int_V \{B_{ikj}(\underline{x},\underline{\xi},t) * \mathring{\sigma}_{ik}(\underline{x},t)\}\, dV(\underline{x}) \quad (5)$$

In the above, tensors G_{ij}, F_{ij}, and B_{ikj} are fundamental solutions of Eq.4 for the displacements, tractions, and strains, respectively. Also, $\mathring{v}_i$, $\mathring{u}_i$, and $\mathring{\sigma}_{ik}$ are initial velocities, initial displacements, and initial stresses, respectively. Finally, c_{ij} is the jump term whose value depends on the field point $\underline{\xi}$ and operation * denotes time convolution. Equation 5 gives the displacement at any point in the solid in terms of the boundary displacements and tractions, the body force, initial values and interior stresses for the general nonlinear transient problem. It should be noticed that all material nonlinearities can be accounted for via the last volume integral in Eq. 5 and that for steady-state or quasi-static problems, the time convolutions degenerate to simple multiplications. Integral equation formulations for cases of structural mechanics discussed previously can be obtained as special cases from Eq. 5.

The direct BEM results from numerical solution of Eq. 5. To that purpose, the boundary and occasionally the interior of the body in question is discretized by surface elements and cells, respectively. These are much the same as the elements used in the FEM. Variation of displacements, tractions and stresses within an element is decribed via shape functions. By allowing field point $\underline{\xi}$ to coincide with all boundary nodes, a system of constraint equations is generated that allows solution of the unknown boundary quantities in terms of the prescribed ones. Both surface and volume integrals are numerically done in local coordinates using Gaussian quadrature schemes. If $\underline{\xi}$ coincides with $\underline{x}$, a singular case results and computation of the resulting Cauchy principal values is combined with computation of the jump terms and done using indirect schemes (11, 21, 27). Otherwise, we have nonsingular cases and simple Gaussian quadrature suffices. Once the boundary solution is finished, then displacements and stresses at selected points in the interior of the body can be computed via Eq.5.

Following this standard numerical implementation process, the resulting system equations in dynamics are either of the form

$$[F]^1\{u\}^n + [G]^1\{t\}^n = [F]^2\{u\}^{n-1} + \ldots + [F]^n\{u\}^1 + [G]^2\{t\}^{n-1} + \ldots + [G]^n\{t\}^1 \tag{6}$$

for a time-stepping approach or

$$[\tilde{F}]\{\tilde{u}\} = [\tilde{G}]\{\tilde{t}\} \tag{7}$$

for a transformed approach. Equation 7 is valid for a sequence of values of the transform parameter. In the above, $\{u\}$ and $\{t\}$ are vectors of the nodal displacements and tractions, respectively, and $[F]$ and $[G]$ are system matrices. The superscripts in Eq. 6 denote time step with n being the current time and the overbars in Eq. 7 denote transformed (Laplace or Fourier) quantities. Following imposition of the boundary conditions, Eqs. 6 and 7 can be written in the form

$$[A]\{x\}^n = [B]\{y\}^n + \{w\}^{n-1} = \{z\}^n \tag{8}$$

and

$$[\tilde{A}]\{\tilde{x}\} = [\tilde{B}]\{\tilde{y}\} = \{\tilde{z}\} \tag{9}$$

respectively. In the above, $\{x\}$ and $\{y\}$ are the unknown and prescribed boundary quantities, respectively, and $\{w\}$ represents the contribution of the previous time steps. Matrices $[A]$ and $[B]$ result from partitioning and subsequent combinations of $[F]$ and $[G]$, and $\{z\}$ is the known right-hand side. In the case of inelastic problems, the incremental character of Eq. 8 is still applicable, except that computation of the contribution of the plasticized regions ($\{w\}^{n-1}$) is done iteratively and requires volume integrations.

Structure of a Typical Program

A simplified flowchart of a typical BEM program for structural analysis is shown in Fig. 1. It should be noted that state-of-the-art at present is three - dimensional material nonlinearities on one hand and three-dimensional transient dynamics on the other hand. Many operations such as pre-and post-processing, numerical integration, assembly of the system matrices, imposition of boundary conditions, matrix reductions and computation of boundary and interior displacements, tractions and stresses are common to all types of structural analysis. The element library, the library of Gaussian integration points, and utility libraries for matrix multiplication and other standard matrix operations must also be made available to the entire program independently of the type of analysis sought.

Static analysis is now standard (19) and resembles FEM-based static analyses. The new element introduced by an inelastic analysis (2) is volume discretization of plasticized

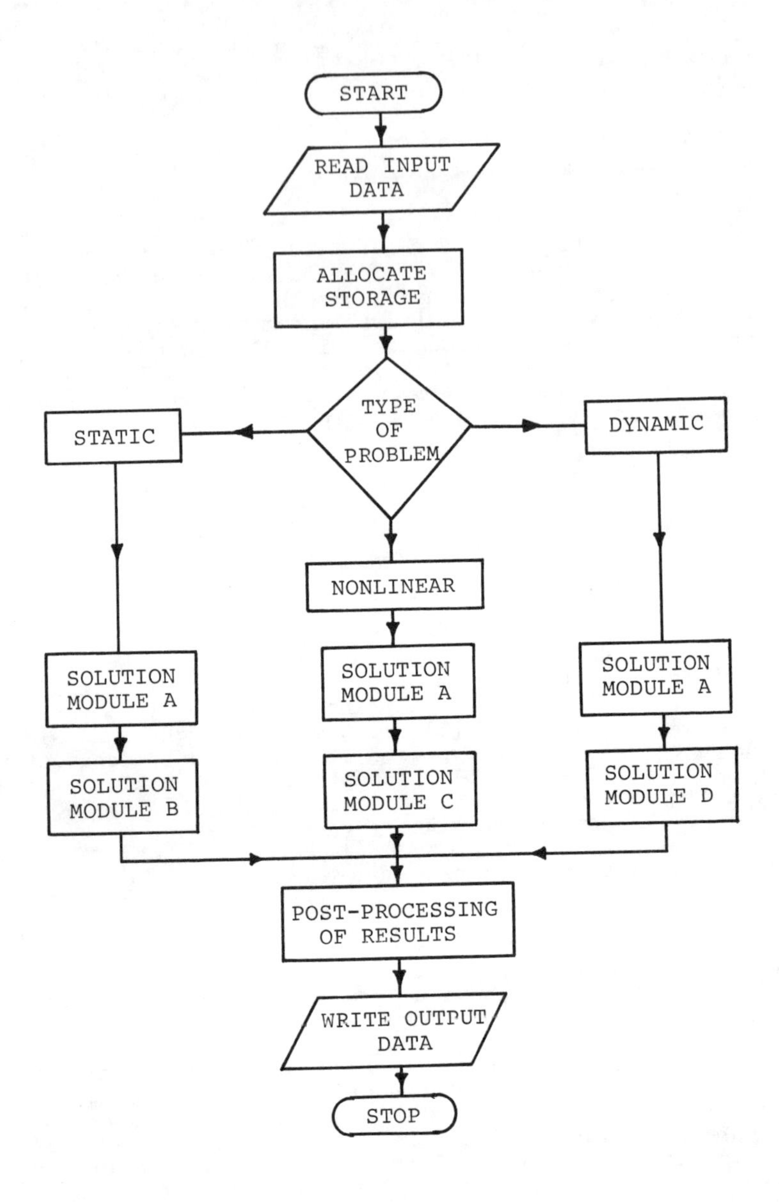
START
READ INPUT DATA
ALLOCATE STORAGE
TYPE OF PROBLEM
STATIC
DYNAMIC
NONLINEAR
SOLUTION MODULE A
SOLUTION MODULE B
SOLUTION MODULE A
SOLUTION MODULE C
SOLUTION MODULE A
SOLUTION MODULE D
POST-PROCESSING OF RESULTS
WRITE OUTPUT DATA
STOP

where

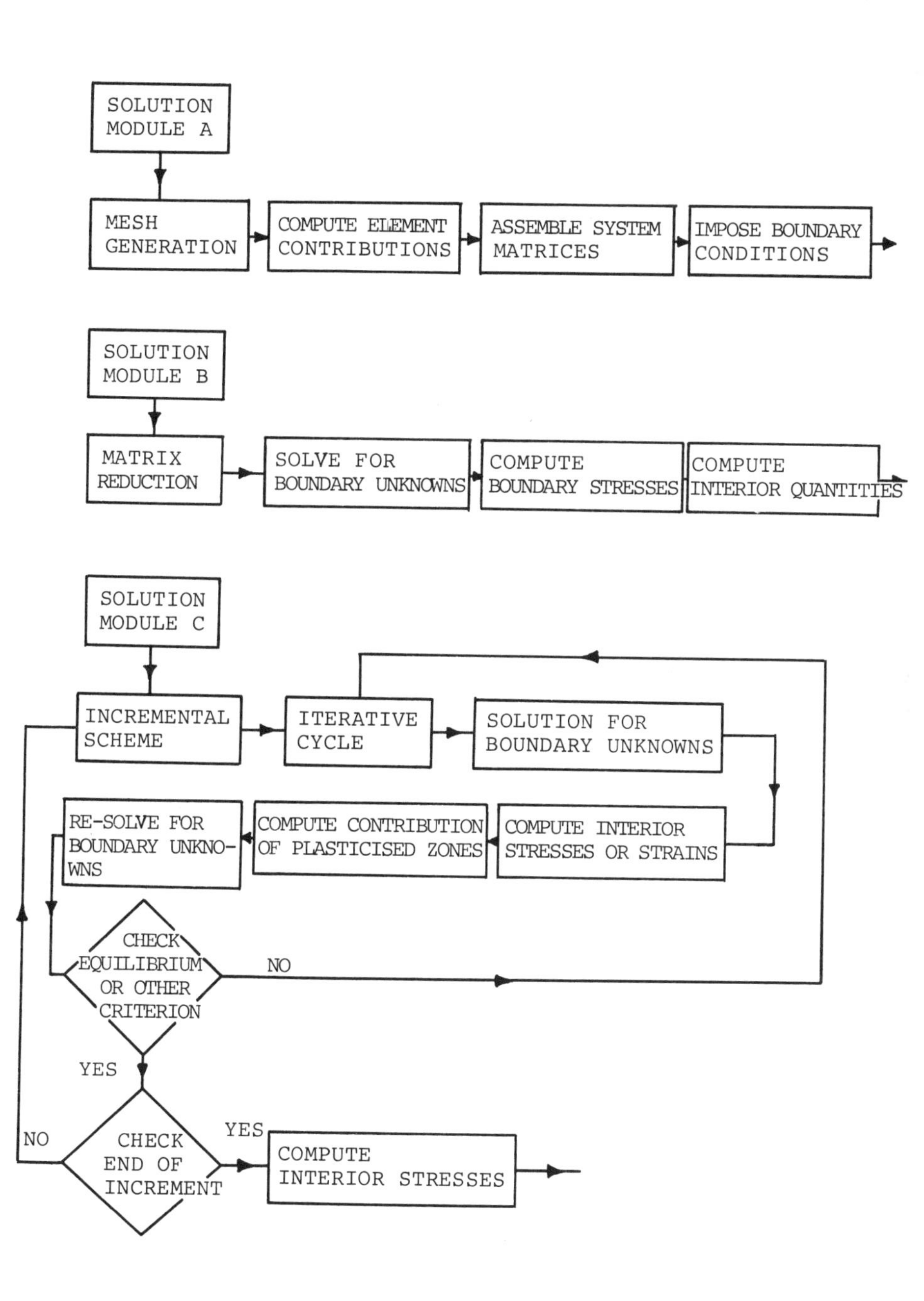
SOLUTION MODULE A
MESH GENERATION
COMPUTE ELEMENT CONTRIBUTIONS
ASSEMBLE SYSTEM MATRICES
IMPOSE BOUNDARY CONDITIONS
SOLUTION MODULE B
MATRIX REDUCTION
SOLVE FOR BOUNDARY UNKNOWNS
COMPUTE BOUNDARY STRESSES
COMPUTE INTERIOR QUANTITIES
SOLUTION MODULE C
INCREMENTAL SCHEME
ITERATIVE CYCLE
SOLUTION FOR BOUNDARY UNKNOWNS
COMPUTE INTERIOR STRESSES OR STRAINS
COMPUTE CONTRIBUTION OF PLASTICISED ZONES
RE-SOLVE FOR BOUNDARY UNKNO-WNS
CHECK EQUILIBRIUM OR OTHER CRITERION
NO
YES
CHECK END OF INCREMENT
NO
YES
COMPUTE INTERIOR STRESSES

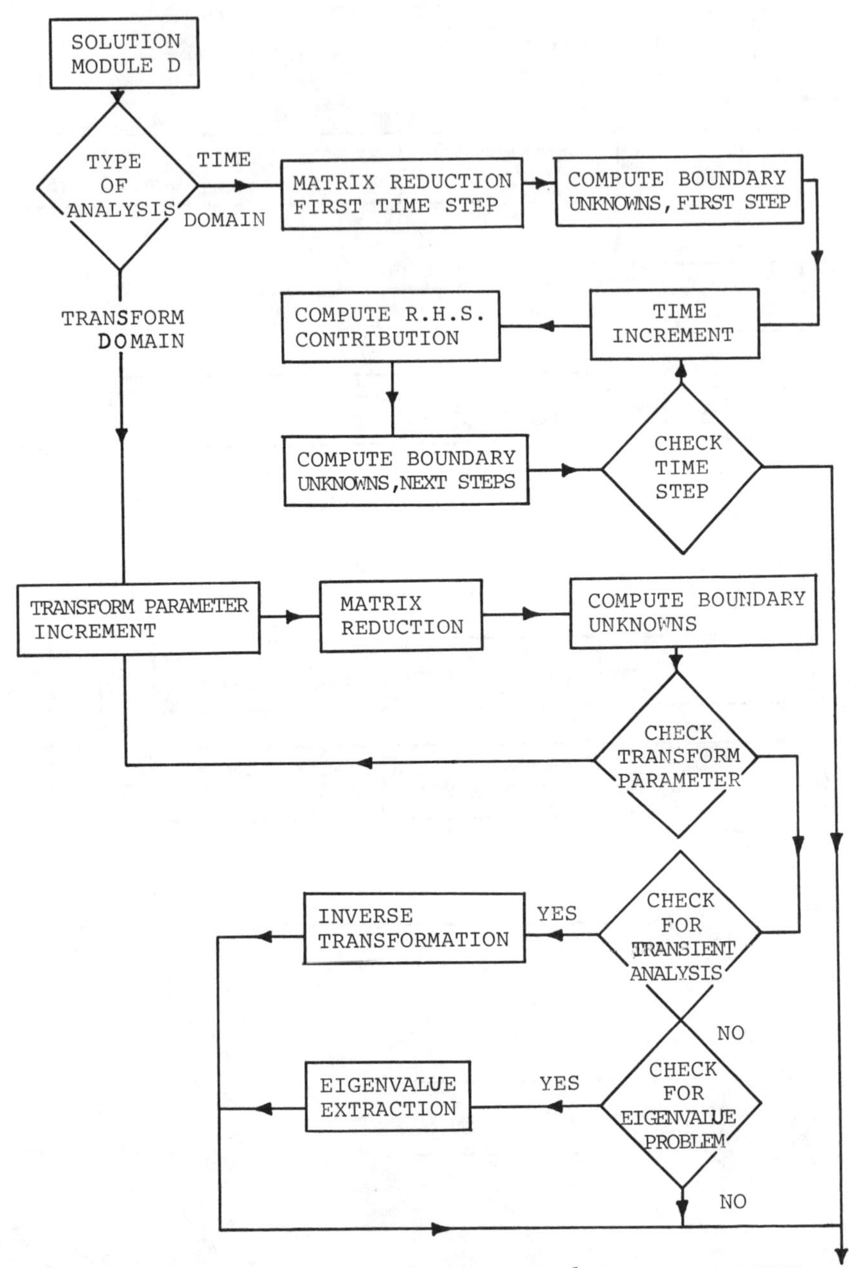

Figure 1. Flowchart of typical BEM general-purpose program.

regions. Implementation is done via standard incremental / iterative schemes. Two basic approaches are used, i.e., either an initial stress or an initial strain technique. In both cases, the iterative cycle begins by computing the contribution of the volume integrals to the right-hand side of the algebraic system of equations using previously known values of the stress or strain at the plasticized interior locations. Boundary solution follows and updated values of the stress or strain are computed from the new boundary information. These update values are used to correct the right hand side and a new boundary solution is obtained. The cycle is repeated until satisfaction of an imposed criterion is achieved, in which case one proceeds with a new load increment. The usual criterion for the iteration cycle is that successive values of the stress or strain at an interior point are within a prescribed tolerance. The end of the loading sequence is prescribed information. As the loading proceeds, plasticity spreads to new regions which must be described by additional discretization.

In the case of dynamic analysis (4,21,27) one must distinguish between time stepping and transform domain approaches. In the latter case there are also two possibilities: Fourier transform or Laplace transform domain cases. The Fourier transform has the advantage that it corresponds to the time harmonic or steady state environment. As such, it results in the classical free vibration problem if no loading is prescribed and upon solution of the resulting homogeneous system of equations the natural frequencies (eigenvalues) and curresponding modal shapes (eigenvectors) are obtained. Inverse numerical transformation of the solution obtained for a spectrum of values of either the frequency or the Laplace transform parameter results in the transient solution. This operation is known as Fourier synthesis in the former case.

The following discussion expounds on a subroutine arrangement that closely resembles the flowchart of Fig. 1. The emphasis is on the dynamic option. The static case can be obtained from the dynamic one as a special case.The inelastic option can be obtained by inserting the iterative scheme in the incremental loop of the time stepping algorithm and ignoring the inertia effects. The inelastic option was previously discussed in detail and no further comments will be made. The subroutine arrangement is commensurate with the programming experience of the author (21) and other programming options obviously exist. For generality, no particular name is given to the subroutines: letters denote major operations and numbers operations within these major tasks.

PROGRAM - The main program. Control variables are read from a data file supplied by the user. Storage space is dynamically allocated to the rest of the program. Subroutines SUBRA01-SUBRJ01 are sequentially called.

SUBRA01 - Data pertaining to geometry of the problem(nodal poinds and element connectivity), material properties and boundary conditions is read from the data file. This information is either placed in common statements and made available to the rest of the program or stored in arrays for subsequent use. Description of the problem can be done automatically through use of a mesh generator.

SUBRB01 - Library of Gaussian points and weights for the Gauss-Legendre quadrature schemes.

SUBRC01 - Library of local coordinates of typical boundary elements.

SUBRD01 - Computes values of the shape functions and their derivatives at the nodes of all boundary elements. This information is useful in computing coordinates of integration points as well as normal and tangent vectors at those points.

SUBRE01 - Subroutine that controls assebly of the system matrices [G] and [F] and of the field variable vectors {t} and {u} during time step n or during the nth value of the transform parameter. Furthermore, imposition of the boundary conditions is also controlled by this subroutine. Mixed boundary value problems can be solved. Subroutines SUBRE02 - SUBRE09 are sequentially called.

SUBRE02 - The surface integration scheme commences here. The subroutine first loops over all nodal points and then over all elements. If a nonsingular case is encountered, SUBRE03 and SUBRE04 are called. Otherwise, SUBRE05 and SUBRE06 are called for singular cases.

SUBRE03 - This subroutine controls the integration error through element subdivision. The number of subelements depends on the distance between field point and element, on the order of the quadrature scheme and on the error tolerance prescribed.

SUBRE04 - Numerical integration of nonsingular cases. The resulting coefficient submatrices are written on tapes. During the first step, the static fundamental solutions are also integrated because these results are needed in computing jump terms and Cauchy principal-values. During the integration process, subroutines SUBRE10 through SUBR E12 are called.

SUBRE05 - The singular element is subdivided into triangles and each triangle is further subdivided into sectors along both radial and angular directions. These subdivisions depend on the position of the singular node on the element.

SUBRE06 - Numerical integration of singular cases. Integration is done using polar coordinates. The comments made in SUBRE04 apply here as well.

SUBRE07 - Assembles the [G] coefficient matrix for time step n or for nth value of the transform parameter by reading from tapes the coefficient submatrices computed in subroutine SUBRE04 and SUBRE06.

SUBRE08 - Assembles the [F] coefficient matrix in the manner described above.

SUBRE09 - In accordance with the boundary conditions, system matrices [G] and [F] are appropriately partitioned into matrix [A], multiplying the unknown displacements and tractions {x} and into matrix [B] multiplying the prescribed values {y}. The latter multiplication is carried out and a right-hand side vector {z} results.

SUBRE10 - This subroutine is called by both SUBRE04 and SUBRE06. It returns the shape functions, their derivatives, the Jacobian determinant and the outward unit normal vector, all evaluated at the integration points for a given type of boundary element.

SUBRE11 - Supplies values of the dynamic fundamental singular solutions to both SUBRE04 and SUBRE06.

SUBRE12 - Supplies values of the static fundamental singular solutions to both SUBRE04 and SUBRE06 during the first time or transform parameter increment only.

SUBRF01 - Out-of-core inversion of matrix [A] by Gauss elimination. For the time domain approach, this is done only once during the first time increment $n = 1$. For the transform domain approach, the unknown displacements and tractions are found by multiplying the inverse of [A] with the right hand side vector {z} for every value of transform parameter.

SUBRG01 - The time-stepping algorithm is implemented in this subroutine. The information from all previous time steps n-1, n-2,..., 2,1 is read from tapes and processed. Subroutine SUBRG02 is called. The current unknown displacements and tractions are found by adding the contributions of all previous time steps $\{z\}^{n-1}$ to the current right hand side $\{z\}^n$ and pre-multiplying by the inverse of $[A]$ for n=1.

SUBRG02 - Computes the contributions of all previous time steps $\{z\}^{n-1}$ and returns it to SUBRG01.

SUBRH01 - The inverse Laplace or Fourier transformations are implemented in this subroutine. Once the entire spectrum of the transform displacements and tractions has been collected, numerical inversion commences. Thus, the displacements and tractions are obtained in discrete form in the time domain.

SUBRH02 - This subroutine contains the fast Fourier transform algorithm that speeds up the inversion process done in SUBRH01.

SUBRI01 - Computes the time history of the stress tensor at prescribed points on the boundary from the time history of the displacements and tractions.

SUBRI02 In case of wave propagation problems, the incident field in the form of displacements, tractions and stresses is read from the data file and superimposed to the scattered fields that are computed in the program.

SUBRJ01 Both input information and information computed by the program are printed and / or plotted in this subroutine. Standard pre-and post-processing subroutine packages can be referenced for completing this task.

There is also a large number of small subroutines that do specialized tasks such as standard matrix and vector operations, computation of minimum and / or maximum distances between nodes and elements, implementation of symmetry, printing and plotting of data, etc. By using-dynamic memory allocation and by storing data on fast access discs instead of storing in core, it is possible to solve up to medium sized three-dimensional problems with less than 200,000 octal words of core memory and without having to use overlays. The time requirements are heavy, however, and depend primarity on the number of time or transform parameter steps used.

General Purpose Programs

Three general purpose programs will be discussed in this section:

(a) BEASY, Danson et al (14) : This program is capable of two-dimensional, axisymmetric and three dimensional potential and stress analysis. It can also do uncoupled thermal analysis by first solving a potential type problem to obtain the temperature and heat flux distributions and then using this information in the stress analysis. Under development (and probably completed by now) are two-dimensional and axi-symmetric elastoplastic stress analysis, two-dimensional solution of the diffusion equation for time-dependent thermal calculations, and solution of the scalar wave equation in three-dimensions.

The element library contains constant, linear and quadratic one-dimensional (line) and two-dimensional (area) elements. The area elements are quadrilaterals, but under development are triangular elements. All these elements are of the non-conforming type, i.e., there are two nodal directories : one for description of the geometry and one for description of the field variables. The former ones are along the edges of an element while the latter ones form a grid inside the element. Although non-conforming elements are easier to program and are ideal for meshes of rapidly changing density, conforming elements give more accurate results and require a much smaller number of nodal points to accomplish this (22). Private communication suggests that BEASY's element library is being enriched with conforming elements.

The program can handle mixed type of boundary conditions. An out-of-core block solver is used for matrix reduction. Once the boundary solution is obtained, computation of the values of the field variables at selected interior points can proceed. Symmetry is handled by simply reflecting the structure about its planes of symmetry and continuing with numerical integration of the fundamental solution - shape function kernel products over the reflected parts. This cuts down on mesh preparation and calculations by a significant amount. A very useful facility is zoning, where the boundary of the structure is divided into a finite number of subregions, each with its own material properties. This allows solution of problems with locally varying material properties.

The program is written in Fortran IV and has typical in put / out put facilities. The user may attach to the program pre-and post-processing packages. Recent versions of the program (7) have a pre-processor to help in data generation and a post-processor for data display.

A typical example solved by BEASY is shown in Fig. 2. It concerns the transient temperature distributions in a turbine disc. The initial temperature of the disc is 295.1^{o} K and values for the thermal conductivity, density and specific heat are 5. $Wm^{-1}\,^{o}K^{-1}$, 8221. kgm^{-3} and 550. $JKg^{-1}\,^{o}K^{-1}$, respectively. Figure 2(a) shows a FEM mesh with 71 isoparametric quadratic elements resulting in 278 nodes. The BEM discretization in Fig. 2(b) has 90 linear elements and 106 nodes and the structure was divided into 18 zones, each with different prescribed values for the heat transfer coeffi-

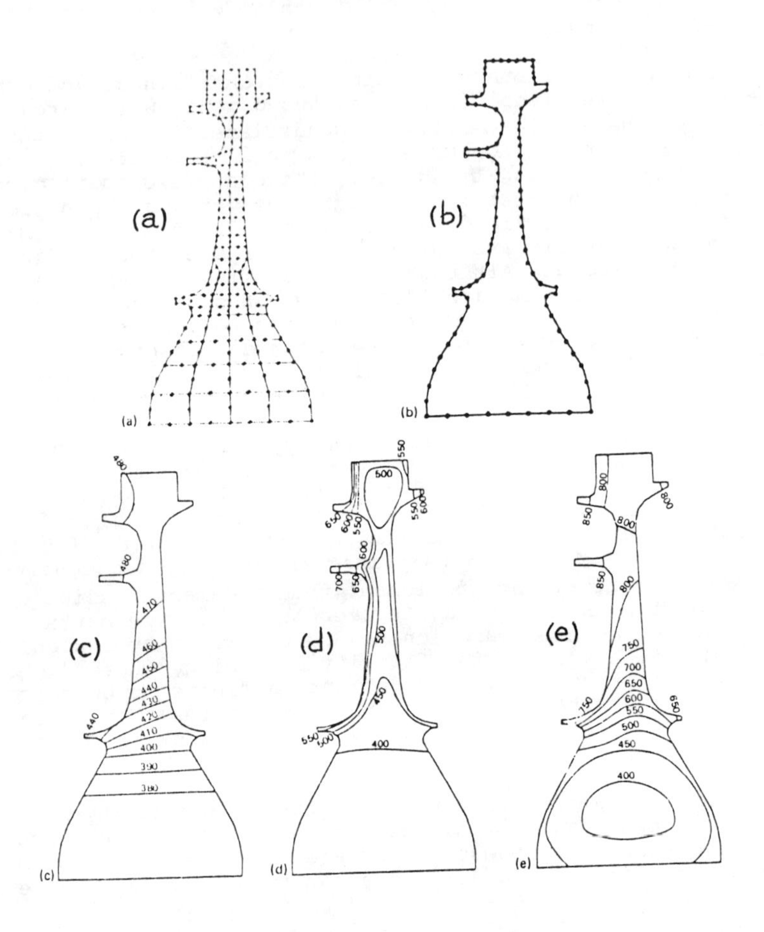

Figure 2. Turbine disc : (a) FEM mesh, (b) BEM mesh
(c), (d), (e) Transient temperature distribution

cient and the temperature of the surrounding gas. Figures 2(c)-(e) show isothermals at typical time intervals obtained by both methods that agree to within plotting accuracy. A critical comparison of BEASY against two standard FEM packages for problems of two-dimensional, axisymmetric, and three-dimensional stress analysis can be found in Brebbia et al (6).

(b) CASTOR, Chaudouet and Devalan (8) : This program is actually composed of four parts, and only one of them,CASTOR3D, is based on the BEM. The remaining three are based on the FEM. CASTOR3D is capable of three - dimensional, static linear stress analysis and three-dimensional, steady-state thermal analysis with a transient option under development. The pre-processor has a module for checking the geometry and a module for drawing the mesh, while the post-processor is capable of plotting the results in detail. The size of the program is about 35,000 Fortran statements and a maximum of 1200 nodal points can be handled. Memory is not dynamically allocated.

The program employs 8 node quadrilateral and 6 node triangular conforming area elements. It is also possible to divide the structure into a maximum of 10 subregions. Generation of nodes, elements and subregions is semi-automatic. The program can also reproduce planar and cyclic symmetry.

Mixed boundary conditions can be handled. Solution at selected interior points follows the boundary solution. In the case of subregions,the surfaces of the subregions act as boundaries.Displacement continuity and traction equilibrium is imposed on those boundaries that are interfaces. The program has re-start capabilities for a structure under multiple loading conditions or for a structure with localized changes in the geometry and boundary conditions. Discontinuities in the tractions manifested at sharp edges or corners are accounted for.

More recent versions of CASTOR3D (9) have grown to 60,000 Fortran statements and have a complete transient thermal analysis option available. Figure 3 shows a heat exchanger under thermal and mechanical loads. Due to symmetry, only a quarter of the structure is discretized into 4 subregions, 104 elements and 279 nodes. Figure 3 (a) shows the boundary conditions for the thermal problem with an additional boundary condition for the external surface being heat transfer by convection with air at 20°C. Figure 3 (b) shows the resulting temperature field, with the maximum temperature being 90° C. The boundary conditions for the stress analysis are shown in Fig. 3(c), while Fig. 3(d) depicts the iso-value curves of equivalent Von-Mises stresses.

(c) BEST3D, Banerjee et al (3): This program is still under development and is expected to reach its mature form by

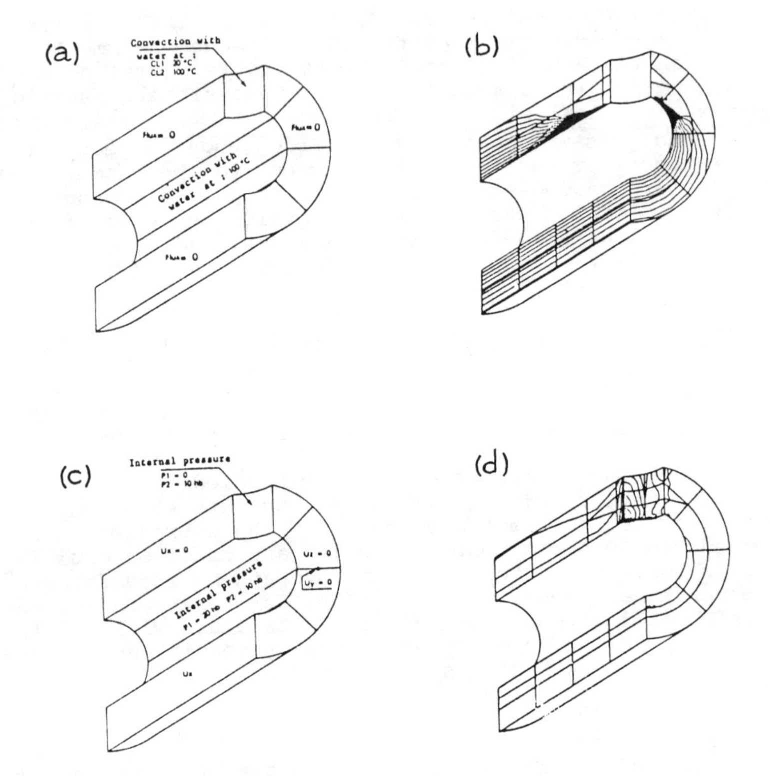

Figure 3. Heat exchanger: (a) Thermal boundary conditions, (b) Temperature distribution, (c) Stress boundary conditions, (d) Stress distribution

the end of 1987. It is a stress analysis system based on the BEM that is capable of elastic, inelastic, periodic dynamic, and transient dynamic analysis of two-and three-dimensional solids. The elastic analysis can handle both isotropic and cross-anisotropic media under thermal, centrifugal and the usual mechanical loads. The inelastic analysis includes isotropic plasticity with variable hardering, kinematic plasticity with multiple yield surfaces, and anisotropic viscoplasticity with thermally sensitive material behavior. Periodic dynamic analysis can be done in either Fourier of Laplace transformed domains, while transient dynamic analysis is either through inverse numerical transformation of the transformed solution or through time-stepping. There is also a provision for solution of eigenvalue problems via the mass matrix concept. The elastic and inelastic analysis portions of the program are complete. The dynamic analysis portions are the ones under development.

The element library has linear elements (2 node line element, 3 node triangle, 4 node quadrilateral, 8 node cell), quadratic elements (3 node line element, 6 node triangle, 8 node quadrilateral, 20 node cell), and an infinite element resulting from modification of the 8 nodes quadrilateral. In addition, a substructuring capability is provided which allows the structure to be modelled as an assembly of several generic modelling regions, each of which acts as a complete structure. Program limits are a total of 2500 nodes, 600 surface elements, 10 generic modelling regions and 20 time points for dynamic analyses.

General mixed boundary conditions are allowed. The solver used by BEST3D operates at the submatrix level, and the system matrix is stored on a direct access file. The decomposition process is Gaussian reduction to upper triangular, submatrix form. Calculation of the solution vector is carried out in a separate subroutine, using the decomposed form of the system matrix from the direct access file. Inelastic analysis requires an iteration process within a load increment to satisty the constitutive relations. Transient analysis is strictly incremental, but evaluation of the load vector requires new surface integrations at each time step.

The program is written in IBM Fortran, although development versions exist in Fortran 77. Data input is format-free. Pre-and post-processing facilities do not exist at present. This development process has spawned a number of special purpose programs for two and three dimensional,inelastic and transient analysis (1,21,26).

BEST3D has been subjected to an extensive validation effort. About fifty simple test cases exist on file. The following two examples are drawn from these cases. First, consider a long column attached to an elastic half-space and loaded in tension. The tip deflection of the beam is the sum of the extension of the beam plus the displacement of the half-space under a distributed load from the base of the column. A typical mesh is shown in Fig. 4, along with the results. It should be noted that an acceptable solution cannot be obtained by modelling the problem as a single region, unless the half-space discretization becomes very detailed. By modelling the beam and half-space as separate generic modelling regions, less than 1% error is committed with the simple mesh utilizing quadratic elements. Consider next an internally pressurized thick cylinder of inner radius a = 1.0 and outer radius b=2.0. Axial deformation of the two front and two back faces is constrained to simulate conditions of plane strain. A Von Mises model for an ideally plastic material is used. Figure 5 (a) shows two typical mesh patterns, while Fig. 5 (b) plots the radial deformation of the cylinder's outer face u_b versus applied pressure p. Both quantities are normalized using the shear modulus G, and the yield stress σ_o. The same problem was solved by modelling the cylinder with two generic modelling regions, but without any appreciable difference in the results.

Special Purpose Programs

A number of special purpose programs that have directly or indirectly appeared in the literature are presented in this section. They are listed according to the reference in which they appear.

(a) BASQUE, Lachat and Watson (19): Probably the first comprehensive program for two-and three-dimensional elastostatics based on the direct BEM. The contribution of these researchers was to introduce shape functions from the FEM to model the field variables (tractions and displacements). Thus, the entire process of mumerical integration of fundamental solution-shape function products over an element could be done using an adaptive numerical approach that allows for error control. The element library has line elements and triangular as well as quadrilateral surface elements, over which the field variables may vary in a linear, quadratic or cubic fashion. Infinite line and quadrilateral elements result from appropriate modification of the corresponding regular elements. Later versions utilize cubic spline functions for representation of the field variables. The program has subregioning facilities and can handle mixed boundary conditions. Its greatest asset is careful control of the integration error, which allows good quality results to be obtained for relatively crude discretizations.

(b) PECET, Doblaré and Alarcón (15): This is a large special purpose program for three-dimensional elastostatic analysis. It is written in Fortran V and consists of about 10,000 statements. It is modular in form. The element library has two surface elements, the 6 node triangle and the 8 node quatrilateral. Piecewise heterogeneous domains can be modelled via subzoning. It can handle mixed boundary conditions as well as body forces such as temperature, self-weight

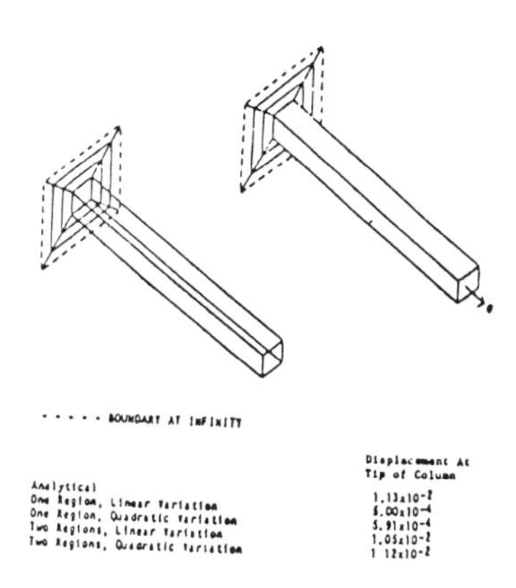

BOUNDARY AT INFINITY
Displacement At Tip of Column
Analytical 1.13x10-2
One Region, Linear Variation 5.00x10-4
One Region, Quadratic Variation 5.91x10-4
Two Regions, Linear Variation 1.05x10-2
Two Regions, Quadratic Variation 1.12x10-2

Figure 4. Column attached to an elastic half-space

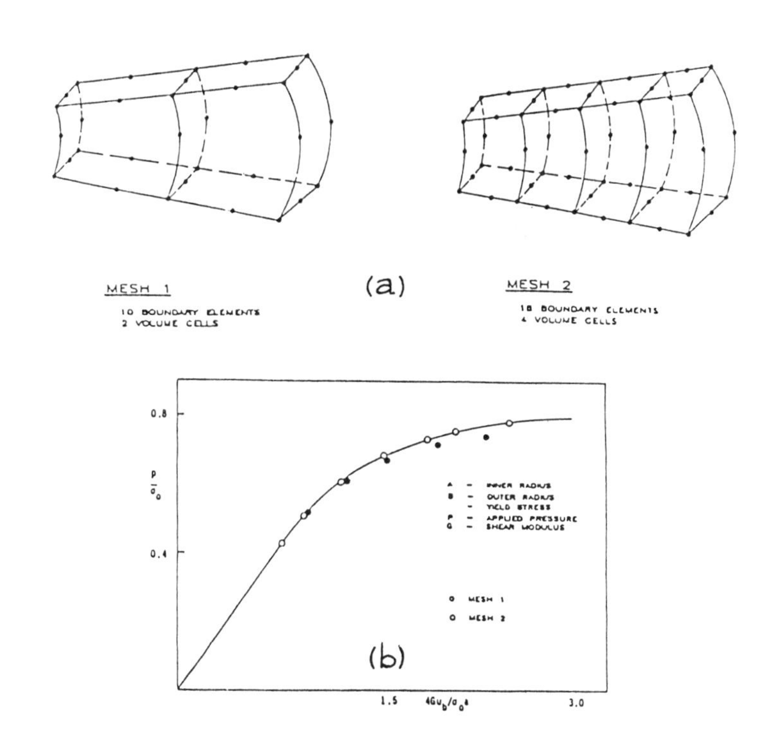

Figure 5. Thick cylinder under radial internal pressure: (a) Discretization, (b) Load-displacement response.

centrifugal forces and point loads without volume integration. Data input is free-format, and there are a few pre-and post-processing capabilities. Typical examples solved involve spheres, cubes, and doloses.

(c) Patterson et al (25): Their program is based on the indirect discrete method of modified Trefftz method. It is essentially a BEM approach that uses non-conforming surface elements. The singular integrations are handled by lifting the source point a small distance above the surface of the element so that a non-singular case results which can be integrated by conventional means. A critical evaluation of this technique can be found in Ref. 22. The program can handle three-dimensional potential and elastostatic problems. It employs quadrilateral area elements and has a sub-regioning facility. Typical design problems that have been analyzed include stress analysis of a circular bar with a spherical cavity and thermal analysis of a glandless motor circulating pump case with a hot neck.

(d) Ingraffea et al (17): This program is for computing stress intensity factors at points along crack fronts in a three-dimensional solid. It operates in an interactive viroment: a pre-processor assists the user in preparing an initial model for the structure containing one or more cracks; a BEM-based code is used for the stress analysis and a post-processor assists the user in modifying the mesh to reflect changing boundary conditions and topology and in displaying the final results. The BEM code utilizes 6 node triangular and 8 node quadrilateral surface elements. Crack elements are produced by appopriate modifications of the above two types of elements. In order to reproduce a crack, domain substructuring must be used so that the two crack surfaces belong to different subdomains. The analysis is quasi-static and the material must exhibit linear elastic material behavior.

(e) Crouch and Starfield (10) : Three BEM-based programs for two-dimensional, linear stress analysis are presented. They are based on the fictitious stress method, the displacement discontinuity method, and the standard direct boundary element method, respectively. The first two methods are actually indirect type boundary element approaches (2). The most appropriate of these programs for problems of geomechanics is the one based on the displacement discontinuity method because it can handle cracks. This method uses the analytical solution to the problem of a constant discontinuity in displacement over a finite line segment in an infinite plane. As a consequence, it is ideally suited to follow the deformation pattern of two crack boundaries as they slide with respect to each other. All three programs employ constant-type line elements and have the capability of handling symmetry. Solution at specified interior points follows boundary solution.

(f) Cruse (13) : Two programs are presented for the stress analysis of linear elastic solids. The first, called BIE-CRX

(12) is for two-dimensional fracture analysis while the second, called BINTEQ (11), is for three-dimensional elastostatics. The former uses line elements with linear variation of the field variables, as well as a crack element, while the latter uses surface elements again with linear variation of the field variables. Both are based on the direct BEM and both have the symmetry feature. Figure 6(a) shows a gas turbine disk rim loaded by uniform applied displacements on the ends to simulate disk strains. Due to symmetry, only the shaded region is discretized, as shown in Fig. 6(b). Using program BINTEQ, the peak stress distribution is plotted along the surface of the shallow notch (curve to the left of peak) and along the surface of the deep notch (curve to the right of peak). Two-dimensional results for the stress intensity factor K_t obtained by BIE-CRX are also shown. A three-dimensional FEM analysis of the same problem gives results of comparable accuracy but at a much higher computing cost.

(g) Rizzo et al (27) : A program based on the direct BEM for time harmonic analysis of elastic wave scattering problems for structures embedded in a three-dimensional homogeneous medium is presented in this reference. The program uses both fundamental solutions for the fullspace and the half-space. In the latter case, it is not necessary to discretize the horizontal free surface of the soil medium. The radiation boundary condition associated with semi-infinite media is automatically satisfied by virtue of the choice of fundamental solutions. Transient response is recovered via a numerical inverse transformation (Fourier synthesis). Special precautions are taken to avoid lack of unique solution to the resulting system of equations at certain fictitious eigenfrequencies by adopting long standing methods from acoustics. The surface elements used are the 6 node triangle and the 8 node quadrilateral. The program can also solve structural vibration type problems.

(h) Beskos and Karabalis (4) ; Beskos and Spyrakos (5) : Two programs based on the direct, time domain BEM for 3 & 2 - dimensional foundation problems are respectively presented in these references. The foundation is assumed massless, resting on a homogeneous medium and under either external dynamic loads or oblique travelling waves. Three general cases are considered: rigid surface foundation, rigid embedded foundation, and flexible surface foundation. Since the fundamental solution corresponding to the infinite space is used, discretization of the surface of the soil around the foundation is generally required. The radiation condition is automatically satisfied. Constant line and area elements are employed in conjunction with a time-stepping algorithm.

(i) Dominguez (16) : A program based on the direct, harmonic BEM is used to analyse three dimensional foundations resting on a homogeneous halfspace. The program can handle both surface and embedded foundations under relaxed and non-

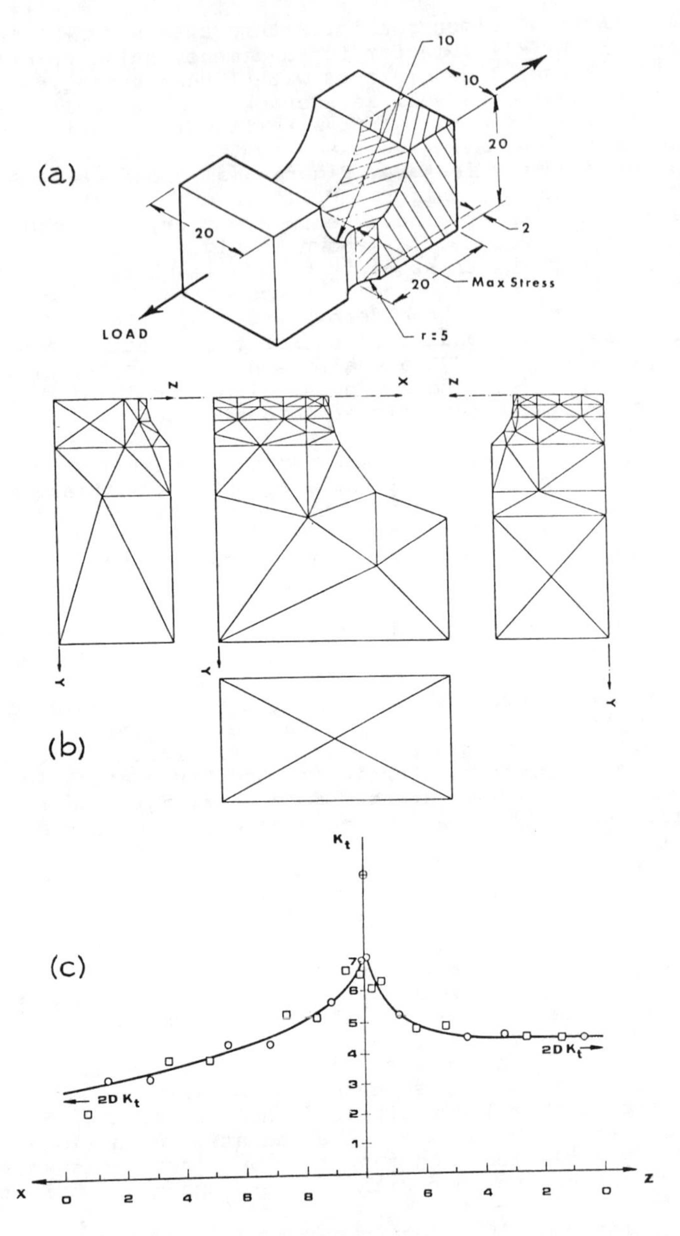

Figure 6. Idealized turbine disk rim: (a) Geometry and loading, (b) boundary discretization, (c) Stress distribution

relaxed boundary conditions. It is tailored to compute the dynamic stiffness coefficients of massless foundations and employs complex number formalism. Constant rectangular surface elements are used and it is necessary to model part of the soil outside the foundation as well as the soil-foundation interface.

(j) Kobayashi et al (18) : A versatile program combining the BEM with the FEM for solving three-dimensional problems of viscoelastodynamics is reported in this paper. The focus here is on soil - sructure interaction type problems: the FEM is for bounded interior domains consisting of the structure and of soils exhibiting non - homogeneous or anisotropic properties, while the BEM is used for the infinite or semi - infinite homogeneous and isotropic exterior domain. Solution is in the Fourier transformed domain and viscoelastic material behavior is accounted for by using complex elastic moduli. The 8 node quadratic surface element is used in BEM part of the program, while typical solid elements are used in the FEM part. The two meshes are combined through enforcement of compatibility and equilibrium conditions at the interfaces of the interior and exterior domains. The program also has some pre - and post - processing capabilities. It has been used to solve an impressive example of a bridge pier built on viscoelastic homogeneous soil with the ground surface subject to a time - harmonic, vertically polarized shear wave of oblique incidence.

(k) Mukherjee and Chandra (24) ; Mukherjee and Banthia (23): Two programs based on the direct BEM have been used for two-dimensional nonlinear problems of mechanics. The former reference focuses on plasticity and viscoplasticity in the presence of large strains and large displacements. The solution algorithm is based on a time-marching scheme using an updated Lagrangian approach. Constant line elements and square cells are used to model the surface and interior of a problem. Typical problems solved are tensile deformation of specimens with cut-outs. The latter reference focuses on problems of creep fructure. A crack is modelled as a thin elliptic cutout, while the plate material is described by an elastic - power law, creep constitutive model. Both stationary and propagating cracks can be considered. The latter case requires remeshing as the crack advances. In addition to the crack element, the same boundary element types previously described are also employed. The program has been used to solve for the stress redistribution near stationary cracks in general two-dimensional solids under constant remote loads.

(l) Wardle and Crotty (28) : A program based on the direct BEM for static two dimensional, non-homogeneous problems encountered in mining applications is described in this reference. The program employs straight line elements and assumes a linear variation of tractions and displacements along each segment. Inhomogeneities are accounted for by sub-dividing the body into piecewise homogeneous sections. Some special problems of mining engineering, such as primitive stresses acting on the system, infinite boundaries, and traction discontinuities at the corners of regions, can easily be handled. Solution at interior points follows the boundary solution. Typical problems solved include finding stresses in the surrounding medium as excavation of tunnels proceeds.

Conclusions

This chapter presented both classes of general purpose and special purpose computer programs based on the BEM for problems of structural mechanics that are available at present. Typical examples of what these programs can solve were also included. In order to familiarize the reader with basic concepts of the BEM that are necessary backgroud for understanding the workings of any BEM-based computer program, a section covering the theory and numerical implementation aspects of the method was included. Furthermore, the structure of a typical BEM program was discussed in detail to further elucidate the concepts behind general purpose program construction and to avoid making this chapter read like a simple list of commercial programs. It is finally the opinion of the author that the present BEM computer codes that are commercially available is the proverbial tip of the iceberg: the next decade will bring into being many new BEM programs as well as programs based on coupling of the FEM and BEM. In other words, the situation is similar to what the FEM was twenty years ago.

Appendix I - References

1. Ahmad, S., 'Linear and Nonlinear Dynamic Analysis by Boundary Element Method,' thesis presented to State University of New York, at Buffalo, New York, in 1986, in partial fulfillment of the requirements for the degree of Doctor of Philosophy.

2. Banerjee, P.K., and Butterfield, R., Boundary Element Methods in Engineering Science, McGraw - Hill, London, 1981.

3. Banerjee, P.K., Wilson, R.B., and Miller, N., 'Development of a Large BEM System for Three-dimensional Inelastic Analysis,' in T.A. Cruse, A.B. Pifko and H. Armen, Editors, Advanced Topics in Boundary Element Analysis, AMD - Vol. 72, ASME, New York, 1985, pp. 1-20.

4. Beskos, D.E. and Karabalis, D.L., 'Dynamic Response of Three-Dimensional Foundations,' Final Report to the National Science Foundation, Part A, Dept. of Civil and Mineral Engineering, University of Minnesota, Minneapolis, Minnesota, 1984.

5. Beskos, D.E. and Spyrakos, C.C., 'Dynamic Response of Two-Dimensional Foundations,' Final Report to the National Science Foundation, Part B, Dept. of Civil and Mineral Engineering, University of Minnesota, Minneapolis, Minnesota, 1984.

6. Brebbia, C.A., Umetani, S. and Trevelyan, J.,' Critical Comparison of Boundary and Finite Element Methods for Stress Analysis,' in C.A. Brebbia and B.J. Noye, Editors BETECH 85, Proceedings of the 1st Boundary Element Technology Conference, Springer-Verlag, Berlin, 1985, pp. 225-256.

7. Brebbia, C.A., and Mercy, A.C.,' Use of Boundary Elements as Computer Aided Design Tool,' in G.A. Keramidas and C.A. Brebbia, Editors, Computational Methods and Experimental Measurments, Vol. II, Springer-Verlag, Berlin, 1986, pp. 809-833.

8. Chaudouet, A., and Devalan, P.,' CA.ST.OR,' in C.A. Brebbia, Editor, Finite Element Systems - A Handbook, Springer - Verlag, Berlin, 1982, pp. 127-146.

9. Chaudouet, A., and Afzali, M., 'CA.ST.OR 3D: Three Dimensional Boundary Element Analysis Code,' in C.A., Brebbia, T. Futagami, and M. Tanaka, Editors, Boundary Elements, Proceedings of the Fifth International Conference, Springer-Verlag, Berlin, 1983, pp. 797-809.

10. Crouch, S.L., and Starfield, A.M., Boundary Element Methods in Solid Mechanics, G. Allen and Unwin, London, 1983.

11. Cruse, T.A., 'An Improved Boundary Integral Equation Method for Three-dimensional Elastic Stress Analysis,' Computers and Structures, Vol. 4, 1974, pp. 741-754.

12. Cruse, T.A., 'Two-dimensional Boundary Integral Equation Fracture Mechanics Analysis,' Applied Mathematical Modelling, Vol. 2, 1978, pp. 287-293.

13. Cruse, T.A., 'Design Applications of BIE Methods,' Earthmoving Industry Conference Paper No. 840761, Peoria, Illinois, Society of Automotive Engineers, Warrendale, Pennsylvania, 1984.

14. Danson, D.J., Brebbia, C.A. and Adey, R.A., 'BEASY, a Boundary Element Analysis System,' in C.A. Brebbia, Editor, Finite Element Systems - A Handbook, Springer-Verlag, Berlin, 1982, pp. 81-98.

15. Doblaré, M. and Alarcón, E., 'A Three-dimensional BIEM Program,' in C.A. Brebbia, Editor, Finite Element

Systems - A Handbook, Springer-Verlag, Berlin, 1982, pp. 325-345.

16. Dominguez, J., 'Dynamic Stiffness of Rectangular Foundations,' Publication No. R78-20, Massachusetts Institute of Technology, Cambridge, Massachusetts, 1978.

17. Ingraffea, A.R., Gerstle, W.H. and Perucchio, R., 'Fracture Analysis with Interactive Computer Graphics,' in D.E. Beskos, Editor, Boundary Element Methods in Structural Analysis, American Society of Civil Engineers New York, 1987, pp. 235-271.

18. Kobayashi, S., Nishimura, N. and Mori, K., 'Application of Boundary Element - Finite Element Combined Method to Three-dimensional Viscoelastodynamic Problems,' in D. Qinghua, Editor, Boundary Elements, Pergamon Press, Oxford, 1986, pp. 67-74.

19. Lachat, J.C., and Watson, J.O., 'Effective Numerical Treatment of Boundary Integral Equations,' International Journal for Numerical Methods in Engineering, Vol. 10, 1976, pp. 991-1005.

20. Manolis, G.D., and Beskos, D.E., 'Dynamic Stress Concentration Studies by Boundary Integrals and Laplace Transform,' International Journal for Numerical Methods in Engineering, Vol. 17, 1981, pp. 573-599.

21. Manolis, G.D., Ahmad, S., and Banerjee, P.K.,' Boundary Element Implementation for Three-dimensional Transient Elastodynamics,' in P.K., Banerjee and J.O. Watson, Editors, Developments in Boundary Element Methods - 4, Elsevier Applied Science Publishers, London, 1986, pp. 29-65.

22. Manolis, G.D., and Banerjee, P.K., 'A Comparison of Conforming versus Non-Conforming Boundary Elements in Elastomechanics,' International Journal for Numerical Methods in Engineering, 1986, pp. 1885-1904.

23. Mukherjee, S., and Banthia, V., 'Non-linear Problems of Fracture Mechanics, in P.K. Banerjee and S. Mukherjee, Editors, Developments in Boundary Element Methods - 3, Elsevier Applied Science Publishers, London, 1984, pp. 87-114.

24. Mukherjee, S., and Chandra, A., 'Boundary Element Formulations for Large Strain - Large Deformation Problems of Plasticity and Viscoplasticity,' in P.K., Banerjee and S. Mukherjee, Editors, Developments in Boundary Element Methods - 3, Elsevier Applied Science Publishers, London, 1984, pp. 27-58.

25. Patterson, C. Sheikh, M.A., and Scholfield, R.P., 'On the Appication of the Indirect Discrete Method for Three-dimensional Design Problems,' in C.A. Brebbia and B.J. Noye, Editors, BETECH 85, Proceedings of the 1st Boundary Element Technology Conference, Springer-Verlag, Berlin, 1985, pp. 221-223.

26. Raveendra, S.T., 'Advanced Development of BEM for Two- and Three-dimensional Nonlinear Analysis,' thesis presented to State University of New York, at Buffalo, New York, in 1984, in partial fulfillment of the requirements for the degree of Doctor of Philosophy.

27. Rizzo, F.J., Shippy D.J. and Rezayat, M., 'Boundary Integral Equations Analysis for a Class of Earth - Structure Interaction Problems,' Final report to the National Science Foundation, Dept. of Engng. Mechanics, University of Kentucky, Lexington, Kentucky, 1985.

28. Wardle, L.J., and Crotty, J.M., 'Two-dimensional Boundary Integral Equation Analysis for Non-homogeneous Mining Applications,' in C.A. Brebbia, Editor, Recent Advances in Boundary Element Methods, Pentech Press, London, 1978, pp. 233-252.

Present Status and Future Developments

by Dimitri E. Beskos*, M. ASCE

An overview of the Boundary Element Method (BEM) and its applications in structural analysis is provided. Advantages and disadvantages of the method are systematically presented. Possible future extensions and developments are also briefly discussed.

Introduction

The Boundary Element Method (BEM) has emerged during the last 15 years as a new powerful computational tool for the numerical solution of a great variety of problems in engineering mechanics in general and structural analysis in particular. A historical review of the BEM was nicely presented in the first chapter of this book by Professor F.J. Rizzo, the man who essentially introduced the method in 1967 through his pioneering paper on elastostatics (3). The next ten chapters of the book dealt with various applications of the BEM to structural analysis problems and served to demonstrate the wide range of applicability of the method and its distinct advantages.

This closing chapter makes an effort on the one hand to systematically review the method and its applications in structural analysis by enumerating its advantages and disadvantages and on the other hand to discuss its possible future extensions and developments. The interested reader can consult another similar article of this author (2) for more details. The comprehensive review article of Atluri (1) dealing with both finite and boundary elements is also another source of information on the subject.

Advantages and Disadvantages

This section provides a systematic enumeration of the advantages and disadvantages of the BEM as compared with the Finite Element Method (FEM), which is presently considered the most popular and widely used numerical method in structural analysis.

The BEM presents some distinct advantages, such as

1) It is a general numerical method with a wide range of applications in structural analysis including linear elastic and elastoplastic torsion of bars, linear elastic analysis of beams, plates and shells as well as general two- and three-dimensional solids under static or dynamic loads, elastoplastic analysis with small or large deformations of two-and three-dimensional structures

---

* Professor, Department of Civil Engineering, University of Patras, Patras 261 10, Greece

under static loads, linear elastic or inelastic fracture analysis, linear stability analysis of beams and plates and dynamic soil-structure and fluid- structure interaction problems. The method also enjoys a firm mathematical basis and constitutes a practical and efficient computational tool for both linear and nonlinear problems.

2) It usually requires only a surface discretization and not a discretization of both the interior and the surface of the domain of interest as it is the case with "domain" type methods like the FEM. This reduction of the spatial dimensions of the problem by one greatly facilitates the data preparation job, permits an easy conduction of mesh refinement studies and leads to a matrix equation of a much smaller order than the one produced by the FEM.

3) The facts that the approximations involved in the method are confined to the surface of the domain and that the influence matrices associated with the BEM consist of dominant elements on or near their main diagonal due to the singular nature of the employed Green's functions are the reasons for obtaining results of higher accuracy than by the FEM, especially in problems involving stress concentrations.

4) The use of the Green's function associated with the problem of interest makes possible the easy treatment of infinite or semi-infinite domains without artificial "box-type" discretizations as it is the case with the FEM.

5) The above advantages of the BEM over the FEM are greatly pronounced in linear problems characterized by three-dimensional (especially infinite or semi-infinite), homogeneous and isotropic domains.

6) The method is capable of providing values of the unknowns in the interior of the domain in a pointwise fashion so that no interelement continuity problems can arise as in the FEM. In addition the computation is restricted to points of interest and does not involve all the mesh points as in the FEM.

7) The BEM can be used for the construction of stiffness matrices for homogeneous very big finite elements (super-elements) or infinite elements which can be used in a finite element formulation characterized by a drastic reduction in the number of the unknowns of the system.

The BEM is also characterized by a number of disadvantages, such as

1) It is associated with nonsymmetric and fully populated influence matrices in contrast to the FEM, which involves symmetric and sparse matrices that require less computational effort for their inversion. The order of the former matrices is, however, much lower than that of the latter and their numerical treatment can be done efficiently through special schemes. Besides, the former

matrices are numerically well conditioned.

2) It is very difficult or practically impossible to obtain the Green's functions for some types of problems, especially for those involving strong anisotropies and inhomogeneities. Of course, approximate fundamental solutions in conjunction with an iterative approach or interior discretization can be employed at an increased computational cost. The FEM is a better choice for problems characterized by rapidly changing physical properties in the domain.

3) The BEM is inefficient as compared to the FEM for problems with a geometry for their mathematical model characterized by one or two spatial dimensions disproportionally small with respect to the others but dimensionally effective, such as those involving the analysis of moderately thick plates and shells or thin narrow strips. The BEM is also inefficient for one-dimensional problems, such as those involving the analysis of beams and frameworks.

4) The presence of known distributed body forces in linear problems or pseudo-incremental body forces in nonlinear problems creates volume integrals which require an internal discretization of the domain. However, this internal discretization is much simpler than that of the FEM, can be restricted to a small portion of the domain in some cases and most importantly, does not lead in any increase in the order of the final system of algebraic equations.

5) There are very few special purpose BEM computer programs presently available as compared to the great number of corresponding FEM programs. This is because the BEM is still in the stage of development, competes against a very efficient and well established technique, the FEM and the architecture of a BEM computer program is not conceptually as easy as the one for the FEM.

Although at a first glance the BEM appears to have few similarities with the FEM, the two methods share a common basis as they can be derived from weighted residuals or variational statements which provide a connection between the BEM and quite a number of other known numerical methods. Thus, from the viewpoint of their common origin and the fact that many advantages of one method correspond to disadvantages for the other method, one can consider the BEM and the FEM not as antagonistic techniques but rather as techniques which can complement each other. The BEM indeed has been successfully combined with the FEM (as well as with some other numerical methods) for the solution of a variety of problems in mechanics thereby creating a hybrid type of method that combines the advantages of the two methods and reduces or eliminates their disadvantages.

Future Developments

The BEM is presently approaching a level of maturity. However, many further developments of a special or general character are still needed. This section serves the purpose of briefly discussing new developments in the field of the BEM that are expected to take place in the near future. The list of these future developments is by no means complete and simply reflects the personal opinion of the author. Thus, the following developments are expected in the near future:

1) Use of computer symbolic manipulation for the automatic generation and synthesis of influence matrices and of adaptive mesh refinement in a way that each step depends on the error information provided by the previous ones.

2) Development of comprehensive general purpose copmuter programs and more special purpose ones to cover all the existing areas of application of the BEM both equipped with increased user-oriented features and graphic display capabilities.

3) Development of computer programs that can take advantage of the special features of the new super-computers (e.g., GRAY-1, -2, CDC Cyber 205), such as vectorization and parallel processing, for the more efficient solution of large order problems.

4) Efficient treatment of the problems of combining the BEM with other numerical methods, especially the FEM and the construction of finite super-elements with the aid of the BEM. There is also a need to resolve the efficiency controversy between the BEM and the FEM by a series of detailed comparison studies for various classes of problems in order to be able to combine them appropriately on the basis of their particular advantages-disadvantages.

5) Construction of exact and approximate Green's functions for certain classes of problems and evaluation of the effectiveness of the latter functions or development of nonconventional BEM's like the generalized boundary methods which do not require employment of Green's functions.

6) Development of efficient schemes for the static and dynamic analysis of plates and thin shells in the elastic and inelastic range.

7) Development of efficient techniques for the treatment of geometric and material nonlinearities in problems dealing with, e.g., plasticity or viscoplasticity under small or large deformations, contact stress analysis, postbuckling behavior and inelastic fracture.

8) Development of dynamic nonlinear analysis techniques for problems, such as, dynamic plasticity, dynamic inelastic contact stress analysis, inelastic creep, dynamic fracture, fatigue and damage, as well as dynamic inelastic stability.

9) Development of efficient methodologies for the treatment of large scale nonlinear static or dynamic problems by domain decomposition techniques.

10) Three-dimensional dynamic soil-structure interaction problems of a linear or a nonlinear character such as those dealing with the dynamics of foundations, the behavior of underground structures and the isolation of structures from surface waves.

11) Three-dimensional dynamic fluid-structure interaction problems dealing with, e.g., liquid storage tanks or vessels, submerged of floating structures, aerospace structures and buildings under wind forces.

12) Complicated problems characterized by a combination of many of the previously mentioned physical aspects, such as those encountered in off-shore platforms or modern space stations.

13) Problems involving structural optimization, optimal structural control and identification studies including the development of nondestructive testing techniques.

14) Development and offering of special graduate level courses and short courses on the BEM by engineering schools and professional engineering societies in conjunction with the production of some application-oriented books for the sake of graduate students and practicing engineers.

Acknowledgements

The author would like to express his thanks to Mrs H. Alexandridis for her conscientious typing of the manuscript.

Appendix.-References

1. Atluri, S.N., "Computational Solid Mechanics (Finite Elements and Boundary Elements): Present Status and Future Directions", Proceedings of the 4th International Conference on Applied Numerical Modelling, H.M. Hsia et al, Eds, Tainan, Taiwan, 1984, pp. 19-37.

2. Beskos, D.E., "Introduction to Boundary Element Methods", Chapter 1 in Boundary Element Methods in Mechanics, D.E. Beskos, Ed., North-Holland, Amsterdam, 1987.

3. Rizzo, F.J., "An Integral Equation Approach to Boundary Value Problems of Classical Elastostatics", Quarterly of Applied Mathematics, Vol. 25, 1967, pp. 83-95.

SUBJECT INDEX

AUTHOR INDEX